Hydrogen Purification and Separation

The purification of hydrogen is necessary to fulfill purity standards of a wide variety of prospective uses, and it is also a key concern regarding the efficient supply of hydrogen. *Hydrogen Purification and Separation* reviews various hydrogen separation methods as well as membranes used in hydrogen separation. It discusses absorption and adsorption methods, as well as novel technologies such as cryogenic methods and plasma-assisted technology, and the related economic assessments and environmental challenges.

- Introduces miscellaneous membrane-assisted processes for hydrogen separation
- Provides different physiochemical absorption methods for hydrogen purification
- Discusses application of sorbents and swing technologies in hydrogen purification
- Uniquely covers hydrogen separation using novel methods
- Includes economic assessments and environmental challenges of hydrogen purification in detail

Part of the multivolume Handbook of Hydrogen Production and Applications, this stand-alone book guides researchers and academics in chemical, environmental, energy, and related areas of engineering interested in development and implementation of hydrogen production technologies.

Mohammad Reza Rahimpour is a Distinguished Professor of Chemical Engineering at Shiraz University, Iran. He earned a PhD in chemical engineering at Shiraz University, in collaboration with the University of Sydney, Australia, in 1988. He leads a pioneering research group specializing in fuel processing technology, with a particular focus on the catalytic conversion of both fossil fuels, such as natural gas, and renewable fuels, like bio-oils derived from lignin, into valuable energy sources. Throughout his academic career, Dr. Rahimpour has specialized in hydrogen production technologies, including steam methane reforming, water electrolysis, photocatalytic hydrogen production, and hydrogen production from both renewable sources and fossil fuels. His extensive expertise and innovative research significantly contribute to advancements in hydrogen production and its practical applications.

Mohammad Amin Makarem earned a PhD in chemical engineering at Shiraz University. His research interests include gas separation and purification, nanofluids, microfluidics, catalyst synthesis, reactor design, and green energy.

Parvin Kiani earned a degree in chemical engineering at Shiraz University. Her research has focused on gas separation, clean energy, and catalyst synthesis.

Hydrogen Purification and Separation

Edited by
Mohammad Reza Rahimpour,
Mohammad Amin Makarem and Parvin Kiani

CRC Press
Taylor & Francis Group
Boca Raton London New York

CRC Press is an imprint of the
Taylor & Francis Group, an **informa** business

Designed cover image: © Shutterstock, Fotogrin

First edition published 2025
by CRC Press
2385 NW Executive Center Drive, Suite 320, Boca Raton FL 33431

and by CRC Press
4 Park Square, Milton Park, Abingdon, Oxon, OX14 4RN

CRC Press is an imprint of Taylor & Francis Group, LLC

ISBN: 978-1-032-46605-7 (hbk)
ISBN: 978-1-032-46606-4 (pbk)
ISBN: 978-1-003-38252-2 (ebk)

DOI: 10.1201/9781003382522

Typeset in Times
by codeMantra

Contents

SECTION I An Overview of Hydrogen Purification and Separation Methods

*Maryam Meshksar, Sina Mosallanezhad, and
Mohammad Reza Rahimpour*

Bilal Kazmi and Syed Ali Ammar Taqvi

SECTION II Hydrogen Separation Using Membranes

*Fatemeh Salahi, Fatemeh Zarei-Jelyani, and
Mohammad Reza Rahimpour*

*Samuel Eshorame Sanni, Denen Ashiekaa Vershima,
Emeka Emmanuel Okoro, and Babalola Aisosa Oni*

*Fatemeh Kavousi, Anna O'Sullivan, Archishman Bose,
and Sudipta De*

*Henry Bryan Trujillo Ruales, Enrico Drioli,
and Adolfo Iulianelli*

SECTION III Absorption and Absorption Methods for Hydrogen Purification

SECTION IV Other Technologies for Hydrogen Purification

Preface

As the world intensifies its focus on sustainable energy solutions, hydrogen emerges as a key player due to its versatility and eco-friendly attributes. The *Handbook of Hydrogen Production and Applications* represents a comprehensive and authoritative exploration of the pivotal role hydrogen plays in addressing the global energy landscape and environmental challenges. This ambitious seven-volume project delves into every facet of hydrogen, ranging from its production using both nonrenewable and renewable resources, purification, and separation processes, to its storage, transportation, and various applications across diverse technologies. With an emphasis on fostering a deeper understanding of hydrogen utilization, particularly in fuel cells, this handbook aims to serve as an invaluable resource for researchers, engineers, and policymakers striving to advance the forefront of clean energy technology. Each volume within this compendium is dedicated to a specific aspect, collectively forming a comprehensive guide that illuminates the multifaceted dimensions of hydrogen production and its myriad applications.

Volume 3, "Hydrogen Purification and Separation", explores the critical processes involved in refining hydrogen for various applications. In Section I, "An Overview of Hydrogen Purification and Separation Methods", readers are introduced to the fundamental concepts underlying hydrogen separation and purification technologies. Economic assessments and environmental challenges associated with these technologies are also addressed, providing a holistic perspective on the considerations driving advancements in this field.

Section II delves into the realm of "Hydrogen Separation Using Membranes", presenting an array of membrane-based technologies. From polymeric and zeolite membranes to ceramic, hybrid, silica, and mixed matrix membranes, this section comprehensively explores the diverse approaches employed in separating hydrogen from mixed gas streams. The discussion extends to hollow fiber and porous inorganic membranes, highlighting their unique characteristics and applications in hydrogen purification.

Turning to Section III, "Absorption and Adsorption Methods for Hydrogen Purification," readers are introduced to alternative purification techniques. The use of ionic liquids and deep eutectic solvents, carbonaceous sorbents, and metal–organic frameworks for hydrogen purification is examined in detail. Additionally, swing technologies for hydrogen purification are explored, providing insights into dynamic purification processes that adapt to varying operating conditions.

The volume concludes with Section IV, which examines "Other Technologies for Hydrogen Purification". Here, plasma-assisted technology emerges as a promising avenue for hydrogen purification, offering innovative solutions to enhance purity levels and efficiency in hydrogen production processes.

This volume aims to serve as a comprehensive guide for researchers, engineers, and policymakers involved in hydrogen purification and separation. By offering insights into a diverse array of purification methods, from membrane-based technologies to absorption and adsorption processes, this volume provides a thorough understanding of the challenges and opportunities in refining hydrogen for various applications. We extend our sincere appreciation to all contributors who have shared their expertise to make this volume possible, and we invite readers to delve into the intricacies of hydrogen purification and separation, paving the way for a sustainable and transformative future in hydrogen utilization.

Mohammad Reza Rahimpour
Mohammad Amin Makarem
Parvin Kiani

Contributors

Ping Ai
Huazhong Agricultural University
Wuhan, China

Ahmed Alengebawy
Huazhong Agricultural University
Wuhan, China

Shailendra Kumar Arya
Panjab University
Chandigarh, India

Sagar Ban
Kathmandu University
Kavre, Nepal

Archishman Bose
University College
Cork, Ireland

Zhonghao Chen
Xi'an Jiaotong–Liverpool University
Suzhou, China

Sudipta De
Jadavpur University
Kolkata, India

Tanmay Jyoti Deka
Queen's University
Belfast, Northern Ireland, UK

Enrico Drioli
Institute on Membrane Technology of
the National Research Council
Università della Calabria
Rende, Italy

Alberto Figoli
Institute on Membrane Technology
National Research Council
Rende, CS, Italy

Nayef Ghasem
Department of Chemical and Petroleum
Engineering
United Arab Emirates University
Al-Ain, UAE

Adolfo Iulianelli
Institute on Membrane Technology of
the National Research Council
Università della Calabria
Rende, Italy

Jyoti Kaushal
Panjab University
Chandigarh, India

Fatemeh Kavousi
University College
Cork, Ireland

Bilal Kazmi
University of Karachi
Karachi, Pakistan

Madhu Khatri
Lovely Professional University
Punjab, India

Maryam Meshksar
Shiraz University
Shiraz, Fars, Iran

Sina Mosallanezhad
Shiraz University
Shiraz, Fars, Iran

A.O. Ogunyinka
Tshwane University of Technology
Pretoria, South Africa

Emeka Emmanuel Okoro
University of Port Harcourt
Choba, Rivers State, Nigeria

Babalola Aisosa Oni
University of Port Harcourt
Choba, Rivers State, Nigeria

Ahmed I. Osman
Queen's University
Belfast, Northern Ireland, UK

Anna O'Sullivan
University College
Cork, Ireland

S.L. Pityana
Tshwane University of Technology
Pretoria, South Africa

A.P.I. Popoola
Tshwane University of Technology
Pretoria, South Africa

O.M. Popoola
Tshwane University of Technology
Pretoria, South Africa

Mohammad Reza Rahimpour
Shiraz University
Shiraz, Fars, Iran

Maryam Takht Ravanchi
Petrochemical Research and
    Technology Company
National Petrochemical Company
Tehran, Iran

Henry Bryan Trujillo Ruales
Institute on Membrane Technology of
    the National Research Council
Università della Calabria
Rende, Italy

E.R. Sadiku
Tshwane University of Technology
Pretoria, South Africa

Fatemeh Salahi
Shiraz University
Shiraz, Fars, Iran

Samuel Eshorame Sanni
Covenant University
Ota, Ogun State, Nigeria

Syed Ali Ammar Taqvi
NED University of Engineering and
    Technology
Karachi, Pakistan

Denen Ashiekaa Vershima
Covenant University
Ota, Ogun State, Nigeria

Jingxue Wang
Qingdao University of Science and
    Technology
Qingdao, People's Republic of China

Yinglong Wang
Qingdao University of Science and
    Technology
Qingdao, People's Republic of China

Pow-Seng Yap
Xi'an Jiaotong–Liverpool University
Suzhou, China

Kexin Yin
Qingdao University of Science and
    Technology
Qingdao, People's Republic of China

Fatemeh Zarei-Jelyani
Shiraz University
Shiraz, Fars, Iran

Jifu Zhang
Qingdao University of Science and
  Technology
Qingdao, People's Republic of China

Mengjin Zhou
Qingdao University of Science and
  Technology
Qingdao, People's Republic of China

Reviewer Acknowledgments

The editors feel obliged to appreciate the dedicated reviewers (listed below) who were involved in reviewing and commenting on the submitted chapters and whose cooperation and insightful comments were very helpful in improving the quality of the chapters and books in this series.

Dr. Fatemeh Haghighatjoo
Department of Chemical Engineering
Shiraz University
Shiraz, Iran

Dr. Maryam Meshksar
Department of Chemical Engineering
Shiraz University
Shiraz, Iran

Ms. Soheila Zandi Lak
Department of Chemical Engineering
Shiraz University
Shiraz, Iran

Mr. Sina Mosallanejad
Department of Chemical Engineering
Shiraz University
Shiraz, Iran

Section I

*An Overview of
Hydrogen Purification and
Separation Methods*

1 Introduction to Hydrogen Separation and Purification Technologies

Maryam Meshksar, Sina Mosallanezhad, and Mohammad Reza Rahimpour

1.1 INTRODUCTION

Concern over greenhouse gas emissions, especially carbon dioxide (CO_2), and the diminishing supply of fossil fuels have made the search for renewable energy sources an urgent and challenging task (Khosravani et al., 2023; Kiani et al., 2022; Mosallanezhad et al., 2023). Based on the findings, transportation was the second largest source of carbon dioxide emissions in 2014, accounting for around 23% of the total. Demand for transportation fuel is predicted to increase by more than 20% in 2023, according to the WEO (Birol, 2007). Reducing transportation emissions is essential.

One of the most prominent energy carriers that might be used for this function is hydrogen (H_2) (Meshksar et al., 2020a; 2022a; Kiani et al., 2023). Besides its role in power generation, hydrogen is used in other industries, including methanol, ammonia, dimethyl ether, and Fischer–Tropsch (F-T) synthesis (Meshksar et al., 2022b). Hydrocarbon reforming, photoelectrolysis, water splitting, biomass gasification, and water electrolysis are some methods used to produce pure hydrogen gas (Salahi et al., 2023).

Statistics from the China Hydrogen Alliance and the China National Petroleum and Chemical Planning Institute indicate that China can produce around 41 million tons of hydrogen annually, yielding 33.42 million tons. Specifically, roughly 12.7 million tons of hydrogen gas are produced annually as a standalone product (synthetic gas devoid of hydrogen) that meets the quality standards for industrial use and may be sold directly as industrial gas (Du et al., 2021). Coal is the primary source of hydrogen, accounting for 63.54% of the total, with 21.24 million tons. A total of 7.08 million tons of H_2 gas, 4.6 million tons of natural gas (NG), and 0.5 million tons of electrolyzed water follow. Biological hydrogen synthesis (Cao et al., 2020), photocatalytic water breakdown utilizing solar energy (Tasleem and Tahir, 2020), and supercritical steam coal (Sun et al., 2021) are all methods for producing hydrogen that is still in the early stages of development and experimentation (Table 1.1). The raw materials used in various processes significantly affect the composition and impurity levels of the hydrogen gas produced. Table 1.2 shows the distribution of several forms of crude H_2.

DOI: 10.1201/9781003382522-2

TABLE 1.1

Techniques for Generating Hydrogen

H_2 Production Method	Technical Feature
Supercritical steam coal	In this technology, supercritical water is used as a medium that provides a homogeneous and high-speed reaction because of its unique physical and chemical properties so that the chemical energy of coal is directly and efficiently converted into hydrogen energy (Sun et al., 2021).
Photocatalytic water decomposition by solar energy	Photocatalyst powders or electrodes can produce photo-generated carriers by absorbing solar energy, decomposing water into H_2 and O_2. The photocatalytic H_2 production can be subdivided mainly into heterogeneous photocatalytic (HPC) H_2 production and photo-electrochemical (PEC) H_2 production (Tasleem and Tahir, 2020).
Biological H_2 production	H_2 is a product of microorganisms' metabolism using biomass and organic wastewater as raw materials. Based on the type of microorganisms and their metabolic mechanisms, the biological H_2 production technology includes water-splitting H_2 production, photo-fermentative H_2 production, dark fermentative H_2 production, and H_2 production combined with photo-fermentation and dark fermentation (Cao et al., 2020).

Source: Adapted from (Du et al., 2021).

Quickly developing novel and highly efficient purification technologies that can create low-cost and high-quality H_2 is crucial to enable the broad use of H_2 energy in transportation. The study looked at the characteristics of traditional technologies for removing hydrogen gas, such as membrane separation, pressure swing adsorption (PSA), and metal hydride separation (Meshksar and Rahimpour, 2023). Improvements in separation efficiency can only be achieved through an ongoing investigation into new and improved adsorption materials, cost-effective membrane materials, poison-proof and energy-efficient metal hydride materials, and creative ways to combine and separate these materials (Barabadi et al., 2023).

1.2 H_2 PURIFICATION TECHNOLOGIES

To link H_2 production and consumption, H_2 purification technology is crucial. The widespread use of H_2 depends on reliable, affordable, and stable H_2 sources and purifying techniques. Physical and chemical methods may predominate in H_2 purification (Aasadnia et al., 2021). The first group includes adsorption methods, including vacuum adsorption, temperature swing adsorption (TSA), and PSA. Also included are membrane separation techniques using inorganic or organic membranes, low-temperature adsorption, and cryogenic distillation. The second group uses a catalytic and separation process using metal hydrides (Figure 1.1). For H_2 generation through centralized large-scale gasification of coal and NG reforming, the most common methods of purification include using PSA after transformation, desulfurization, and decarbonization. These processes are used when the H_2 supply is more than 10,000 Nm^3/h. Low operating costs and long service life are the defining characteristics of

TABLE 1.2

Composition of Different Types of Crude H_2

Component (%)

Process	H_2 (%)	CO (%)	CO_2 (%)	CH_4 (%)	N_2 (%)	Ar (%)	Total Sulfur (%)	H_2O (%)	O_2 (%)	Others (%)
Coal gasification [27]	25–35	35–45	15–25	0.1–0.3	0.5–1	–	0.2–1	15–20	–	–
NG reforming [28]	70–75	10–15	10–15	1–3	0.1–0.5	–	–	–	–	–
Methanol reforming [29]	75–80	0.5–2	20–25	–	–	–	–	–	–	–
Coke oven gas [30]	45–60	5–10	2–5	25–30	2–5	–	0.01–0.5	–	0.2–0.5	2–5
Methanol purge gas [31]	70–80	4–8	5–10	2–8	5–15	0.1–2	–	–	–	–
Synthetic ammonia tail gas [32]	60–75	–	–	–	15–20	–	–	1–3	10–15	–
Biomass gasification [33]	25–35	30–40	10–15	10–20	1	–	0.2–1	–	0.3–1	–

Source: Adapted from Du et al., 2021.

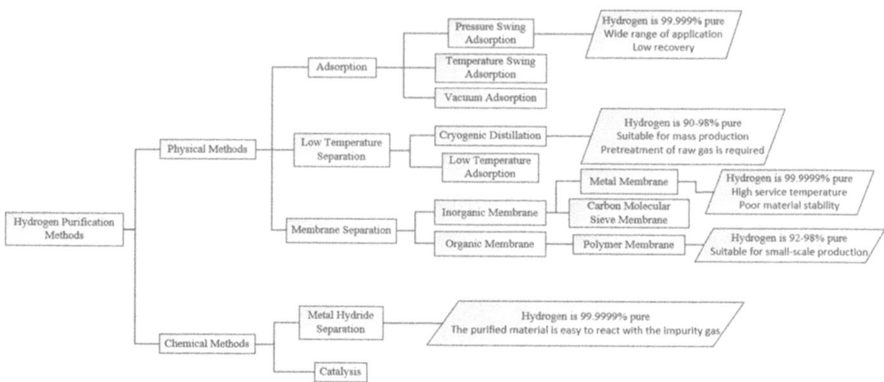

FIGURE 1.1 Classification of H_2 purification technologies. (Adapted from Du et al., 2021.)

the long-standing PSA technology. However, fuel cell automobiles have decreased recovery rate and yield efficiency due to the H_2 produced by traditional PSA techniques, which usually have many contaminants. On top of that, it could be more financially feasible.

Cryogenic distillation is possible for large-scale manufacturing due to the modest, precise impurity elimination criteria, such as limited CO levels to $\leq 0.2\,$ppm. However, the method typically yields H_2 with a purity of 85%–99%, which does not meet the requirements for the application. Improved H_2 recovery efficiency in the centralized by-product mode is possible with flexible methods that handle different pollutants and an H_2 supply range from 1,000 to 10,000 Nm³/h. For example, one may combine an organic membrane with a PSA process to produce methanol purge gas. In contrast, the gas from the refinery's coke ovens and by-products is obtained via a two-stage or multi-stage PSA process. Sulfide, formaldehyde (HCHO), and formic acid (HCOOH) are contaminants that low-temperature adsorption may effectively remove. However, this method makes heavy energy use and is only practical for limited, low-volume uses of cold sources (Schorer et al., 2019). Gas sources with a high concentration of inert components may be successfully separated using the metal hydride and Pd membrane separation processes.

On the other hand, the process of H_2 recovery involves exposing purified materials to polluted gas, which reduces the purification efficacy. Novel membrane technologies, such as ionic liquid (IL) membranes (Meshksar et al., 2020c), electrochemical H_2 pump membranes (Bernardo et al., 2020), and carbon molecular sieve membranes (CMSMs) (Hamm et al., 2017), are the subject of present-day scientific investigation. A lot of focus is currently on these technologies. The large-scale adoption of these technologies is challenging to forecast with any degree of certainty.

1.2.1 PSA TECHNIQUE

The cyclic pressure fluctuations used in PSA gas separation and purification use different adsorption capabilities of other gases. Both the adsorbent type and the particular technical procedure determine the PSA's separating effect. Figure 1.2 shows the typical PSA system's flow diagram.

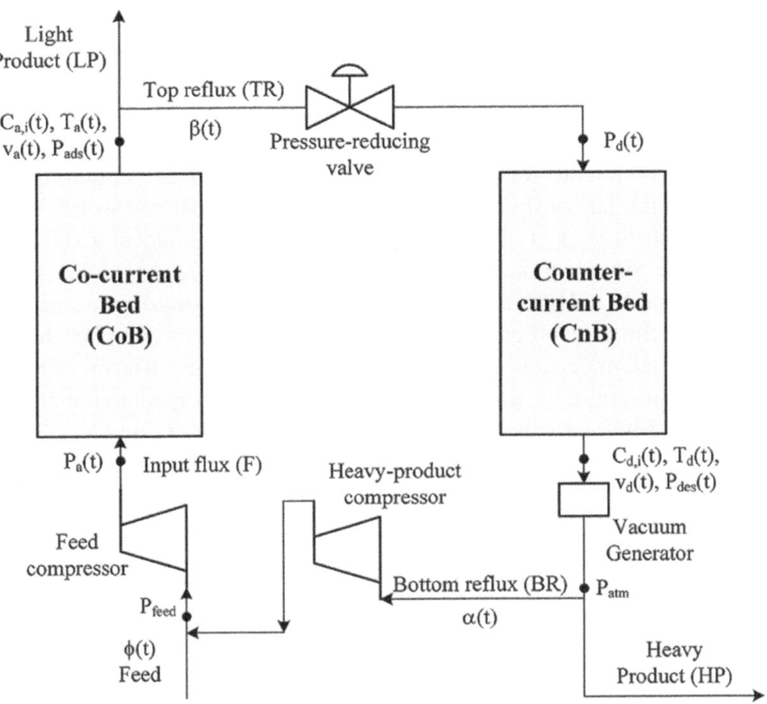

FIGURE 1.2 Classical PSA system. (Adapted from Agarwal et al., 2010.)

When separating and purifying PSA, H_2 is the way to go because of its much higher static capacity than other gas molecules like CO_2, CO, and CH_4 (Hamm et al., 2017). The industrial usage of H_2 has already been very successful for air products, air liquids, and other well-known gas companies throughout the globe.

Some of the most common adsorbents are activated carbon, silica gel, alumina, and zeolite molecular sieves. While most studies focused on removing CO_2, there were several reports of improvements and developments in these adsorbents for other pollutants. The method of CO_2 removal in experimental devices employing PSA technology was investigated by Lively et al. (2012) using hollow fibers as an adsorbent material. Additional optimization is necessary since the produced H_2 had a recovery rate of 88.1% and a purity of 99.2%. Using palm shell charcoal's high CO_2 absorption capacity, Shamsudin et al. (2019) obtained a recovery rate of 88.43% and an H_2 purity of about 100%. He et al. (2020) used a dip-coated Ni foam framework to build a structured activated carbon system for rapid pressure swing adsorption (RPSA). The adsorption rate constant K was found to be $0.0029\,s^{-1}$ at a flow rate of 200 mL/min and a pressure of 0.4 MPa. The value is almost double that of traditional adsorbents. In the presence of H_2, the material showed improved CO_2 adsorption efficiency. Another study by Kuroda et al. (2018) purified biomass-derived H_2 using hydroxyl aluminum silicate clay (HAS-Clay), showing a solid capacity to absorb CO_2. The adsorption and separation of H_2S is another potential use of this adsorbent.

Innovative adsorption materials may benefit significantly from metal–organic frameworks (MOFs), freshly synthesized materials with highly adjustable properties and structures (Yousefi et al., 2023). To remove CO_2 contaminants from tail gas, Agueda et al. (2015) employed UTSA-16 as an adsorbent and simulated the PSA steam methane reforming process. With a 93%–96% recovery rate and 2–2.8 moles per kilogram per hour yield, the results showed that the H_2 purity ranged from 99.99% to 99.999%.

Innovative compounds that may efficiently remove many contaminants from hydrogen gas simultaneously are now the focus of scientific research. Using an ion exchange approach, Brea et al. (2019) synthesized a NaX molecular sieve from raw materials while simultaneously including CaX and MgX molecular sieves. The adsorption of the $H_2/CH_4/CO/CO_2$ gas combination was simulated, and the results show that these three adsorbents can yield H_2 with a purity level higher than 99.99%. The use of the CaX molecular sieve showed the best efficiency in terms of H_2 recovery rate and yield. Additionally, a study by Banu et al. (2013) compared the effectiveness of four distinct MOF adsorbents: UiO-66(Zr)-Br, UiO-67(Zr), and Zr-Cl$_2$AzoBDC. The results showed that UiO-66(Zr)-Br was the most successful in purifying the H_2 produced during methane steam reforming. A novel adsorbent, Cu-AC-2, was created by Relvas et al. (2018) to treat a mixture of gases, namely H_2, CH_4, CO, and CO_2. H_2 had a purity level of over 99.97%, while CO_2 was below 0.17 ppm.

To increase the efficacy of H_2 purification, it is essential to improve and simplify the PSA process. Ahn et al. (2012) utilized the two- and four-bed PSA systems to remove H_2 gas from coal, with nitrogen (N_2) serving as the principal impurity. The four-bed PSA procedure outperformed the two-bed PSA technique, leading to a 71%–85% recovery rate and a 96%–99.5% hydrogen purity. Abdeljaoued et al. (2018) created a theoretical model for a four-bed PSA system. They used a four-bed PSA system for a twelve-step experimental investigation at room temperature. Hydrogen produced by ethanol steam reforming was the subject of an examination for the removal of contaminants. There was hope that future upgrades would raise the rate of H_2 recovery to above 75% and reduce the CO concentration to less than 20 ppm. Moon et al. (2018) examined the eight-bed PSA technique in their study. This method outperformed the four-bed PSA process by about 11%, with a maximum H_2 recovery rate of 89.7% and an H_2 purity of 99.99%. The five-step one-bed PSA cycle and the six-step two-bed PSA cycle were also modeled by Zhang et al. (2019). Next, they evaluated each model according to H_2 yield, recovery rate, and purity parameters.

Compared to the one-bed PSA process, the two-bed approach demonstrated an 11% recovery rate and 1 mol/kg·h yield. Li et al. (2019) investigated how PSA was affected by adsorption pressure and time. The purification of hydrogen gas produced by methane steam reforming was accomplished using a six-step, two-bed PSA technique. The produced H_2 was 80% pure and had a purity level of over 99.95%. An increase in the adsorption pressure was required to keep the H_2 purity constant in the presence of increasing concentrations of CH_4 pollutants. A four-bed PSA device using 5 Å molecular sieve adsorbents was created by Yáñez et al. (2020) for their investigation. The synthesized ammonia tail gas has the following composition: H_2:N_2:CH_4:Ar = 58:25:15:2, and this apparatus was designed to filter H_2 from it. Reports indicated a recovery rate of 55.5%–75.3% and an H_2 purity of 99.25%–99.97%.

One important topic to research is how to optimize the process flow utilizing classical PSA to increase the recovery rate and purity of H_2. One method to clean adsorbents with a high capacity for removing impurities is vacuum pressure swing adsorption or VPSA. This procedure revitalizes the adsorbents. You et al. (2012) demonstrated that, when run under the same conditions, the VPSA and PSA techniques may produce H_2 with similar degrees of purity. Nonetheless, compared with PSA, the VPSA approach demonstrated a recovery rate that was about 10% higher.

Furthermore, an experiment was carried out by Lopes et al. (2012) on rapid vacuum pressure swing adsorption (RVPSA). The results showed that, in comparison to PSA, RVPSA enhanced the hydrogen yield by around 410%. Golmakani et al. (2017) contrasted the PSA, VPSA, and TSA procedures in their study. Regarding recovery rate and cost-effectiveness, they discovered that the VPSA method produced hydrogen for fuel cells relatively quickly. They looked at three different procedures and found that VPSA was the best. Furthermore, they created a VPSA model with four beds and sixteen stages (Golmakani et al., 2020). Scientists looked at how N_2 affected the process's efficiency in order to speed up VPSA's H_2 recovery and reduce energy use.

Various pollutants and the demand for H_2 consumption are being addressed by developing or using novel PSA devices. An unknown PSA device, Sour PSA, has been created and studied by air products to collect acidic pollutants, such as CO_2 and sulfides in H_2. The compact pressure swing adsorption (CPSA) system was developed by Majlan et al. (2009), which continually uses a high circulation rate to generate H_2 without adsorbent renewal. Furthermore, it can enhance the purity of H_2 to 99.999% by reducing the CO concentration in the $H_2/CO/CO_2$ gas mixture from 4,000 to 1.4 ppm and the CO_2 concentration from 5% to 7 ppm. To separate CO, CO_2, H_2O, and H_2 from the input gas at the right conditions, Zhu et al. (2018) presented a seven-step, two-bed elevated-temperature pressure swing adsorption (ET-PSA) system. Using this process, H_2 was successfully generated with a purity level of 99.9991% and a recovery rate of 99.6%. Gases like CO and CO_2 may be removed from H_2 using this procedure.

1.2.2 Membrane Separation Technique

Environmental and chemical engineers rely heavily on porous membranes for various processes, including gas/solid, gas/solid, liquid/liquid, solid/liquid, and liquid/liquid/gas separation, purification, and concentration. The sizes of the particles and the techniques employed for separation determine the membrane processes (Zafarnak et al., 2023; Roostaie et al., 2020). Membrane filtering is a more inexpensive alternative to distillation for separating isotropic liquids. In contrast to more conventional filtering technologies, membrane filtration is both safe and efficient, and it has many advantages, such as:

   i. Reduced sludge production
  ii. Superior effluent quality
 iii. Reduced energy need

The most common filtration techniques in membrane operations are dead-end filtering, tangential flow filtration, and cross-flow filtration. These two techniques are shown in Figure 1.3. Referring to Figure 1.3a, cross-flow filtration involves a

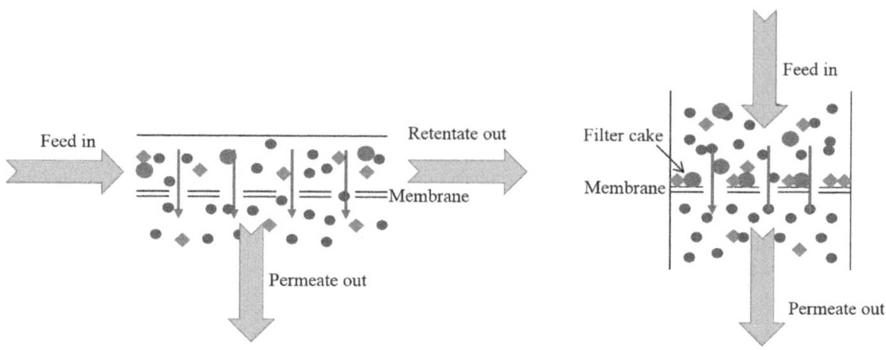

FIGURE 1.3 Filtration schematics: (a) Cross-flow and (b) dead-end membrane. (Adapted from Meshksar et al., 2020b.)

tangential feed stream flow toward the membrane surface. Hence, only a portion of the gas stream and its impurities are collected on the other side of the membrane. The purification process is improved by recirculating the already passed stream. All of the fluid is pushed beyond the membrane surface in dead-end filtration. After passing through the membrane, any impurities with sizes more significant than their pores are gathered and stored behind the membrane module (Meshksar et al., 2020b).

Some advantages and disadvantages are associated with each of these flow arrangements. Dead-end separation membranes are easier and cheaper to build than cross-flow filtering. Separation in laboratories is often accomplished using the dead-end membrane technique, a batch procedure that involves slowly adding the feed solution to the membrane medium. The main problems with dead-end membrane filtration are membrane fouling and concentration polarization. Because of their sweeping effects and high passing flow shear rates, cross-flow filtration systems are less likely to get fouled, but they are more expensive and labor-intensive (Meshksar et al., 2020b).

The effectiveness of the membrane in separating and purifying H_2 is fundamentally dependent on its performance. The two most prevalent types of membranes today are metal and polymer. Furthermore, newer types of membranes, such as CMSM, MOF, and nanomaterial membranes, may show better separation performance. Therefore, the following study tests and evaluates the effectiveness of several membrane materials in H_2 purification.

1.2.2.1 Metal Membranes

Catalysis on the thick metal membranes converts hydrogen gas into protons and electrons. Protons can cross the metal barrier and rejoin with electrons on the other side to form H_2. Only hydrogen gas can travel selectively via the metal membrane, which blocks the passage of O_2, N_2, CO_2, and other gases. The remarkable H_2 passivity, high hydrogenolysis, and self-catalytic hydrogenolysis process resistance of palladium (Pd) membranes make them a popular metal membrane material (Rahimpour et al., 2017). However, Pd membranes are subject to hydrogen embrittlement at low temperatures and have high manufacturing costs.

Other metal components, such as Cu, Ni, Ag, Y, Au, may be added to the Pd membrane to create a Pd alloy membrane. At the same time as it increases the rate of hydrogen permeability, it widens the Pd lattice, which solves the problem of hydrogen embrittlement. Nayebossadri et al. (2019) examined how well Pd, PdCu, and PdAg membranes separated hydrogen gas from NG. At different concentrations, the researchers measured how well these membranes worked. After comparing the three membranes, it was found that the PdAg membrane had the best H_2 permeability. A bilayer bcc-PdCu alloy membrane was created by Zhao et al. (2020) by electroplating Pd and Cu onto a ceramic support membrane in alternating patterns. The membrane was permeable to hydrogen gas and resistant to very low temperatures. As a result, it shows promise as a material for room-temperature hydrogen gas separation.

It is possible for Pd alloy membranes and pure Pd membranes to self-sustain. Their thickness is limited to tens to hundreds of micrometers so that they can withstand mechanical stresses adequately. Lower H_2 permeability rate and higher expenses result from too thick membranes. So, it's possible to make a supported Pd composite membrane by applying a Pd or Pd alloy membrane to a porous material's surface. With support, the Pd membrane may be made more mechanically strong with less Pd and thinner overall. This improves the total cost and H_2 permeation rate across the membrane, as mentioned before. A nanoscale Pd membrane was coated onto poly-benzimidazole-4,40-(hexafluoroisopropylidene)-bis(benzoic acid) (PBI-HFA) by Kong et al. (2017) using the vacuum electroless plating (VELP) procedure. The Pd/PBI-HFA composite membrane exhibited good selectivity for H_2/N_2 and H_2/CO_2 and successfully prevented the entry of CO. A Pd membrane and a NaY molecular sieve were used by Kiadehi and Taghizadeh (2019) on porous stainless steel substrates. An H_2/N_2 selectivity of 736 was observed at 450°C when the H_2 and N_2 mixture was infiltrated into the Pd/NaY/PSS composite membrane.

Furthermore, Iulianelli et al. (2019) used the metal vapor synthesis approach to produce a thin membrane of supported Pd_{70}-Cu_{30}/-Al_2O_3. The membrane displayed selectivity values of 1,800 for H_2/N_2 and 6,500 for H_2/CO_2 at 400°C and 50 kPa. An Al_2O_3 membrane containing Pd was created by Huang et al. (2020) by applying a surface coating of the naturally occurring mineral Nontronite-15A on porous Al_2O_3. Compared to earlier composite membranes, this method maintains a high H_2 permeability while reducing production costs.

When compared to other metals, Pd has a lower hydrogen permeability. New evidence points to unique body-centered cubic (bcc) lattice structures in vanadium group metals, such as V, Nb, and Ta. Compared to Pd, these metals have better mechanical strength and H_2 permeability, but aren't good at dissociating or adsorbing H_2 (Liguori et al., 2020). However, developing a surface layer of condensed oxide prevents the entry of H_2. The thin membrane's low hydrogen permeability results from this, even though the vanadium group metals in its lattice structure have a high hydrogen permeability.

Furthermore, hydrogen embrittlement is more common in these metals compared to Pd. To solve this problem, symmetrical composite membranes covered the vanadium group metals with a delicate Pd layer. The combined capabilities of the Pd membrane's H_2 adsorption and dissociation capability and the vanadium group metals' H_2 penetration capabilities decreased total cost.

1.2.2.2 Polymer Membranes

Gas permeability rate variations are crucial to the working mechanism of polymer membrane separation. Polyamide, polysulfone (PSF), and polyimide (PI) are three of the most popular polymer membrane materials utilized today (Yáñez et al., 2021).

Superior mechanical strength, thermal stability, permeability, and selectivity are the hallmarks of an ideal polymer membrane material. Highly porous polymer membranes generally have low selectivity, but the inverse is also true. Researchers have developed mixed matrix membranes (MMMs) to improve the performance of polymer membranes, which are limited by the trade-off between selectivity and permeability. These membranes incorporate inorganic materials such as zeolite, silicon dioxide, and carbon molecular sieves (CMS) (Strugova et al., 2018). Rezakazemi et al. (2012) constructed PDMS/4A MMMs by incorporating 4A zeolite nanoparticles onto a polydimethylsiloxane (PDMS) substrate. The experimentally produced MMMs outperformed the pure PDMS membrane regarding H_2/CH_4 selectivity and H_2 permeability. The Matrimid® 5218-DDR MMM was created by Peydayesh et al. (2017) by incorporating Deca-dodecanol 3R (DDR) zeolite into a Matrimid® 5218 PI substrate. Because of this, the H_2/CH_4 selectivity increased by 189%, while the H_2 permeability increased by 100%.

The effectiveness of polymer membranes may be further improved by introducing polymer mixing. In comparison to a PSF or PI membrane alone, the PSF/PI membrane developed by Hamid et al. (2019) improved the selectivity for H_2 over CO_2 (4.4). It increased the capacity to let H_2 gas flow through. The membrane was able to purify 80% of the H_2 gas. On the other hand, the PSF/PI membrane proved to have more consistent chemical and physical properties, resulting in a novel polymer membrane that performed very well.

A hollow fiber membrane's mechanical performance and specific surface area are superior compared to constructing a standard plate membrane. These findings also guide the construction of gas separation membranes. A bilayer hollow fiber membrane was created by Naderi et al. (2019). The outer selective layer comprised a mixture of polybenzimidazole (PBI) and sulfonated polyphenylene sulfone (sPPSU). The inner support layer, on the other hand, was PSF. At 90°C and 14 bar, the membrane showed a hydrogen permeability of 16.7 GPU and a hydrogen to carbon dioxide (H_2/CO_2) selectivity of 9.7, according to the results of the tests. Consequently, the membrane demonstrated desirable properties for separating H_2 and CO_2 under hot conditions.

Membranes made from the aforementioned polymers are highly selective toward hydrogen gas. Also, researchers have developed separation membranes with a high selectivity for CO_2, allowing them to remove CO_2 from H_2 effectively (Chen et al., 2020). To achieve negative selectivity, polymer membranes need a particular affinity for CO_2, which is necessary since CO_2 has a larger molecular diameter than H_2. By combining a poly(4-methyl-1-pentene) (PMP) substrate with a MOF composed of MIL 53(Al), Abedini et al. (2014) were able to create an MMM. The MMMs outperformed the pure PMP membranes regarding thermal stability and negative selectivity for CO_2/H_2 in experimental trials.

At the same time, MMMs achieved negative selectivity that exceeded the Robeson upper limit when the input pressure grew. MMMs composed of polyvinylamine and a covalent organic framework (COF) were developed by Cao et al. (2016). These MMMs showed a 15 CO_2/H_2 selectivity and 396 GPU CO_2 penetration rate. Salim et al. (2018) also created novel oxidation-resistant membranes. The crosslinked polyvinyl alcohol-polysiloxane substrate contains quaternary ammonium hydroxide, fluoride, and tetrafluoroborate in these membranes. The CO_2/H_2 selectivity of the membranes is more than 100, and their CO_2 permeation rate is 100 GPU. These membranes are expected to purify hydrogen gas for use in fuel cell automobiles. Nanocomposite membranes improved CO_2 penetration rate and CO_2/H_2 selectivity when graphene oxide (GO) was added to polydisperse monosulfide (PDMS) (Nigiz and Hilmioglu, 2020). The selectivity of CO_2 over H_2 rose from 7.1 to 11.7 at a concentration of 0.5% GO and 0.2 MPa transmembrane pressure, while the permeability of CO_2 reached 3,670 Barrer.

Furthermore, ZIF-8-TA nanoparticles were created by Chen et al. (2020) by a hydrophilic modification of ZIF-8 with tannic acid (TA). The nanoparticles were combined with hydrophilic polyvinyl amine to make a MMM. The CO_2 permeability was found to be 987 GPU and the CO_2/H_2 selectivity to be 31 when tested at a feed pressure of 0.12 MPa. These results show that the CO_2 from H_2 separation process was successful.

1.2.2.3 Carbon-Based Membranes

Amorphous microporous carbon molecular sieve (CMS) membranes are typically made by carbonizing or pyrolyzing polymer precursors in a vacuum or inert gas. Polymer precursors include phenolic resins, polyfurfuryl alcohols, polyimide, and its derivatives (Li et al., 2015). A thorough research was carried out to enhance the selectivity and permeability of CMS membranes. Llosa Tanco et al. (2021) created composite alumina-CMS membranes (Al-CMSMs) using tubular porous alumina as a support structure. When tested at 30°C, these membranes outperformed Robeson's upper limit for polymer membranes regarding gas separation efficiency for H_2 and CH_4. Hydrogen gas, combined with methane gas, has a purity level of 99.4%. Xu et al. (2020) created highly selective CMSMs by heating polyetherketone-cardo polymers to a point where they thermally disintegrated. A permeability of 5,260 Barrer was recorded for the carbonized CMS membranes, which were heated to 700°C. There were 311, 142, and 75 selectivity values for H_2/CH_4, H_2/N_2, and H_2/CO, respectively. When carbonization was carried out at 900°C, the H_2/CH_4 selectivity reached 1,859.

There has been a lot of talk about graphene-based membranes for gas separation since they are new carbon-based membranes. The atomically thin materials, graphene and GO, are solid, mechanically and chemically stable. The optimum membrane material is graphene because of its thin thickness, allowing for maximal permeation flow and minimum transmission resistance. This is because the relationship between the thickness and permeability of a membrane is inversely proportional. However, only some graphene-based materials are suitable for gas separation since they need more holes. Achieving high porosity in graphene sheets and uniformly scattered nanopores of acceptable size and shapes are the main goals of the work, which aims to increase graphene-based membranes efficiency in gas separation (Yang et al., 2021). Two PG monolayers, one carrying -graphyne H_2 (-GYH) and the other -graphyne N_2 (-GYN), were simulated by Sang et al. (2017) to separate H_2

from a gas mixture that also included N_2, H_2O, CH_4, CO_2, and CO. The GYN membranes were the best choice for removing hydrogen gas from a combination because of their high selectivity and hydrogen permeability. The effective H_2 separation from CH_4 and CO_2 was established by Hanada et al. (2017a) using two-dimensional nanomaterials that resembled graphene (g-C_3N_4). According to theoretical studies, membranes subjected to 2.5% and 5% biaxial strains might increase their pore areas, hence increasing H_2 permeability. The H_2/CO_2 and H_2/CH_4 selectivity would be unaffected by this improvement. Also, Wei et al. (2018) utilized density functional theory (DFT) to analyze the effectiveness of 3N-PG and 6N-PG monolayers, which are composed of N_2 and PG membranes, in removing hydrogen from a gas mixture that also included CO, N_2, and CH_4.

In addition, it was shown that 3N-PG and 6N-PG monolayers have higher H_2 permeability than PG membranes. This finding lends credence to the idea that these monolayers might be a game-changing material for hydrogen purification. In their investigation on nano-PG (NPG) membranes, Sun et al. (2019) found that the H_2 permeability reached 106 GPU, which is much higher than polymer membranes. At the same time, the H_2/CH_4 selectivity reached 225, which is similar to what is seen in polymer membranes. In addition, when tested under identical separation conditions and purification standards, NPG membranes are more cost-effective than polymer membranes. Modified alumina tubes were used to create GO nanocomposite membranes by Zeynali et al. (2019). These membranes were stable, with significant H_2/CO_2 and H_2/N_2 selectivity and permeability to H_2.

On top of that, they were less expensive than Pd membranes. To study the path of gas molecules through nanographene C_{216}, Liu et al. (2018) ran a simulation. They found that hydrogen molecules can diffuse across C_{216} membranes with a 0.65 eV barrier. Compared to PG and polymer membranes, H_2's selectivity for CO, O_2, NO, N_2, CO_2, and H_2O peaked at 1,033.

1.2.2.4 MOF Membranes

Additionally, it was demonstrated that monolayers composed of 3N-PG and 6N-PG showed enhanced permeability to H_2 in comparison to PG membranes. This finding underscores the potential of these monolayers as an innovative and novel substance for the purification of H_2. Concurrently, the H_2/CH_4 selectivity attained a magnitude of 225, analogous to the selectivity observed in polymer membranes. In addition, when compared to polymer membranes under identical separation conditions and purification criteria, NPG membranes demonstrate outstanding cost-effectiveness.

In addition, their cost-effectiveness was superior to Pd membranes. A simulation was performed by Liu et al. (2018) to examine the path followed by gas molecules as they traversed nanographene C_{216}. The results of their research indicated that H_2 molecules can penetrate C_{216} membranes at a diffusion barrier of 0.65 eV. The maximal selectivity of H_2 toward NO, O_2, CO, N_2, CO_2, and H_2O was 1,033, which was greater than the selectivity of both PG and polymer membranes.

1.2.2.5 Ionic Liquid Membranes

ILs are highly suitable for developing into membrane separation media due to their favorable characteristics, which encompass nonvolatility, thermal stability, nonflammability, and liquid state at 25°C. In mass transmission between the gas and liquid phases, gas solubility in the liquid phase is a crucial parameter. Determining the gas solubility in ILs confers a substantial benefit due to the scarcity of gas solubility data regarding ILs available in the literature. The solubility of a gas in a particular IL is deemed inadequate, thus the IL is not suitable for gas absorption. In recent years, an increasing quantity of measurements documenting the equilibrium solubility of CO_2 in diverse ILs has become accessible to many researchers (Meshksar et al., 2020c).

1.2.3 METAL HYDRIDE SEPARATION TECHNIQUE

The metal hydride separation method entails hydrogen purification by utilizing H_2 storage alloys, which reversibly absorb and discharge H_2. H_2 storage alloys facilitate the decomposition of H_2 molecules into H atoms. This process is accomplished by implementing reduced temperature and increased pressure conditions. Following this, metal hydrides are generated through various mechanisms such as diffusion, phase transition, and combination reactions, whereas impurity vapors become entrapped within the metal particles. As a result of a decrease in pressure and an increase in temperature, impurity gases are evacuated from the metal particles, leading to the liberation of H_2 from the crystal lattice. Depending on the predominant element present, H_2 storage alloys can be classified as rare earth, titanium, zirconium, or magnesium alloys. Based on the atomic ratio of the principal components, these alloys may also be classified as AB_5-type alloys, AB_2-type alloys, AB-type alloys, or A_2B-type alloys (Xiao et al., 2017). The performance of the H_2 storage alloys directly influences the effectiveness of H_2 purification. As a result, further enhancing these alloys' chemical stability and tolerance could result in improved operational capabilities. Dunikov et al. (2016) separated the H_2/CO_2 mixture utilizing two AB_5-type alloys. The effectiveness of the low-pressure $LaNi_{4.8}Mn_{0.3}Fe_{0.1}$ alloy in purifying H_2 from a mixture comprising 59% H_2 was discovered. It was determined that 94% of H_2 was recovered through purification, indicating that it is suitable for operating PEMFCs. Yang et al. (2017) investigated the $LaNi_{4.3}Al_{0.7}$ H_2 storage alloy under cyclic conditions in an atmosphere with a high concentration of CO. At temperatures exceeding 363 K, the alloy's hydrogen storage capacity progressively decreases. However, it maintains a comparatively swift rate of motion, rendering it viable for a wide range of applications involving hydrogen purification and separation.

Moreover, Hanada et al. (2017b) examined the effect of CO_2 on the H_2 absorption capacity of AB_2-type alloys. Producing metal hydrides appropriate for H_2 storage and purification was the objective. The results indicated that the incorporation of Fe and Co into the alloys resulted in an improvement in their resistance to CO_2. However, the presence of Ni caused an unfavorable effect. Zhou et al. (2019) identified nano-VTiCr and MgH_2 as catalysts facilitating the reaction with low-pressure H_2. Moreover, this catalyst is capable of being repurposed within a gas mixture. Therefore, the substance demonstrated the ability to separate and purify H_2.

1.2.4 CRYOGENIC DISTILLATION

Cryogenic distillation is a method for purifying and separating H_2 by capitalizing on the fluctuating volatility of various input gas components. In contrast to CH_4 and other light hydrocarbons (HCs), the volatility of H_2 is comparatively elevated. Consequently,

HCs, CO, N_2, and others condense at temperatures lower than H_2. The aforementioned method is frequently utilized to separate H_2 and HC. While the low-temperature separation method ensures a significant rate of H_2 recovery, its adaptability to different input gases poses challenges. Therefore, CO_2, H_2O, and other contaminants must be extracted from the supply gases before separating to prevent equipment blockage at low temperatures. The practical operation of gas compressors and refrigeration equipment is additionally accompanied by significant energy consumption and expenditures. While most impurities dissolve at low temperatures, certain substances endure in the gaseous state as saturated vapor. Consequently, the direct extraction of highly purified H_2 becomes a challenging endeavor (Song et al., 2019).

1.3 CONCLUSION

The industrial utilization of hydrogen necessitates adherence to existing standards regarding impurity levels. As hydrogen produced by various processes (e.g., reforming and gasification) contains many contaminants, this chapter focuses on extracting these specific impurities. PSA is a universal method applicable to eliminating the majority of pollutants. Gas pollutants like CO, N_2, CH_4, and H_2O are commonly eliminated via H_2-permeable membranes. In contrast, CO_2-permeable membranes are designed exclusively for CO_2 removal. Due to the sensitivity of metal hydrides to H_2O, CO, and CO_2, they can be utilized to extract Ar, N_2, etc. Nevertheless, the current array of H_2 purification methods is limited in scope, rendering it unfeasible to meet the hydrogen impurity level regulations for particular applications, such as fuel cell vehicles, with a single separation and purification technique. Given the many distinct H_2 sources, employing two or more H_2 purification systems is advisable.

REFERENCES

Aasadnia, M., Mehrpooya, M. & Ghorbani, B. 2021. A novel integrated structure for hydrogen purification using the cryogenic method. *Journal of Cleaner Production*, 278, 123872.

Abdeljaoued, A., Relvas, F., Mendes, A. & Chahbani, M. H. 2018. Simulation and experimental results of a PSA process for production of hydrogen used in fuel cells. *Journal of Environmental Chemical Engineering*, 6, 338–355.

Abedini, R., Omidkhah, M. & Dorosti, F. 2014. Hydrogen separation and purification with poly (4-methyl-1-pentyne)/Mil 53 mixed matrix membrane based on reverse selectivity. *International Journal of Hydrogen Energy*, 39, 7897–7909.

Agarwal, A., Biegler, L. T. & Zitney, S. E. 2010. Superstructure-based optimal synthesis of pressure swing adsorption cycles for precombustion CO2 capture. *Industrial & Engineering Chemistry Research*, 49, 5066–5079.

Agueda, V. I., Delgado, J. A., Uguina, M. A., Brea, P., Spjelkavik, A. I., Blom, R. & Grande, C. 2015. Adsorption and diffusion of H2, N2, CO, CH4 and CO2 in Utsa-16 metal-organic framework extrudates. *Chemical Engineering Science*, 124, 159–169.

Ahn, S., You, Y.-W., Lee, D.-G., Kim, K.-H., Oh, M. & Lee, C.-H. 2012. Layered two-and four-bed PSA processes for H2 recovery from coal gas. *Chemical Engineering Science*, 68, 413–423.

Banu, A.-M., Friedrich, D., Brandani, S. & DÜ Ren, T. 2013. A multiscale study of MOFs as adsorbents in H2 PSA purification. *Industrial & Engineering Chemistry Research*, 52, 9946–9957.

Barabadi, A., Makarem, M. A. & Meshksar, M. 2023. CO2 capture with ionic liquid membrane. In: Makarem, M. A. (ed.) *Reference Module in Earth Systems and Environmental Sciences*. Netherlands: Elsevier.

Bernardo, G., Araújo, T., Da Silva Lopes, T., Sousa, J. & Mendes, A. 2020. Recent advances in membrane technologies for hydrogen purification. *International Journal of Hydrogen Energy*, 45, 7313–7338.

Birol, F. 2007. World energy prospects and challenges. *Asia-Pacific Review*, 14, 1–12.

Brea, P., Delgado, J., Águeda, V. I., Gutiérrez, P. & Uguina, M. A. 2019. Multicomponent adsorption of H2, CH4, CO and CO2 in zeolites NaX, CaX and MgX. Evaluation of performance in PSA cycles for hydrogen purification. *Microporous and Mesoporous Materials*, 286, 187–198.

Cao, L., Iris, K., Xiong, X., Tsang, D. C., Zhang, S., Clark, J. H., Hu, C., Ng, Y. H., Shang, J. & Ok, Y. S. 2020. Biorenewable hydrogen production through biomass gasification: A review and future prospects. *Environmental Research*, 186, 109547.

Cao, X., Qiao, Z., Wang, Z., Zhao, S., Li, P., Wang, J. & Wang, S. 2016. Enhanced performance of mixed matrix membrane by incorporating a highly compatible covalent organic framework into poly(vinylamine) for hydrogen purification. *International Journal of Hydrogen Energy*, 41, 9167–9174.

Chen, F., Dong, S., Wang, Z., Xu, J., Xu, R. & Wang, J. 2020. Preparation of mixed matrix composite membrane for hydrogen purification by incorporating Zif-8 nanoparticles modified with tannic acid. *International Journal of Hydrogen Energy*, 45, 7444–7454.

Du, Z., Liu, C., Zhai, J., Guo, X., Xiong, Y., Su, W. & He, G. 2021. A review of hydrogen purification technologies for fuel cell vehicles. *Catalysts*, 11, 393.

Dunikov, D., Borzenko, V., Blinov, D., Kazakov, A., Lin, C. Y., Wu, S. Y. & Chu, C. Y. 2016. Biohydrogen purification using metal hydride technologies. *International Journal of Hydrogen Energy*, 41, 21787–21794.

Golmakani, A., Fatemi, S. & Tamnanloo, J. 2017. Investigating PSA, VSA, and TSA methods in SMR unit of refineries for hydrogen production with fuel cell specification. *Separation and Purification Technology*, 176, 73–91.

Golmakani, A., Nabavi, S. A. & Manović, V. 2020. Effect of impurities on ultra-pure hydrogen production by pressure vacuum swing adsorption. *Journal of Industrial and Engineering Chemistry*, 82, 278–289.

Hamid, M. A. A., Chung, Y. T., Rohani, R. & Junaidi, M. U. M. 2019. Miscible-blend polysulfone/polyimide membrane for hydrogen purification from palm oil mill effluent fermentation. *Separation and Purification Technology*, 209, 598–607.

Hamm, J. B., Ambrosi, A., Griebeler, J. G., Marcilio, N. R., Tessaro, I. C. & Pollo, L. D. 2017. Recent advances in the development of supported carbon membranes for gas separation. *International Journal of Hydrogen Energy*, 42, 24830–24845.

Hanada, N., Asada, H., Nakagawa, T., Higa, H., Ishida, M., Heshiki, D., Toki, T., Saita, I., Asano, K. & Nakamura, Y. 2017a. Effect of CO2 on hydrogen absorption in Ti-Zr-Mn-Cr based AB2 type alloys. *Journal of Alloys and Compounds*, 705, 507–516.

Hanada, N., Asada, H., Nakagawa, T., Higa, H., Ishida, M., Heshiki, D., Toki, T., Saita, I., Asano, K., Nakamura, Y., Fujisawa, A. & Miura, S. 2017b. Effect of CO2 on hydrogen absorption in Ti-Zr-Mn-Cr based AB2 type alloys. *Journal of Alloys and Compounds*, 705, 507–516.

He, B., Liu, J., Zhang, Y., Zhang, S., Wang, P. & Xu, H. 2020. Comparison of structured activated carbon and traditional adsorbents for purification of H2. *Separation and Purification Technology*, 239, 116529.

Huang, Y., Liu, Q., Jin, X., Ding, W., Hu, X. & Li, H. 2020. Coating the porous Al2O3 substrate with a natural mineral of Nontronite-15A for fabrication of hydrogen-permeable palladium membranes. *International Journal of Hydrogen Energy*, 45, 7412–7422.

Iulianelli, A., Ghasemzadeh, K., Marelli, M. & Evangelisti, C. 2019. A supported Pd-Cu/Al2O3 membrane from solvated metal atoms for hydrogen separation/purification. *Fuel Processing Technology*, 195, 106141.

Khosravani, H., Meshksar, M., Koohi- Saadi, M., Taghadom, K. & Rahimpour, M. R. 2023. Synthesis, characterization, and application of bio-templated Ni-Ce/Al2O3 catalyst for clean H2 production in the steam reforming of methane process. *Journal of the Energy Institute*, 108, 101203.

Kiadehi, A. D. & Taghizadeh, M. 2019. Fabrication, characterization, and application of palladium composite membrane on porous stainless steel substrate with NaY zeolite as an intermediate layer for hydrogen purification. *International Journal of Hydrogen Energy*, 44, 2889–2904.

Kiani, P., Meshksar, M., Rahimpour, M. R. & Iulianelli, A. 2022. CO2 utilization in methane reforming using La-doped SBA-16 catalysts prepared via pH adjustment method. *Fuel*, 322, 124248.

Kiani, P., Rahimpour, H. R. & Rahimpour, M. R. 2023. Chapter 4- Steam reforming process for syngas production. In: Rahimpour, M. R., Makarem, M. A. & Meshksar, M. (eds.) *Advances in Synthesis Gas: Methods, Technologies and Applications*. Netherlands: Elsevier.

Kong, S. Y., Kim, D. H., Henkensmeier, D., Kim, H.-J., Ham, H. C., Han, J., Yoon, S. P., Yoon, C. W. & Choi, S. H. 2017. Ultrathin layered Pd/PBI-HFA composite membranes for hydrogen separation. *Separation and Purification Technology*, 179, 486–493.

Kuroda, S., Nagaishi, T., Kameyama, M., Koido, K., Seo, Y. & Dowaki, K. 2018. Hydroxyl aluminium silicate clay for biohydrogen purification by pressure swing adsorption: Physical properties, adsorption isotherm, multicomponent breakthrough curve modelling, and cycle simulation. *International Journal of Hydrogen Energy*, 43, 16573–16588.

Li, H., Liao, Z., Sun, J., Jiang, B., Wang, J. & Yang, Y. 2019. Modelling and simulation of two-bed PSA process for separating H2 from methane steam reforming. *Chinese Journal of Chemical Engineering*, 27, 1870–1878.

Li, P., Wang, Z., Qiao, Z., Liu, Y., Cao, X., Li, W., Wang, J. & Wang, S. 2015. Recent developments in membranes for efficient hydrogen purification. *Journal of Membrane Science*, 495, 130–168.

Liguori, S., Kian, K., Buggy, N., Anzelmo, B. H. & Wilcox, J. 2020. Opportunities and challenges of low-carbon hydrogen via metallic membranes. *Progress in Energy and Combustion Science*, 80, 100851.

Liu, Y., Liu, W., Hou, J., Dai, Y. & Yang, J. 2018. Coronoid nanographene C216 as hydrogen purification membrane: A density functional theory study. *Carbon*, 135, 112–117.

Lively, R. P., Bessho, N., Bhandari, D. A., Kawajiri, Y. & Koros, W. J. 2012. Thermally moderated hollow fiber sorbent modules in rapidly cycled pressure swing adsorption mode for hydrogen purification. *International Journal of Hydrogen Energy*, 37, 15227–15240.

Llosa Tanco, M. A., Medrano, J. A., Cechetto, V., Gallucci, F. & Pacheco Tanaka, D. A. 2021. Hydrogen permeation studies of composite supported alumina-carbon molecular sieves membranes: Separation of diluted hydrogen from mixtures with methane. *International Journal of Hydrogen Energy*, 46, 19758–19767.

Lopes, F. V., Grande, C. A. & Rodrigues, A. E. 2012. Fast-cycling VPSA for hydrogen purification. *Fuel*, 93, 510–523.

Majlan, E. H., Daud, W. R. W., Iyuke, S. E., Mohamad, A. B., Kadhum, A. A. H., Mohammad, A. W., Takriff, M. S. & Bahaman, N. 2009. Hydrogen purification using compact pressure swing adsorption system for fuel cell. *International Journal of Hydrogen Energy*, 34, 2771–2777.

Meshksar, M., Afshariani, F. & Rahimpour, M. R. 2020a. Chapter 10- Solar reformers coupled with PEMFCs for residential cogeneration and trigeneration applications. In: Basile, A. & Spazzafumo, G. (eds.) *Current Trends and Future Developments on (Bio-) Membranes*. Netherlands: Elsevier.

Meshksar, M., Farsi, M. & Rahimpour, M. R. 2022a. Effect of Ni active site position and synthesis route on activity, stability, and morphology of Ce promoted Ni/Al2O3 catalyst for clean H2 production. *Journal of Environmental Chemical Engineering*, 10, 108471.

Meshksar, M., Farsi, M. & Rahimpour, M. R. 2022b. Hollow sphere Ni-based catalysts promoted with cerium for steam reforming of methane. *Journal of the Energy Institute*, 105, 342–354.

Meshksar, M. & Rahimpour, M. R. 2023. Chapter 9- Ionic liquid membranes for syngas purification. In: Rahimpour, M. R., Makarem, M. A. & Meshksar, M. (eds.) *Advances in Synthesis Gas: Methods, Technologies and Applications*. Netherlands: Elsevier.

Meshksar, M., Roostaee, T. & Rahimpour, M. R. 2020b. Chapter 10- Membrane technology for brewery wastewater treatment. In: Basile, A. & Comite, A. (eds.) *Current Trends and Future Developments on (Bio-) Membranes*. Netherlands: Elsevier.

Meshksar, M., Sedghamiz, M. A., Zafarnak, S. & Rahimpour, M. R. 2020c. Chapter 13- CO2 separation with ionic liquid membranes. In: Rahimpour, M. R., Farsi, M. & Makarem, M. A. (eds.) *Advances in Carbon Capture*. Netherlands: Woodhead Publishing.

Moon, D.-K., Park, Y., Oh, H.-T., Kim, S.-H., Oh, M. & Lee, C.-H. 2018. Performance analysis of an eight-layered bed PSA process for H2 recovery from IGCC with pre-combustion carbon capture. *Energy Conversion and Management*, 156, 202–214.

Mosallanezhad, S., Haghighatjoo, F. & Rahimpour, H. R. 2023. Fossil fuels storage technologies and challenges. In: Rahimpour, M. R. (ed.) *Reference Module in Earth Systems and Environmental Sciences*. Netherlands: Elsevier

Naderi, A., Chung, T.-S., Weber, M. & Maletzko, C. 2019. High performance dual-layer hollow fiber membrane of sulfonated polyphenylsulfone/Polybenzimidazole for hydrogen purification. *Journal of Membrane Science*, 591, 117292.

Nayebossadri, S., Speight, J. D. & Book, D. 2019. Hydrogen separation from blended natural gas and hydrogen by Pd-based membranes. *International Journal of Hydrogen Energy*, 44, 29092–29099.

Nigiz, F. U. & Hilmioglu, N. D. 2020. Enhanced hydrogen purification by graphene-poly(dimethyl siloxane) membrane. *International Journal of Hydrogen Energy*, 45, 3549–3557.

Peydayesh, M., Mohammadi, T. & Bakhtiari, O. 2017. Effective hydrogen purification from methane via polyimide Matrimid® 5218- Deca-dodecasil 3R type zeolite mixed matrix membrane. *Energy*, 141, 2100–2107.

Rahimpour, M., Samimi, F., Babapoor, A., Tohidian, T. & Mohebi, S. 2017. Palladium membranes applications in reaction systems for hydrogen separation and purification: A review. *Chemical Engineering and Processing: Process Intensification*, 121, 24–49.

Relvas, F., Whitley, R. D., Silva, C. & Mendes, A. 2018. Single-stage pressure swing adsorption for producing fuel cell grade hydrogen. *Industrial & Engineering Chemistry Research*, 57, 5106–5118.

Rezakazemi, M., Shahidi, K. & Mohammadi, T. 2012. Hydrogen separation and purification using crosslinkable Pdms/zeolite A nanoparticles mixed matrix membranes. *International Journal of Hydrogen Energy*, 37, 14576–14589.

Roostaie, T., Meshksar, M. & Rahimpour, M. R. 2020. Chapter 15- Biofuel reforming in membrane reactors. In: Iulianelli, A. & Basile, A. (eds.) *Current Trends and Future Developments on (Bio-) Membranes*. Netherlands: Elsevier.

Salahi, F., Zarei- Jelyani, F., Meshksar, M., Farsi, M. & Rahimpour, M. R. 2023. Application of hollow promoted Ni-based catalysts in steam methane reforming. *Fuel*, 334, 126601.

Salim, W., Vakharia, V., Chen, K. K., Gasda, M. & Ho, W. S. W. 2018. Oxidatively stable borate-containing membranes for H2 purification for fuel cells. *Journal of Membrane Science*, 562, 9–17.

Sang, P., Zhao, L., Xu, J., Shi, Z., Guo, S., Yu, Y., Zhu, H., Yan, Z. & Guo, W. 2017. Excellent membranes for hydrogen purification: Dumbbell-shaped porous γ-graphynes. *International Journal of Hydrogen Energy*, 42, 5168–5176.

Schorer, L., Schmitz, S. & Weber, A. 2019. Membrane based purification of hydrogen system (MEMPHYS). *International Journal of Hydrogen Energy*, 44, 12708–12714.

Shamsudin, I., Abdullah, A., Idris, I., Gobi, S. & Othman, M. 2019. Hydrogen purification from binary syngas by PSA with pressure equalization using microporous palm kernel shell activated carbon. *Fuel*, 253, 722–730.

Song, C., Liu, Q., Deng, S., Li, H. & Kitamura, Y. 2019. Cryogenic-based CO2 capture technologies: State-of-the-art developments and current challenges. *Renewable and Sustainable Energy Reviews*, 101, 265–278.

Strugova, D. V., Zadorozhnyy, M. Y., Berdonosova, E. A., Yablokova, M. Y., Konik, P. A., Zheleznyi, M. V., Semenov, D. V., Milovzorov, G. S., Padaki, M., Kaloshkin, S. D., Zadorozhnyy, V. Y. & Klyamkin, S. N. 2018. Novel process for preparation of metal-polymer composite membranes for hydrogen separation. *International Journal of Hydrogen Energy*, 43, 12146–12152.

Sun, J., Feng, H., Xu, J., Jin, H. & Guo, L. 2021. Investigation of the conversion mechanism for hydrogen production by coal gasification in supercritical water. *International Journal of Hydrogen Energy*, 46, 10205–10215.

Sun, C., Zheng, X. & Bai, B. 2019. Hydrogen purification using nanoporous graphene membranes and its economic analysis. *Chemical Engineering Science*, 208, 115141.

Tasleem, S. & Tahir, M. 2020. Current trends in strategies to improve photocatalytic performance of perovskites materials for solar to hydrogen production. *Renewable and Sustainable Energy Reviews*, 132, 110073.

Wei, S., Zhou, S., Wu, Z., Wang, M., Wang, Z., Guo, W. & Lu, X. 2018. Mechanistic insights into porous graphene membranes for helium separation and hydrogen purification. *Applied Surface Science*, 441, 631–638.

Xiao, J., Tong, L., Yang, T., Bénard, P. & Chahine, R. 2017. Lumped parameter simulation of hydrogen storage and purification systems using metal hydrides. *International Journal of Hydrogen Energy*, 42, 3698–3707.

Xu, R., He, L., Li, L., Hou, M., Wang, Y., Zhang, B., Liang, C. & Wang, T. 2020. Ultraselective carbon molecular sieve membrane for hydrogen purification. *Journal of Energy Chemistry*, 50, 16–24.

Yáñez, M., Ortiz, A., Gorri, D. & Ortiz, I. 2021. Comparative performance of commercial polymeric membranes in the recovery of industrial hydrogen waste gas streams. *International Journal of Hydrogen Energy*, 46, 17507–17521.

Yáñez, M., Relvas, F., Ortiz, A., Gorri, D., Mendes, A. & Ortiz, I. 2020. PSA purification of waste hydrogen from ammonia plants to fuel cell grade. *Separation and Purification Technology*, 240, 116334.

Yang, E., Alayande, A. B., Goh, K., Kim, C.-M., Chu, K.-H., Hwang, M.-H., Ahn, J.-H. & Chae, K.-J. 2021. 2D materials-based membranes for hydrogen purification: Current status and future prospects. *International Journal of Hydrogen Energy*, 46, 11389–11410.

Yang, F. S., Chen, X. Y., Wu, Z., Wang, S. M., Wang, G. X., Zhang, Z. X. & Wang, Y. Q. 2017. Experimental studies on the poisoning properties of a low-plateau hydrogen storage alloy LaNi4.3Al0.7 against CO impurities. *International Journal of Hydrogen Energy*, 42, 16225–16234.

You, Y.-W., Lee, D.-G., Yoon, K.-Y., Moon, D.-K., Kim, S. M. & Lee, C.-H. 2012. H2 PSA purifier for CO removal from hydrogen mixtures. *International Journal of Hydrogen Energy*, 37, 18175–18186.

Yousefi, S., Meshksar, M., Rahimpour, H. R. & Rahimpour, M. R. 2023. Chapter 11- MOF mixed matrix membranes for syngas purification. In: Rahimpour, M. R., Makarem, M. A. & Meshksar, M. (eds.) *Advances in Synthesis Gas: Methods, Technologies and Applications*. Netherlands: Elsevier.

Zafarnak, S., Meshksar, M., Rahimpour, H. R. & Rahimpour, M. R. 2023. Chapter 12-Membrane technology for syngas production. In: Rahimpour, M. R., Makarem, M. A. & Meshksar, M. (eds.) *Advances in Synthesis Gas: Methods, Technologies and Applications.* Netherlands: Elsevier.

Zeynali, R., Ghasemzadeh, K., Sarand, A. B., Kheiri, F. & Basile, A. 2019. Experimental study on graphene-based nanocomposite membrane for hydrogen purification: Effect of temperature and pressure. *Catalysis Today*, 330, 16–23.

Zhang, N., Xiao, J., Bénard, P. & Chahine, R. 2019. Single-and double-bed pressure swing adsorption processes for H2/CO syngas separation. *International Journal of Hydrogen Energy*, 44, 26405–26418.

Zhao, C., Sun, B., Jiang, J. & Xu, W. 2020. H2 purification process with double layer bcc-PdCu alloy membrane at ambient temperature. *International Journal of Hydrogen Energy*, 45, 17540–17547.

Zhou, C., Fang, Z. Z., Sun, P., Xu, L. & Liu, Y. 2019. Capturing low-pressure hydrogen using VTiCr catalyzed magnesium hydride. *Journal of Power Sources*, 413, 139–147.

Zhu, X., Shi, Y., Li, S. & Cai, N. 2018. Elevated temperature pressure swing adsorption process for reactive separation of CO/CO2 in H2-rich gas. *International Journal of Hydrogen Energy*, 43, 13305–13317.

2 Economic Assessments and Environmental Challenges of Hydrogen Separation and Purification Technologies

Bilal Kazmi and Syed Ali Ammar Taqvi

2.1 INTRODUCTION

Hydrogen (H_2) has recently been recognized as a potentially useful form of renewable energy that has the potential to not only assist in meeting the growing demand for energy around the world (Administration, 2019). But it also aids in resolving the socioeconomic issues that many nations are currently grappling with. H_2, thanks to its adaptability, sustainability, and absence of emissions, has the potential to be a solution to the problem of rising energy consumption, thereby contributing to the reduction of emissions of greenhouse gases (GHGs), the creation of new jobs, and the acceleration of economic growth (Mehrpooya et al., 2021; Bhaskar et al., 2022).

H_2 has a number of potential advantages, one of the most important of which is its ability to assist in the transition to a low-carbon economy (Chen, Li and Chung, 2012). As the world aims toward attaining sustainability, there is an increasing need for renewable energy sources that are ecologically friendly (Mueller-Langer et al., 2007). H_2 is being recognized as a critical resource for economies attempting to minimize their carbon emissions and achieve their environmental aims (Suleman, Dincer and Agelin-Chaab, 2016). The employment of H_2 as an environmentally friendly source of energy possesses the potential to foster socioeconomic progress by means of the generation of new employment possibilities and accelerating the rate of economic development. The emergence of a hydrogen-based economy needs substantial investments in the areas of research, infrastructure, and the production process (Mueller-Langer et al., 2007). This investment has the potential to generate new job openings in a variety of industries, such as engineering, construction, and transportation. The establishment of H_2 economy has the potential to help drive economic growth by encouraging the development of new industries (Shao et al., 2009). These new sectors include H_2 fuel cell automobiles and H_2 generators powered by H_2. H_2 offers solutions to the socioeconomic problems that plague many nations, particularly in more isolated or rural areas. These solutions come on top of the economic

　　　　　　　　　　　　　　　DOI: 10.1201/9781003382522-3

advantages offered by H_2. The inability of many nations to obtain dependable and reasonably priced sources of energy can stifle economic expansion and impede social progress. H_2 is one potential answer to these problems with energy access; this is especially true in more rural places where connecting to the grid can be difficult or expensive. H_2 may be produced locally using renewable energy sources, offering a dependable and sustainable source of energy that can help support economic development and improve quality of life. H_2 can be used to support and improve the quality of life (Levin and Chahine, 2010).

The expansion of industry and the human population around the world have contributed to a rise in the demand for energy, which in turn has placed a pressure on the energy infrastructure that is already in place. In this light, the liquefaction of H_2 has the potential to play a key part in the solution to the energy issue (Dokhani, Assadi and Pollet, 2022; Wu, Lan and Yao, 2023). H_2 is seen as a source of energy that is both clean and renewable, and it possesses the potential to supplant fossil fuels. On the other hand, H_2 is almost always created in the gaseous state, which makes it challenging to transport and difficult to store effectively. When H_2 is cooled to a temperature of $-253°C$ as part of the liquefaction process, the volume of the H_2 is reduced by a factor of 700. Because of this, it is now much simpler to transport and store significant amounts of H_2. Compressing H_2 gas, chilling it with a heat exchanger, and expanding it through a turbine are the three steps that make up the process that is known as the Claude Cycle (Yang and Li, 2023). This cycle can be used to liquefy H_2, which is a useful by-product of other processes. This process consumes a lot of energy and calls for a large amount of electricity to be carried out (Yang and Li, 2023). However, the energy that is necessary to liquefy H_2 can be produced from renewable sources such as wind or solar power, which makes the process favorable to the environment.

When it has been reduced to a liquid state, H_2 is ready to be shipped by pipelines, ships, or trucks to a variety of destinations where it will be used as a fuel. Because of its versatility as a fuel for automobiles, power plants, and industrial activities, H_2 is an attractive option for usage as an energy source (Peschel, 2020). The liquefaction of H_2 can potentially contribute to addressing the present energy crisis by facilitating the transportation and storage of H_2 (Hren et al., 2023). H_2 gas liquefaction may be able to store and transport this vital component of a sustainable energy system. H_2 liquefaction could help meet the global demand for clean, eco-friendly energy. H_2 must be purified before liquefaction to remove impurities and pollutants that could affect the process and product quality. Water, oxygen, nitrogen, and other contaminants in hydrogen gas may affect product quality. Pressure swing adsorption (PSA) is one method for purifying H_2 gas. Asgari et al. (2014) used membrane separation and cryogenic distillation. PSA is a common method used to purify H_2 gas (Li et al., 2016). This process requires the utilization of H_2 gas, which is used to go through a bed of adsorbent material that exhibits selective adsorption properties toward impurities such as oxygen, nitrogen, and water (Mellek and Kaya, 2020). Subsequent to the adsorption process, it is probable to restore the adsorbent material to its initial state by decreasing pressure or increasing temperature, which encourages the discharge of the purified H_2 gas. The utilization of membrane separation is a technique employed for the purification of H_2 gas (Lei et al., 2021).

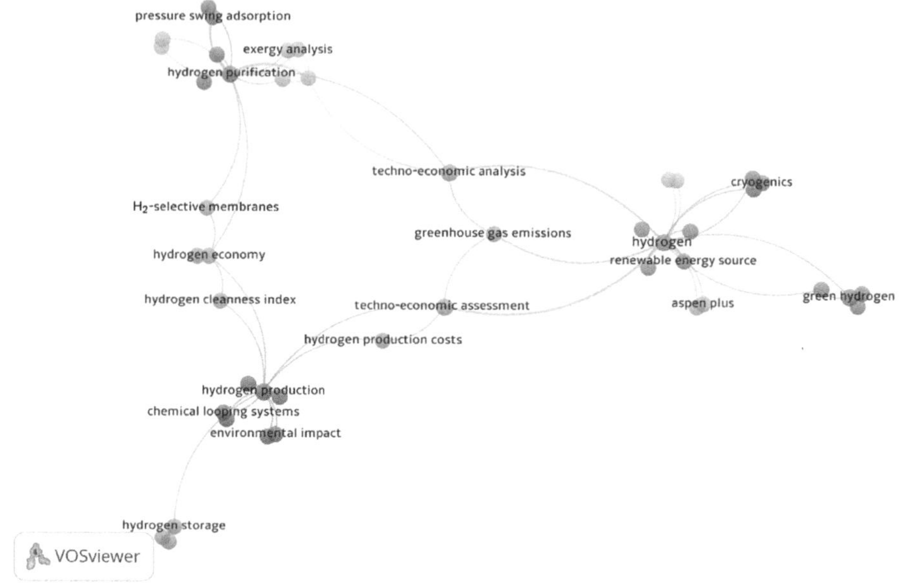

FIGURE 2.1　Overview of the potential networks of hydrogen purification and separation technologies.

The process involves the utilization of a membrane that exhibits selective permeability toward H_2 gas, which facilitates the separation of H_2 from other gases, such as oxygen and nitrogen. This results in the restricted passage of pure H_2 through the membrane (He, 2017). The cryogenic distillation process is a relatively complicated and costly process utilized for the purification of H_2 gas (Du and Liu, 2021). This method requires the cooling of H_2 gas to extremely low temperatures, leading to the condensation and afterward removal of all impurities, including water and other gases. Subsequently, the H_2 gas that passes through purification is captured and might need further treatments to accomplish liquefaction (Bitter and Tashvigh, 2022). A network-based linking aspect of the H_2 purification technologies is shown in Figure 2.1.

Conventional technologies are commonly employed for the purpose of hydrogen purification. The selection of technology is dependent upon the specific requirements of the utilization, as each technology possesses its own set of advantages and disadvantages (as outlined in Table 2.1). These technologies have been subject to extensive studies and improvements over time, making them reliable and beneficial to produce superior-grade hydrogen for distinct applications in industry.

Hence, this chapter offers an insightful perspective on the economic and environmental potential of conventional H_2 purification technologies. It is imperative to consider these factors when assessing the feasibility of utilizing such technologies in the generation of the purified form of H_2. The selection of technology is dependent on the specific requirements of the application, in addition to the accessibility and expenditure of resources such as raw materials, energy, and labor. The increasing

TABLE 2.1

Advantages and Disadvantages of H$_2$ Purification Technologies (Silva et al., 2013; Lei et al., 2021; Chen and Ahn, 2022)

Technology	Advantages	Disadvantages
Pressure swing adsorption (PSA)	– Relatively low capital cost – High energy efficiency – Can handle a wide range of feed gas compositions	– Operating cost can be high – May require frequent replacement of adsorbent material – Limited scalability
Membrane separation	– Low capital and operating cost – Simple design and operation – Can handle varying feed gas compositions	– Low separation efficiency – Susceptible to fouling and degradation – Limited scalability
Cryogenic Separation	– High separation efficiency – Can handle large volumes of gas – Can recover valuable by-products	– High capital and operating cost – High energy consumption – Requires special equipment and handling of cryogenic fluids – Limited scalability
Amine scrubbing	– Low capital cost – Can remove impurities and CO_2 – Can handle varying feed gas compositions	– High operating cost – High water usage and waste generation – Susceptible to amine degradation and corrosion – Limited scalability

demand for H$_2$ is predicted to be supported by a reduction in environmental effects and an improvement in economic feasibility of H$_2$ production, owing to the utilization of renewable energy sources and technological advancements.

2.2 ECONOMIC AND ENVIRONMENTAL ASPECT OF H$_2$ PURIFICATION TECHNOLOGIES

The utilization of conventional H$_2$ purification technologies is crucial in ensuring the generation of purified H$_2$ that can be efficiently applied to diverse applications. The utilization of these technologies necessitates a comprehensive evaluation of their economic and environmental implications.

One of the main economic considerations in the production of H$_2$ is the cost of the purification process (Peramanu, Cox and Pruden, 1999). The utilization of conventional H$_2$ purification technologies is crucial in ensuring the generation of purified H$_2$ that can be efficiently applied to diverse applications. The utilization of these technologies necessitates a comprehensive evaluation of their economic and environmental implications (Zafanelli et al., 2023). These methods are superior to cryogenic distillation and other large-scale methods in terms of cost and infrastructure requirements, making them ideal for production on a smaller scale. Although cryogenic distillation is the most effective method for purifying H_2, it also comes with the highest cost due to its large initial investment and ongoing maintenance expenses (Naquash et al., 2023). The utilization of cryogenic distillation is often seen in circumstances involving extensive production, whereby the technology's higher efficiency allows its comparatively higher costs (Aasadnia, Mehrpooya and Ghorbani, 2021). Absorption and adsorption technologies are considered to be economically viable; however, they necessitate substantial quantities of absorbent or adsorbent material and energy for their functioning. The selection of technology is contingent upon the particular demands of the application, in addition to the accessibility and expense of essential resources such as raw materials, energy, and labor.

The environmental implications of conventional H_2 purification technologies are contingent upon various factors, such as the specific type of technology employed and the energy source utilized during the purification process (Nordio et al., 2021; Hren et al., 2023). GHG emissions, energy consumption, and trash generation are the key environmental challenges related to H_2 manufacturing (Cownden, Mullen and Lucquiaud, 2023). In comparison to methods like cryogenic distillation, which uses a lot of energy and produces a lot of pollutants, PSA and membrane separation technologies have negligible environmental effects (Li et al., 2016; Bitter and Tashvigh, 2022). These technologies have a greater potential for production on a smaller scale, which leads to the minimizing of the environmental impacts linked to the production of H_2. The implementation of absorption and adsorption technologies is linked to the minimum adverse environmental effects due to their utilization of non-hazardous substances and absence of waste by-products. The environmental ramifications of said technologies are contingent upon the particular materials employed. The environmental impact of cryogenic distillation is noteworthy owing to the substantial energy consumption involved in the process, leading to considerable emissions of GHGs. The utilization of sustainable energy sources, such as wind and solar power, has the potential to considerably mitigate the ecological consequences of cryogenic distillation.

Hence, the economic and environmental prospects of conventional H_2 purification technologies must be considered when evaluating their viability for use in the production of high-quality H_2. The choice of technology depends on the specific requirements of the application, as well as the availability and cost of raw materials, energy, and labor. As the demand for H_2 continues to grow, the use of renewable energy sources and advancements in technology are expected to further decrease the environmental impacts and increase the economic viability of H_2 production. The details of the conventional H_2 purification technologies with respect to economics and environment from a process system engineering perspective are listed in Table 2.2.

2.2.1　Economic Prospect of PSA for H_2 Purification

PSA is a well-known H_2 purification technology that has been widely used in various industries due to its economic feasibility. Lui et al. (2020) explained that the processes of cleaning, reforming, gas shift reaction, and separation yield hydrogen with a purity of 99.9% from syngas. Warm or hot gas cleaning can enhance the thermal efficiency of combined cycle gasification systems. The PSA technique,

TABLE 2.2

Process System Engineering-Based Assessment of H_2 Purification Technologies (Naquash et al., 2023; Zafanelli et al., 2023)

Technology	Process Complexity	Process Control	Safety Considerations	Process Flexibility	Capital Cost	Operating Cost	Energy Efficiency	Environmental Impact	Ease of Maintenance	Scalability	Contaminant Removal Efficiency	Purified H_2 Purity
Pressure swing adsorption (PSA)	Moderate	Relatively easy to control	Low risk of explosion or fire	Can be adjusted for varying feed gas compositions	Moderate	High	Moderate	Low	Moderate	High	High	Up to 99.999%
Membrane separation	Low	Relatively easy to control	Low risk of explosion or fire	Limited ability to adjust for varying feed gas compositions	Low	Low	High	Low	High	Low to moderate	Low to moderate	Up to 99.5%
Cryogenic separation	High	Relatively difficult to control	High risk of explosion or fire	Limited ability to adjust for varying feed gas compositions	High	High	High	Moderate	Low	High	High	Up to 99.999%
Amine scrubbing	Low to moderate	Relatively easy to control	Low risk of explosion or fire	Can be adjusted for varying feed gas compositions	Low	High	Low to moderate	High	High	Low	Low to moderate	Up to 99%

a traditional method for hydrogen separation, involves utilizing an adsorbent bed to absorb impurities present in syngas at high pressure and subsequently releasing them as the pressure is reduced. The PSA process necessitates a hydrogen concentration of 70 mol% in the incoming gas stream, with a desired efficiency of 99.99%. PSA systems differ in terms of adsorption capacity, flow rate, regeneration process, and material composition. Generally, the capital cost of a PSA system is lower than other purification technologies such as cryogenic distillation and membrane separation due to its simpler design and fewer components (He, 2017). Moreover, PSA systems require less power than other purification technologies, which results in lower operating costs.

One of the key economic advantages of PSA is its scalability. PSA technology can be easily scaled up or down to meet the specific H_2 demand of the industry (Lee et al., 2021). When it comes to producing high-purity H_2, PSA systems come in two sizes: those designed for use in a laboratory and those designed for use in a chemical plant. PSA technology is more cost-effective than alternatives that demand extensive customization to match the unique needs of each business because of its scalability. When compared to alternative methods of H_2 purification, PSA technology also provides substantial energy savings (Naquash et al., 2023). The primary source of energy consumption in a PSA system is attributed to the compressors utilized for the purpose of pressurizing and depressurizing the adsorbent beds (Silva et al., 2013). Nonetheless, the energy consumption associated with this purification method is notably lower in comparison to other purification technologies like cryogenic distillation or membrane separation. These alternative methods necessitate substantial energy inputs to uphold low temperatures or high-pressure differentials (Chen and Ahn, 2022).

In addition, the utilization of PSA technology enables uninterrupted operation, thereby reducing the frequency of system shutdowns and enhancing the overall efficacy of H_2 synthesis. The uninterrupted functioning of the PSA system yields a dependable and uniform provision of high-grade hydrogen gas. This is a crucial requirement for industrial processes that necessitate a steady stream of H_2 (Li et al., 2016). PSA technology offers an economic benefit by facilitating the removal of impurities that can be reused or exchanged. Impurities such as carbon monoxide (CO), carbon dioxide (CO_2), and methane (CH_4) can be recovered from the adsorbent material and used for other industrial processes (Mellek and Kaya, 2020). This can result in additional revenue streams for the industry and further reduce the overall operating cost of the PSA system. The cost of the adsorbent material plays a significant role in the overall cost of the PSA system. The cost of the adsorbent material can vary depending on its quality, durability, and availability. However, the cost of the adsorbent material can be minimized by recycling it after the end of its lifespan, or by regenerating it using steam or other methods. Another economic advantage of PSA is its ability to operate at a lower pressure range compared to other purification technologies. PSA can effectively purify H_2 streams with a purity level of 99.9% or higher at a pressure range of 5–20 bar, which is typically found in industrial H_2 production processes (Kakiuchi et al., 2023). Lower operating pressure not only reduces energy costs, but also results in a lower level of compressor maintenance costs.

Hence, PSA technology offers several economic advantages that make it a preferred choice for H_2 purification in many industries. The scalability, energy efficiency, continuous operation, recovery of impurities, and maturity of PSA technology make it a cost-effective solution for industries that require high-purity H_2.

2.2.2 Economic Prospect of Membrane-Based Process for H_2 Purification

Membrane-based H_2 purification technology is gaining attention due to its promising economic benefits (Zhao et al., 2020). Compared to other conventional H_2 purification technologies such as PSA and cryogenic separation, membrane-based systems have lower capital and operational costs. Falco et al. (2013) also add that developing membrane technology is crucial to the success of membrane-assisted water gas shift reactors over PSA or chemical absorption procedures.

The main economic advantage of membrane-based H_2 purification technology is its simplicity and efficiency in separating H_2 from other gases (Bitter and Tashvigh, 2022). Chuah and Lee (2020) also emphasized that smaller plant footprint and greater energy efficiency are two benefits of membrane-based processes for H_2 purification over conventional methods such as cryogenic distillation and PSA; these attributes contribute to reduced production costs. Nevertheless, the issue of scalability continues to be a significant obstacle for graphene-based membranes. Subsequent investigations should center on assessing their efficacy in practical industrial environments. The membrane material used in the system is relatively inexpensive, and the process requires less energy than other methods, reducing operational costs (Bitter and Tashvigh, 2022). Bitter and Tashvigh (2022) also stated that membrane separation technology is the most sustainable and promising method for purifying H_2 on account of its continuous operation, minimal energy consumption, and carbon footprint, as well as its easy maintenance. Fabricated from polybenzimidazole (PBI), these membranes possess exceptional chemical, thermal, and mechanical stabilities in addition to a high intrinsic H_2/CO_2 selectivity. Enhancements in membrane separation technology may render PBI a viable solution for the efficient production of H_2. Another economic advantage of membrane-based H_2 purification technology is its scalability (He, 2017). Schorer, Schmitz and Weber (2019) emphasized that membrane-based hydrogen purification allows simultaneous compression up to 200 bar, making transit and storage easier. Thus, membrane-based hydrogen purification could dramatically lower hydrogen purification costs, which are currently costly due to high-efficiency losses during compression, cooling, or the multi-stage process. It also says that low CAPEX for the EHP system is feasible due to the significant reductions of system costs that result from recent design improvements and market introductions of various electrochemical conversion systems such as hydrogen fuel cells. The membrane-based purification of hydrogen system has promising economic prospects. It is relatively easy to increase the production capacity of the system by adding more membranes in parallel (Spallina et al., 2019). The modular design of the system allows for easy maintenance and replacement of the membranes, reducing downtime and repair costs (He, 2017). Shao et al. (2009) stated that membrane technology is cost-effective for hydrogen purification due to its energy efficiency, smaller unit cost, and ease of operation. Economic factors favor polymeric membranes over alternatives. However, membrane technology cannot provide the ultrapure hydrogen (>99.99%) needed for fuel cell application, limiting its economic viability.

Membrane-based H_2 purification technology has advanced, but still faces cost hurdles. The membranes' costs can add up quickly, especially if high-performance components are required. An additional factor that can raise maintenance costs is that membranes have a limited life span due to fouling, corrosion, and other factors (Singla et al., 2022). Despite these challenges, membrane-based H_2 purification technology has strong economic possibilities as new materials and designs are developed to increase efficiency and membrane durability.

2.2.3 ECONOMIC PROSPECTS OF CRYOGENIC-BASED PROCESS FOR H_2 PURIFICATION

One of the costliest methods for purifying H_2 is based on the use of cryogenic technology. With the decreasing temperature of the supply gas to extremely low levels (about −253°C), contaminants can be liquefied and subsequently distilled away, leaving pure H_2. This cooling method is both energy-intensive and costly because of the quantity of power it consumes.

Ishaq, Dincer and Crawford (2021) explained that H_2 purification using cryogenics is a very costly option as it requires high energy consumption. Due to its high hydrogen recovery efficiency, the operating cost of the cryogenic process can be very competitive with PSA or membrane processes despite its large initial investment. Thus, cryogenic processes have good economic potential, especially when hydrogen recovery efficiency is high. It's capable of removing a wide range of contaminants, allowing for purities as high as 99%. These contaminants include water, nitrogen, carbon monoxide, and carbon dioxide (Dash, Chakraborty and Elangovan, 2023). This high purity is particularly essential for applications that require high-quality H_2, such as fuel cells (Zhang et al., 2023). Cryogenic-based technology also has the benefit of being an established technique. The energy and industrial gas industries have been using the technology for a long time (Zhang et al., 2021). Therefore, there is a large body of knowledge and competence in cryogenic technology, which may reduce expenses related to implementation.

Cryogenic-based technology offers excellent H_2 purification results, but its high energy consumption makes it difficult to attain cost competitiveness, especially when compared to alternative methods of H_2 purification, such as PSA and membrane-based technology. Equipment and maintenance costs, as well as the price of electricity and other energy sources, can have a considerable impact on the final price tag. Consequently, while cryogenic-based H_2 purification technology is effective and capable of producing highly pure H_2, it is still a costly technology due to its high energy requirements. Therefore, the economic viability of this technique is conditional on a number of variables, such as the accessibility of cheap energy sources, the size of the operation, and the required purity.

2.2.4 ECONOMIC PROSPECTS OF AMINE-BASED PROCESS FOR H_2 PURIFICATION

The process of purifying H_2 gas with amine-based technology includes the use of amine compounds, specifically monoethanolamine (MEA) and polyethyleneimine (PEI), to selectively capture impurities such as carbon dioxide from the gas (Luo et al., 2022). Zou et al. (2022) stated that amine-based techniques remove CO_2 from hydrogen streams and are widely employed in industry. It highlights that amine-based methods have been integrated with other technologies to purify hydrogen and reduce carbon emissions. These features suggest that amine-based technologies may have good economic prospects in H_2 purification, although the document does not analyze or project. Furthermore, the H_2 purification technology based on amines can be configured to function across a spectrum of pressures and temperatures, rendering it appropriate for diverse use cases (Ghannadi and Dincer, 2024). As an illustration, the utilization of this substance can facilitate the purification of H_2 gas for application in fuel cell automobiles or the production of electrical energy in power stations.

The potential economic outcomes of utilizing amine-based technology for hydrogen purification are uncertain. Amine-based processes exhibit comparatively lower capital and operational expenses in contrast to alternative purification methodologies (Ochienga, Peters and Slagle, 2013), such as PSA and cryogenic distillation. The technique is also advantageous because the amine chemicals employed are cheap and easily accessible. In addition, the amine-based H_2 purification process demonstrated favorable outcomes. It's designed for manufacturing H_2 with a purity level of up to 99.9% (Spallina et al., 2019). As a result, less H_2 is wasted during purification, bringing down the cost of producing H_2.

Furthermore, the amine-based H_2 purification technology is not only cost-effective but also has a proven track record of high performance in various industrial contexts. This suggests that there is a substantial amount of knowledge available to assist in the design and operation of amine-based H_2 purification systems, which reduces the risk and cost associated with their adoption. Table 2.3 provides a comprehensive summary of the H_2 purification methods and their overall economic potential.

2.2.5 ENVIRONMENTAL ASPECTS OF THE CONVENTIONAL H_2 PURIFICATION TECHNOLOGIES

Concern for the environment and the drive for reducing the global carbon footprint have prompted investigators to investigate the environmental consequences of H_2 purification methods (Silva et al., 2013). As the globe strives for a greener tomorrow, efforts are being made to cut down on carbon emissions, cut back on power usage, and increase the uptake of renewable energy (Mehmeti and Angelis-dimakis, 2018).

Since H_2 is a versatile, clean-burning fuel with many potential applications, including transportation and electricity generation, its manufacture has been heralded as an important move toward a more sustainable energy future (Dash, Chakraborty and Elangovan, 2023). The method utilized to purify H_2 and the source of H_2 are two major factors in determining the process's environmental impact. Ahmed et al.

TABLE 2.3

Comparison of Various H$_2$ Purification Technologies in Terms of Economic Parameters (Yao et al., 2017; Bernardo et al., 2020)

Technology	Capital Cost	Operating Cost	Energy Cost	Maintenance Cost	Capital Cost ($/kg H$_2$/day)	Operating Cost ($/kg H$_2$)
Pressure swing adsorption (PSA)	High	Low to moderate	High	Low to moderate	0.25–1.50	0.15–0.25
Membrane separation	Low	Low to moderate	Low	Low	1.00–5.00	0.20–0.50
Cryogenic separation	High	High	High	Low to moderate	3.00–10.00	0.30–0.50
Amine scrubbing	Low to moderate	High	Low to moderate	High	2.00–6.00	0.20–0.40

(Forruque et al., 2021) stated that numerous traditional approaches exhibit exorbitant energy expenditures, restricted efficacy, and significant emissions of GHGs. PSA is an energy-intensive process that yields minimal H$_2$ recovery in H$_2$ PSA purification systems. Environmental and technical concerns are raised regarding the purification of H$_2$ via amine and activated carbon adsorption, which are also addressed. This is why it's essential for assessing the implications of various H$_2$ purification methods on the environment. Using this analysis, the most eco-friendly and persistent preference can be pushed and adopted throughout a wide range of areas. Taking into account the environmental effects of H$_2$ purification methods assures that the gas is produced and used in an environmentally friendly manner.

2.2.5.1 Environmental Aspects of Pressure Swing Adsorption-Based H$_2$ Purification Process

While PSA technology offers several economic advantages for H$_2$ purification, there are also some environmental concerns associated with its use. These concerns include the generation of waste, the use of energy, and the potential for GHG emissions.

PSA technology has a negative impact on the environment due to the garbage it produces. The efficacy of PSA systems for purifying H$_2$ declines over time as the adsorbent material used in them becomes saturated with contaminants (Bernardo et al., 2020). When this happens, the adsorbent material must be changed out, and the old one is thrown away. It can be expensive and time-consuming to properly dispose of used adsorbent material. PSA technology's energy use is another area of environmental concern (Osman et al., 2022). PSA technology is more energy-efficient than other methods of H$_2$ purification, but it still consumes a lot of power (Capa et al., 2023). Compressors used to pressurize and depressurize the adsorbent beds have a high energy footprint that adds to emissions of GHGs and other forms of air pollution. The electricity utilized to run the compressors may also contribute to power plant emissions. If the contaminants removed by the adsorbent material aren't put to good use, PSA technology could likewise contribute to global warming.

CO recovered during H_2 purification can be used in other industrial processes, but if it is not properly utilized, it may be released into the atmosphere and contribute to global warming.

2.2.5.2 Environmental Aspects of Membrane-Based H_2 Purification Process

When compared to alternative H_2 purification methods, membrane-based systems are widely regarded as being more eco-friendly (Bernardo et al., 2020). Some environmental worries remain, however, and new technology hasn't solved them all.

The membranes' production method is a major area of interest. Membranes are notably energy-intensive to manufacture due to the polymers or ceramics they commonly employ (Ramdin, De Loos and Vlugt, 2012). GHG emissions and other environmental consequences may result from producing these materials. Alternative membrane materials that are less harmful to the environment and use less energy are the focus of current research. Waste membrane disposal is another area of concern. Over time, membranes become clogged with contaminants and must be replaced. Disposing of waste membranes in a suitable manner can be costly and may necessitate the use of certain handling techniques. To mitigate the adverse environmental impacts, the membranes employed in the H_2 purification process can be recycled or repurposed (Salim and Ho, 2018). The energy consumption associated with the operation of the membrane-based H_2 purification process is a source of concern (Lei et al., 2021). The energy efficiency enhancements compared to earlier H_2 purification technologies do not render this process entirely self-sufficient in terms of energy. The use of energy in power plants can intensify global warming due to the generation of GHGs (Shao et al., 2009). Employing renewable energy sources to fuel the membrane-based H_2 purification process can mitigate this concern. Lastly, there is the matter of contamination resulting from the H_2 production process. The production of H_2 itself generates emissions, while the purifying procedure based on membranes does not. Steam methane reforming is a common method for generating H_2 that can result in the release of GHGs. Employing sustainable techniques such as electrolysis fueled by renewable energy sources helps mitigate this issue.

There are some environmental problems with membrane-based H_2 purification technology, despite the fact that this technology is typically regarded as being environmentally favorable (Nenoff et al., 2006). Issues with membrane production, used-membrane disposal, energy use, and H_2 generation-related pollutants are all causes for worry (He, 2017). These issues can be reduced by the use of renewable energy sources, renewable H_2 production methods, alternate membrane materials, and the recycling and reuse of used membranes.

2.2.5.3 Environmental Aspects of Cryogenic-Based H_2 Purification Process

Cryogenic H_2 purification technology is frequently employed because of its reliability and efficiency. There are, however, environmental issues that must be resolved.

The immense amount of energy needed to operate the cryogenic equipment is a major issue with the cryogenic-based H_2 purification method. As a result, there is an increase in consumption of energy and emissions of harmful gases. Furthermore, cryogenic equipment uses numerous refrigerants, which may damage the ozone

TABLE 2.4

Comparison between Conventional H_2 Purification Technologies Based on Environment Factors (Levin and Chahine, 2010; Bernardo et al., 2020)

Environmental Factors	Pressure Swing Adsorption	Membrane	Cryogenic	Amine
Energy efficiency	Moderate	High	Low	High
Carbon footprint	Moderate	Low	High	Low
Water consumption	Low	Low	High	Moderate
Chemical usage	Moderate	Low	High	High
Waste generation	Low	Low	Moderate	Moderate
Land use	Low	Low	Moderate	Moderate
Air pollution	Low	Low	Low	Low
Noise pollution	Low	Low	Low	Low
Health and safety	Low	Low	Low	Low

layer and speed up climate change if not managed appropriately. Renewable energy and alternative refrigerants can help alleviate these problems. Cryogenic equipment handling poses safety risks, which is another environmental worry. Accidents and spills during the handling and storage of cryogenic materials pose a serious threat to environmental quality and must be avoided at all costs (Mehrpooya et al., 2021). Implementing appropriate training, safety equipment, and emergency response strategies can help alleviate these worries. The management of waste products resulting from the cryogenic H_2 purification process is an additional area of concern. These waste products may be contaminated and contain hazardous compounds. The incorrect disposal of these wastes presents risks to human health and the ecological system. Implementing appropriate disposal systems is necessary to relieve these concerns. The cryogenic H_2 purification method relies heavily on water for chilling (Zafanelli et al., 2023). In regions with limited water availability, such utilization of water poses a potential threat to the local water reservoir and ecological balance. This issue can be mitigated through meticulous water management and the implementation of alternative cooling technologies.

Finally, there is the issue of pollution caused by the creation of H_2. Indirect emissions are produced during the generation of H_2, but not during the cryogenic purification process (Naquash et al., 2023). Steam methane reforming is a commonly used technique for producing H_2, although it results in the release of GHGs. The application of environmentally friendly techniques such as electrolysis fueled by renewable energy sources can assist in mitigating this issue. Table 2.4 presents a comparative analysis of various conventional H_2 purification systems in terms of their environmental factors.

2.3 ECONOMIC AND ENVIRONMENTAL OPPORTUNITIES AND CHALLENGES FOR H_2 PURIFICATION TECHNOLOGY

The technology used to purify H_2 addresses hazards as well as advantages from an economic and environmental perspective. Zhang et al. (2021) stated that H_2 production methods have pros and cons. Technology, economy, energy use, and cost affect viability. The only way to objectively assess the economic and environmental benefits of hydrogen generating systems is to understand each step of the process. Some particular occasions require onsite production. Besides the many raw components, hydrogen production systems are difficult and energy-intensive. To guide the hydrogen production industry's growth, impartial technology comparisons are crucial. Finally, research should aim toward green, efficient, convenient, safe, and affordable hydrogen production systems.

Manufacturing and using H_2 purification technology poses environmental concerns, such as the release of GHGs, air pollutants, and wastewater. Ahmed et al. (Forruque et al., 2021) explained that H_2 purification technology can expand the use and market penetration of renewable energy sources, reduce six gigatonnes of CO_2 emissions, meet 18% of final energy demand, create 30 million jobs, and power 20%–25% of the transportation sector. However, the study notes that H_2 generation has many obstacles, including sustainability, cost-effectiveness, and nonrenewable energy use. The health of humans and the ecosystem are both at risk from these pollutants. Another major environmental concern is how the production of H_2 from fossil fuels may contribute to global warming. H_2 purification technologies, on the other hand, offer the chance to cut down on GHG emissions and boost air quality. By offering a clean energy source for transportation, power generation, and industrial operations, these technologies can play a crucial part in the shift to a low-carbon economy. Wu, Lan and Yao (2023) highlighted that H_2 has great potential as a clean energy carrier. Despite substantial advances in H_2 purification procedures, operational and technical obstacles in achieving a steady-state and stable H_2 economy have prompted researchers to develop more efficient processes. To highlight the future of this major energy source, a scientific techno-economic study for all H_2 purification approaches is crucial. Hydrogen purification technology has economic and environmental benefits, but its excessive use of fossil fuels may restrict its sustainability. The details of the opportunities and challenges of each technology are presented in Table 2.5.

2.3.1 Environmental Remedies for H_2 Purification Processes

Environmental remedies for enhancing various H_2 purification technologies resulting in the high energy expenditures of various purification processes and the difficulty of purifying fossil fuel-derived hydrogen due to its high impurity concentrations (Zou et al., 2022). It includes a range of strategies to reduce their environmental impact. Here are some examples:

2.3.1.1 PSA Technology
- Employing adsorbent materials with a high degree of selectivity for H_2 in order to enhance efficiency and minimize waste.
- Enhancing the regeneration process to minimize energy use and emissions.
- Reducing the release of emitted pollutants from the process by enhancing system design and maintenance.

2.3.1.2 Membrane Technology
- Advancing the development and utilization of more effective and long-lasting membranes to minimize material wastage and enhance performance.

TABLE 2.5

Outlook on the Challenges and Opportunities of Various Conventional H$_2$ Purification Technologies (Levin and Chahine, 2010)

Technology	Opportunities	Challenges
• Pressure swing adsorption (PSA)	– Can produce high-purity H$_2$ – Suitable for small-scale applications – Can be integrated with other processes	– High operating costs – Frequent adsorbent material replacement may be necessary – Limited scalability for large-scale applications
• Membrane separation	– Simplicated design and operation – Low capital and operating expenses – Appropriate for minor to medium-scale applications	– Poor efficiency in separating – Not scalable for use on a massive scale – Vulnerable to fouling and deterioration
• Cryogenic separation	– High separation efficiency – Can produce high-purity H$_2$ – Can recover valuable by-products	– High operating and capital expenditures – Specialized equipment and cryogenic fluid handling are required – Scalability limitations for large-scale applications
• Amine scrubbing	– Cost-effective initial investment – Eliminates impurities and carbon dioxide – Appropriate for small- to medium-sized operations	– High operating expenses – High water consumption and refuse production – Vulnerable to corrosion and amine degradation – Scalability constraints for large-scale applications

- Optimizing process conditions to decrease the energy needed for gas compression and permeation.
- Implementing recycling initiatives for used membranes to mitigate waste.

2.3.1.3 Amine Technology

- Employing alternate, less harmful solvents to mitigate the ecological consequences of the procedure.
- Optimizing the design and operation of the method to minimize the discharge of pollutants.
- Adopting a waste management initiative to mitigate the ecological consequences of garbage produced by the process.

2.3.1.4 Cryogenic Technology

- Using more efficient heat exchangers to reduce energy consumption and emissions.
- Minimizing the loss of H$_2$ during the process through improved system design and maintenance.
- Developing and using alternative refrigerants with lower environmental impact.

In order to increase the environmental performance of H_2 purification technologies, it is crucial to consistently assess and enhance the design and operation of the process. Additionally, it is important to explore and adopt novel materials and technologies that can minimize waste and emissions.

2.3.2 Life Cycle Analysis of the H_2 Purification Technologies

Given that H_2 is a versatile, clean-burning fuel with many potential applications, including transportation and electricity generation, its manufacture has been heralded as an important move toward a more sustainable energy future. The method utilized to purify H_2 and the source of H_2 are two major factors in determining the process's environmental impact.

The environmental impact of H_2 purification can only be understood by looking at the technology as a whole, from the mining and processing of raw materials through its final disposal. A LCA is useful in this context. An LCA is a procedure that takes into account all stages of a product's or process's potential influence on the environment (Salkuyeh, Saville and MacLean, 2018). Energy use, carbon dioxide emissions, pollution of air and water, and depletion of natural resources are just some of the environmental concerns taken into account. LCA can be used to determine the impact on the environment of H_2 purification processes at every stage, from resource collection to waste disposal (Li et al., 2020). This information can then be used to make decisions about which technology to use and how to optimize the process for minimum environmental impact.

LCA has been used to evaluate the environmental feasibility of various H_2 purification technologies. LCA studies have shown that the environmental impact of PSA and membrane-based H_2 purification technologies is lower than that of cryogenic distillation technology (Valente, Iribarren and Dufour, 2017). This is because PSA and membrane-based technologies consume less energy and require less equipment compared to cryogenic distillation (Dawood, Anda and Shafiullah, 2020). Apart from the LCA, various environmental factors need to be taken into account for H_2 purification technologies. These factors include the origin of the H_2, the accessibility of sustainable energy for powering the process, and the feasibility of carbon capture and storage to alleviate emissions (Khojasteh Salkuyeh, Saville and MacLean, 2017). In conclusion, a sustainable and environmentally responsible energy future cannot be achieved without considering the environmental impacts of H_2 purification technology. The LCA findings for several H_2 purification processes are discussed below.

2.3.2.1 Pressure Swing Adsorption (PSA)

When compared to alternative H_2 purification technologies, PSA's environmental impact is minimal. Production of the adsorbent materials results in significant energy consumption and carbon dioxide emissions, which are the principal environmental implications of PSA. The negative effects of PSA on the environment can be greatly mitigated, however, by increasing the process' energy efficiency and switching to renewable energy sources.

TABLE 2.6

Life Cycle Assessment Based Factors of Various H_2 Purification Technologies (Bhandari, Trudewind and Zapp, 2014; Hren et al., 2023)

Technology	Impact Categories Considered	Life Cycle Stages Considered
Pressure swing adsorption (PSA)	Possible effects on the environment include acidification, eutrophication, human toxicity, and the formation of photochemical ozone	Raw material extraction, manufacturing, transportation, use, and end-of-life
Membrane separation	Possible effects on the environment include acidification, eutrophication, human toxicity, and the formation of photochemical ozone	Raw material extraction, manufacturing, transportation, use, and end-of-life
Cryogenic separation	Possible effects on the environment include acidification, eutrophication, human toxicity, and the formation of photochemical ozone	Raw material extraction, manufacturing, transportation, use, and end-of-life
Amine scrubbing	Possible effects on the environment include acidification, eutrophication, human toxicity, and the formation of photochemical ozone	Raw material extraction, manufacturing, transportation, use, and end-of-life

2.3.2.2　Membrane-Based Purification

Membrane-based H_2 purification has a small carbon footprint compared to alternative methods. Energy consumption for pumping H_2 through the membrane and waste disposal of spent membranes are the principal environmental consequences of membrane-based purification. However, the environmental impact can be greatly mitigated through the use of renewable energy sources and the recycling of old membranes.

2.3.2.3　Cryogenic-Based Purification

Since a lot of power is needed to chill and liquefy H_2, cryogenic-based H_2 purification has a significant effect on the environment. GHG emissions from power generation and the usage of nonrenewable resources are the principal negative effects of cryogenic purification on the environment. The environmental impact, however, can be greatly mitigated with the implementation of renewable energy sources and enhancements to energy efficiency.

2.3.2.4　Amine-Based Purification

Aqueous purification with amines is inferior to alternative techniques. The disposal of used amine solution and the energy required to regenerate the amine solution are the primary environmental concerns associated with amine-based purification. However, the environmental repercussions can be significantly alleviated by implementing strategies such as enhancing energy efficiency and harnessing renewable energy resources (Table 2.6).

2.4　CONCLUSION

Producing high-purity H_2 gas for use in a wide range of industrial settings relies on the availability of H_2 purification technology. The widespread implementation of these technologies, however, faces a number of economic and environmental hurdles.

High initial and ongoing expenses, energy consumption, and routine maintenance are some of the economic challenges of H_2 purification technology. Cryogenic process has relatively high initial and ongoing costs due to the complexity of their equipment and process control systems. Cryogenic and PSA processes have significant operational costs due to their high energy consumption. Energy usage, GHG emissions, and garbage generation are some of the environmental difficulties posed by H_2 purification technology. H_2 purification technologies have an impact on the process's carbon footprint due to their energy requirements, and the usage of nonrenewable energy sources can lead to the exhaustion of scarce materials. It is possible to address these issues by looking into ways to better the economy and the environment. For instance, advancements in material science and process engineering can lead to the development of more efficient and cost-effective H_2 purification technologies. The use of renewable energy sources such as solar and wind power can also help reduce the carbon footprint of H_2 purification processes. In addition, waste streams generated from H_2 purification processes can be repurposed or recycled, such as using CO_2 captured from amine absorption as a feedstock for the production of chemicals. Overall, addressing the economic and environmental challenges of H_2 purification technologies is crucial for their widespread adoption and integration into the future energy mix.

REFERENCES

Aasadnia, M., Mehrpooya, M. and Ghorbani, B. (2021) 'A novel integrated structure for hydrogen purification using the cryogenic method', *Journal of Cleaner Production*, 278. doi: 10.1016/j.jclepro.2020.123872.

Administration, U. S. E. I. (2019) 'International energy outlook 2019 with projections to 2050', *Choice Reviews Online*. doi: 10.5860/CHOICE.44-3624.

Asgari, M. et al. (2014) 'Designing a commercial scale pressure swing adsorber for hydrogen purification', *Petroleum and Coal*, 56(5), pp. 552–561.

Bernardo, G. et al. (2020) 'Recent advances in membrane technologies for hydrogen purification', *International Journal of Hydrogen Energy*, 45(12), pp. 7313–7338. doi: 10.1016/j.ijhydene.2019.06.162.

Bhandari, R., Trudewind, C. A. and Zapp, P. (2014) 'Life cycle assessment of hydrogen production via electrolysis: A review', *Journal of Cleaner Production*, 85, pp. 151–163. doi: 10.1016/j.jclepro.2013.07.048.

Bhaskar, A. et al. (2022) 'Decarbonizing primary steel production: Techno-economic assessment of a hydrogen based green steel production plant in Norway', *Journal of Cleaner Production*. doi: 10.1016/j.jclepro.2022.131339.

Bitter, J. H. and Tashvigh, A. A. (2022) 'Recent advances in polybenzimidazole membranes for hydrogen purification', *Industrial & Engineering Chemistry Research*, 61(18), pp. 6125–6134.

Capa, A. et al. (2023) 'Process simulations of high-purity and renewable clean H2 production by sorption enhanced steam reforming of biogas', *ACS Sustainable Chemistry and Engineering*, 11(12), pp. 4759–4775. doi: 10.1021/acssuschemeng.2c07316.

Chen, H. Z., Li, P. and Chung, T. S. (2012) 'PVDF/ionic liquid polymer blends with superior separation performance for removing CO 2 from hydrogen and flue gas', *International Journal of Hydrogen Energy*, 37(16), pp. 11796–11804. doi: 10.1016/j.ijhydene.2012.05.111.

Chen, Y. and Ahn, H. (2022) 'Optimization strategy for enhancing the product recovery of a pressure swing adsorption through pressure equalization or co-current depressurization: A case study of recovering hydrogen from methane', *Industrial and Engineering Chemistry Research*. doi: 10.1021/acs.iecr.2c04654.

Chuah, C. Y. and Lee, J. (2020) 'Graphene-based membranes for H 2 separation : Recent progress and future perspective', *Membranes*, 10(11), pp. 1–30. doi: https://doi.org/10.3390/membranes10110336.

Cownden, R., Mullen, D. and Lucquiaud, M. (2023) 'Towards net-zero compatible hydrogen from steam reformation—Techno-economic analysis of process design options', *International Journal of Hydrogen Energy*. doi: 10.1016/j.ijhydene.2022.12.349.

Dash, S. K., Chakraborty, S. and Elangovan, D. (2023) 'A brief review of hydrogen production methods and their challenges', *Energies*, 16(3), p. 1141. doi: 10.3390/en16031141.

Dawood, F., Anda, M. and Shafiullah, G. M. (2020) 'Hydrogen production for energy: An overview', *International Journal of Hydrogen Energy*, 45(7), pp. 3847–3869. doi: 10.1016/j.ijhydene.2019.12.059.

Dokhani, S., Assadi, M. and Pollet, B. G. (2022) 'Techno-economic assessment of hydrogen production from seawater', *International Journal of Hydrogen Energy*. doi: 10.1016/j.ijhydene.2022.11.200.

Du, Z. and Liu, C. (2021) 'A review of hydrogen purification technologies for fuel cell vehicles', *Catalysts*, 11, pp. 1–17. https://doi.org/10.3390/catal11030393.

Falco, M. D. E. et al. (2013) 'Palladium-based membranes for hydrogen separation: preparation, economic analysis and coupling with a water gas shift reactor', in *Handbook of Membrane Reactors*. Elsevier, pp. 456–486. doi: 10.1533/9780857097347.2.456.

Forruque, S. et al. (2022) 'Sustainable hydrogen production : Technological advancements and economic analysis', *International Journal of Hydrogen Energy*, 47, pp. 37227–37255. doi: 10.1016/j.ijhydene.2021.12.029.

Ghannadi, M. and Dincer, I. (2024) 'A new system for hydrogen and methane production with post-combustion amine-based carbon capturing option for better sustainability', *Energy Conversion and Management*, 301(September 2023), pp. 1–13. doi: 10.1016/j.enconman.2023.118026.

He, X. (2017) 'Techno-economic feasibility analysis on carbon membranes for hydrogen purification', *Separation and Purification Technology*, 186, pp. 117–124. doi: 10.1016/j.seppur.2017.05.034.

Hren, R. et al. (2023) 'Hydrogen production, storage and transport for renewable energy and chemicals: An environmental footprint assessment', *Renewable and Sustainable Energy Reviews*. doi: 10.1016/j.rser.2022.113113.

Ishaq, H., Dincer, I. and Crawford, C. (2021) 'A review on hydrogen production and utilization : Challenges and opportunities', *Journal of Hydrogen Production*, 47, pp. 26238–26263. doi: 10.1016/j.ijhydene.2021.11.149.

Kakiuchi, T. et al. (2023) 'Modeling and optimal design of multicomponent vacuum pressure swing adsorber for simultaneous separation of carbon dioxide and hydrogen from industrial waste gas', *Adsorption*, 29(1), pp. 9–27. doi: 10.1007/s10450-022-00371-x.

Khojasteh Salkuyeh, Y., Saville, B. A. and MacLean, H. L. (2017) 'Techno-economic analysis and life cycle assessment of hydrogen production from natural gas using current and emerging technologies', *International Journal of Hydrogen Energy*, 42(30), pp. 18894–18909. doi: 10.1016/j.ijhydene.2017.05.219.

Lee, H. et al. (2021) 'Comparative techno-economic analysis for steam methane reforming in a sorption-enhanced membrane reactor: Simultaneous H2 production and CO2 capture', *Chemical Engineering Research and Design*, 171, pp. 383–394. doi: 10.1016/j.cherd.2021.05.013.

Lei, L. et al. (2021) 'Carbon molecular sieve membranes for hydrogen purification from a steam methane reforming process', *Journal of Membrane Science*. doi: 10.1016/j.memsci.2021.119241.

Levin, D. B. and Chahine, R. (2010) 'Challenges for renewable hydrogen production from biomass', *International Journal of Hydrogen Energy*, 35(10), pp. 4962–4969. doi: 10.1016/j.ijhydene.2009.08.067.

Li, B. et al. (2016) 'Pressure swing adsorption / membrane hybrid processes for hydrogen puri fi cation with a high recovery', *Frontiers of Chemical Science and Engineering*, 10(2), pp. 255–264. doi: 10.1007/s11705-016-1567-1.

Li, G. et al. (2020) 'Life cycle assessment and techno-economic analysis of biomass-to-hydrogen production with methane tri-reforming', *Energy*, 199, p. 117488. doi: 10.1016/j.energy.2020.117488.

Lui, J. et al. (2020) 'A critical review on the principles, applications, and challenges of waste-to-hydrogen technologies', *Renewable and Sustainable Energy Reviews*, 134(January), p. 110365.

Luo, J. et al. (2022) 'Effective adsorption of Ultra-dilute CO_2 over Polyethyleneimine-based adsorbent for H_2 purification', *Separation and Purification Technology*, 299(June), pp. 1–10. doi: https://doi.org/10.1016/j.seppur.2022.121686.

Mehmeti, A. and Angelis-dimakis, A. (2018) 'Life cycle assessment and water footprint of hydrogen production methods : From conventional to emerging technologies', *Environments - MDPI*, 5(24), pp. 1–19. doi: 10.3390/environments5020024.

Mehrpooya, M. et al. (2021) 'Conceptual design and evaluation of an innovative hydrogen purification process applying diffusion-absorption refrigeration cycle (exergoeconomic and exergy analyses)', *Journal of Cleaner Production*, 316(July). doi: 10.1016/j.jclepro.2021.128271.

Mellek, Y. and Kaya, A. (2020) 'Simulation and optimization for hydrogen purification performance of vacuum pressure swing adsorption', *Energy Procedia*, 158, pp. 1917–1923.

Mueller-Langer, F. et al. (2007) 'Techno-economic assessment of hydrogen production processes for the hydrogen economy for the short and medium term', *International Journal of Hydrogen Energy*, 32(16), pp. 3797–3810. doi: 10.1016/j.ijhydene.2007.05.027.

Naquash, A. et al. (2023) 'Separation and purification of syngas-derived hydrogen: A comparative evaluation of membrane- and cryogenic-assisted approaches', *Chemosphere*, 313(November 2022). doi: 10.1016/j.chemosphere.2022.137420.

Nenoff, T. M. et al. (2006) 'Membranes for hydrogen purification : An important step toward a hydrogen- based economy', 31(October), pp. 735–744. doi: 10.1557/mrs2006.186.

Nordio, M. et al. (2021) 'Techno-economic evaluation on a hybrid technology for low hydrogen concentration separation and purification from natural gas grid', *International Journal of Hydrogen Energy*, pp. 23417–23435. doi: 10.1016/j.ijhydene.2020.05.009.

Ochienga, R., Peters, C. and Slagle, J. (2013) 'Amine-based gas-sweetening processes prove economically more viable than the Benfield HiPure process'. Bryan Research & Engineering Inc, Bryan, Texas, US.

Osman, A. I. et al. (2022) *Hydrogen production, storage, utilisation and environmental impacts: a review, Environmental Chemistry Letters*. Springer International Publishing. doi: 10.1007/s10311-021-01322-8.

Peramanu, S., Cox, B. G. and Pruden, B. B. (1999) 'Economics of hydrogen recovery processes for the purification of hydroprocessor purge and off-gases', *International Journal of Hydrogen Energy*, 24(5), pp. 405–424. doi: 10.1016/S0360-3199(98)00105-0.

Peschel, A. (2020) 'Industrial perspective on hydrogen purification, compression, storage, and distribution', *Fuel Cells*, 20(4), pp. 385–393. doi: 10.1002/fuce.201900235.

Ramdin, M., De Loos, T. W. and Vlugt, T. J. H. (2012) 'State-of-the-art of CO_2 capture with ionic liquids', *Industrial and Engineering Chemistry Research*, 51, pp. 8149–8177. doi: 10.1021/ie3003705.

Salim, W. and Ho, W. S. W. (2018) 'Hydrogen purification with CO2-selective facilitated transport membranes', *Current Opinion in Chemical Engineering*, 21, pp. 96–102. doi: 10.1016/j.coche.2018.09.004.

Salkuyeh, Y. K., Saville, B. A. and MacLean, H. L. (2018) 'Techno-economic analysis and life cycle assessment of hydrogen production from different biomass gasification processes', *International Journal of Hydrogen Energy*, 43(20), pp. 9514–9528. doi: 10.1016/j.ijhydene.2018.04.024.

Schorer, L., Schmitz, S. and Weber, A. (2019) 'Membrane based purification of hydrogen system', *Journal of Hydrogen Production*, 44, pp. 2–8. doi: 10.1016/j.ijhydene.2019.01.108.

Shao, L. et al. (2009) 'Polymeric membranes for the hydrogen economy: Contemporary approaches and prospects for the future', *Journal of Membrane Science*, 327(1–2), pp. 18–31. doi: 10.1016/j.memsci.2008.11.019.

Silva, B. et al. (2013) 'H2 purification by pressure swing adsorption using CuBTC', In: *Separations Division 2013- Core Programming Area at the 2013 AIChE Annual Meeting: Global Challenges for Engineering a Sustainable Future*, AiChe, vol. 118, pp. 744–756. http://dx.doi.org/10.1016/j.seppur.2013.08.024.

Singla, S. et al. (2022) 'Hydrogen production technologies - Membrane based separation, storage and challenges', *Journal of Environmental Managment*, 302(August 2021), pp. 1–17. doi: 10.1016/j.jenvman.2021.113963.

Spallina, V. et al. (2019) 'Techno-economic assessment of an integrated high pressure chemical-looping process with packed-bed reactors in large scale hydrogen and methanol production', *International Journal of Greenhouse Gas Control*, 88(February), pp. 71–84. doi: 10.1016/j.ijggc.2019.05.026.

Suleman, F., Dincer, I. and Agelin-Chaab, M. (2016) 'Comparative impact assessment study of various hydrogen production methods in terms of emissions', *International Journal of Hydrogen Energy*, 41(19), pp. 8364–8375. doi: 10.1016/j.ijhydene.2015.12.225.

Valente, A., Iribarren, D. and Dufour, J. (2017) 'Life cycle assessment of hydrogen energy systems: A review of methodological choices', *International Journal of Life Cycle Assessment*, 22(3), pp. 346–363. doi: 10.1007/s11367-016-1156-z.

Wu, N., Lan, K. and Yao, Y. (2023) 'An integrated techno-economic and environmental assessment for carbon capture in hydrogen production by biomass gasification', *Resources, Conservation and Recycling*, 188(May 2022). doi: 10.1016/j.resconrec.2022.106693.

Yang, J. and Li, Y. (2023) 'Study on performance comparison of two hydrogen liquefaction processes based on the Claude cycle and the Brayton refrigeration cycle', *Processes*, 11(3), p. 932.

Yao, J. et al. (2017) 'Techno-economic assessment of hydrogen production based on dual fluidized bed biomass steam gasification, biogas steam reforming, and alkaline water electrolysis processes', *Energy Conversion and Management*, 145, pp. 278–292. doi: 10.1016/j.enconman.2017.04.084.

Zafanelli, L. F. A. S. et al. (2023) 'A novel cryogenic fixed-bed adsorption apparatus for studying green hydrogen recovery from natural gas grids', *Separation and Purification Technology*. doi: 10.1016/j.seppur.2022.122824.

Zhang, N. et al. (2021) 'Optimization of pressure swing adsorption for hydrogen purification based on Box-Behnken design method', *International Journal of Hydrogen Energy*, 46(7), pp. 5403–5417. doi: 10.1016/j.ijhydene.2020.11.045.

Zhang, X. et al. (2023) 'Modeling study on a two-stage hydrogen purification process of pressure swing adsorption and carbon monoxide selective methanation for proton exchange membrane fuel cells', *International Journal of Hydrogen Energy*, 48(64), pp. 25171–25184. doi: 10.1016/j.ijhydene.2023.01.138.

Zhao, C., et al. (2020) 'H2 purification process with double layer bcc-PdCu alloy membrane at ambient temperature', *International Journal of Hydrogen Energy*, 45, pp. 17540–17547.

Zou, C. et al. (2022) 'Industrial status, technological progress, challenges, and prospects of hydrogen energy', *Natural Gas Industry*, 9, pp. 427–447. doi: 10.3787/j.issn.1000- 09762022.04.001.

Section II

Hydrogen Separation Using Membranes

3 Polymeric Membranes for Hydrogen Separation

*Fatemeh Salahi, Fatemeh Zarei-Jelyani,
and Mohammad Reza Rahimpour*

3.1 INTRODUCTION

The increasing global demand for energy that is both "clean" and efficient has increased the level of preparedness to implement the proposed "hydrogen economy" as a sustainable resolution to the evolving energy crisis (Zarei-Jelyani et al., 2023a; Salahi et al., 2023a; Mosallanezhad et al., 2023). Humans must overcome the numerous scientific and technological barriers that stand between our current hydrogen (H_2) generation, application, and storage capabilities and those needed for a competitive, sustainable hydrogen economy since global energy consumption is predicted to nearly double by 2050 and our current fossil fuel reserves are under increasing pressure from the political, economic, and environmental domains (Hemminger et al., 2008). Despite the presence of numerous complex technological obstacles, the achievement of the complete potential of a hydrogen economy is contingent upon the mitigation of two economic obstacles: the expense of fuel cells and the expense of hydrogen production. Since then, significant collaborative endeavors have been employed to ensure the secure and efficient generation and storage of hydrogen gas. Hydrogen gas is a widely utilized element in various industries, including metallurgical, chemical, petrochemical, pharmaceutical, and textile (Salahi et al., 2023b). It is utilized in large-scale operations to generate an extensive product range, including raw chemical materials such as hydrogen peroxide, methanol, and ammonia (Zarei-Jelyani et al., 2023b). Nevertheless, the extensive manufacturing of H_2 for these sectors frequently necessitates a capital investment of considerable magnitude for the separation and purification procedures, thereby substantially escalating the cost of H_2 (Ockwig and Nenoff, 2007).

The most common technique for producing hydrogen is steam reforming of light hydrocarbons, primarily methane (CH_4), and this is the basis for several H_2 membrane separation systems (Gunardson and Gunardson, 1998). The procedure, commonly referred to as steam-methane reforming (SMR), comprises a duo of fundamental stages (Salahi et al., 2023a; Meshksar et al., 2022). During the first stage of the reformation process, the chemical reaction between methane (CH_4) and an excess amount of steam (H_2O) results in the formation of carbon monoxide (CO) and hydrogen (H_2) at a temperature of approximately 820°C. The production of further H_2 can be achieved through the water-gas shift (WGS) reaction, which involves the reaction of CO with H_2O (Salahi et al., 2023b; Zarei-Jelyani et al., 2022).

DOI: 10.1201/9781003382522-5

$$CH_4 + H_2O \leftrightarrow CO + 3H_2 \tag{3.1}$$

$$CO + H_2O \leftrightarrow CO_2 + H_2 \tag{3.2}$$

Theoretically, the SMR process produces 4 moles of H_2 and 1 mole of CO_2 for every mole of CH_4 consumed. However, achieving this ideal yield is uncommon in practical applications.

$$CH_4 + 2H_2O \leftrightarrow CO_2 + 4H_2 \tag{3.3}$$

The composition of the H_2 product before purification is contingent upon the specific shift process employed. In a high-temperature shift reactor, it is customary for the composition of the product stream to consist of 73.9% hydrogen, 17.7% carbon dioxide, 6.9% methane, and 1.0% carbon monoxide. A secondary shift process, which entails a shift reaction at a lower temperature range of 190°C–210°C, is frequently employed. The outcome of this process is a product composition consisting of 74.1% H_2, 18.5% CO_2, 6.9% CH_4, and 0.1% CO (Molburg and Doctor, 2003). In the field of gaseous hydrogen separations, membrane compositions encompass a broad spectrum of elements from the periodic table, including organic polymers, metallic alloys inorganic oxides, and various composites such as composites and metal–organic frameworks (Lin, 2001).

Irrespective of the technique employed for the generation of H_2, there will always be a requirement for a cost-efficient and effective approach to isolate it from other less desirable species. At present, there exist three primary methods, either used alone or in conjunction, for the purification of H_2, namely: (i) pressure swing adsorption (PSA), (ii) fractional/cryogenic distillation, and (iii) membrane separation (Sircar and Golden, 2000; Stocker et al., 1998; Bredesen et al., 2004). Although PSA and cryogenic or fractional distillation systems are currently utilized in commercial settings, they tend to lack cost-effectiveness and require significant energy input for the separation and purification of hydrogen. Furthermore, it can be argued that both of these techniques do not offer satisfactory levels of purity required for specific applications within the context of the hydrogen economy. Among the available techniques, membrane separation has emerged as a highly promising method owing to its notable advantages such as low energy requirements, potential for uninterrupted operation, significantly reduced capital expenditure, operational simplicity, and ultimately, cost-effectiveness (Ockwig and Nenoff, 2007).

Since 1950, membranes have been investigated for use in various gas separation applications, including the elimination of oxygen from air, helium from natural gas, and hydrogen from petroleum refinery gas. However, it took almost a quarter of a century for cost-effective commercial membranes to be created that could effectively carry out these separations. One of the initial widespread uses of gas separation membranes was in the recovery of hydrogen. Hydrogen separations encompass a range of processes, such as modifying the H_2/CO ratio in synthesis gas, which is a blend of hydrogen and carbon monoxide typically derived from natural gas steam reforming or coal gasification. Other applications include extracting hydrogen from diverse hydrocarbon streams and purging gases in ammonia production and other petrochemical operations (Zolandz and Fleming, 1992; Koros and Mahajan, 2000).

Polymeric membranes are a kind of membrane that is often employed in hydrogen separation. The present chapter presents a comprehensive survey of polymeric membranes designed for hydrogen separation.

3.2 HISTORY OF POLYMERIC MEMBRANE FOR HYDROGEN SEPARATION

Organic polymers had their roots in the 1840s when nitrocellulose and polyisoprene were discovered and vulcanized (natural rubber) marking significant milestones in the field's rich history. The widespread and diverse utilization of polymeric materials in contemporary society has had a significant influence on various aspects of the global economy and large-scale industry. Therefore, it can be inferred that these materials should be evaluated and utilized as a prospective foundation of the envisioned hydrogen-based economy. The application of membrane separations for gas separations has been under scrutiny since the 1950s. However, it was only in the mid-1970s that DuPont first implemented the utilization of small-diameter hollow fibers as a practical membrane for gas separation (Weller, 1950; Strathmann et al., 1992). Polymer separation membranes are widely used in various industrial applications. However, the recovery, separation, and purification of H_2 are considered to be precious yet challenging targets for these membranes. Conventional techniques employed for these segregations include modification of the H_2/CO ratio in synthesis gas, eliminating H_2 from hydrocarbon streams, and ultimately, purging H_2 from ammonia production and other extensive-scale/commercial petrochemical procedures (Koros and Mahajan, 2000).

Although DuPont's hollow fibers were innovative, their low permeance was insufficient to enable economically viable gas separations. A new advancement is introduced in gas separation through the utilization of spiral-wrapped cellulose acetate membranes (Separex) for separating H_2 and natural gas. With the advancement of transport properties in polymeric H_2 separation membranes, their utilization on commercial scales has become increasingly prevalent in various recovery processes, including the reclamation of H_2 from recycled refinery gas (Ockwig and Nenoff, 2007).

Proton exchange membrane (PEM)-based fuel cells are a prominent polymer-based technology utilized for energy production. These fuel cells can directly and efficiently convert the chemical energy of H_2 into electrical energy, while significantly reducing the emission of greenhouse gases including hydrocarbons, CO_2, CO, SO_x, and NO_x (Farrauto et al., 2003). To widely use fuel cells, a sufficiently widespread distribution of hydrogen must be established. Although polymer-based membranes for H_2 separation can achieve high levels of H_2 purity, the attainment of the highest purities for numerous applications necessitates the use of inorganic membranes, palladium composites, or other advanced separation techniques (Shindo et al., 2014).

3.3 MEMBRANES MADE OF DENSE POLYMERS

Dense-type polymeric membranes are classified as glass or rubber. Glassy polymeric membranes exhibit superior selectivity but lower flux, while rubbery polymeric membranes demonstrate higher flux but lower selectivity. Kluiters (2004)

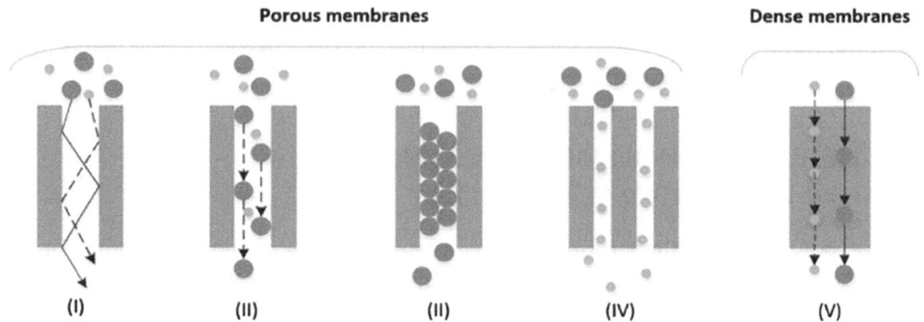

FIGURE 3.1 Transport mechanism for porous membranes: (i) Knudson diffusion; (ii) surface diffusion; (iii) capillary condensation; (iv) molecular sieving; and (v) solution diffusion (Vermaak et al., 2021).

claims that the working temperature for polymer membranes is 100°C. One notable advantage is their capacity to withstand high-pressure drops, coupled with their affordability. Polymer membranes exhibit several drawbacks, namely restricted mechanical robustness, heightened susceptibility to swelling and compression, and susceptibility to specific chemicals like hydrochloric acid, sulfur oxides, and carbon dioxide.

Polymeric membranes can be classified into two categories, namely porous and non-porous. The transport of hydrogen in these membranes can take place through five distinct diffusion processes (as shown in Figure 3.1). If the polymeric membrane exhibits porosity, the process of diffusion is influenced by the size of the pores and the diffusing gas molecules. This can result in four distinct mechanisms, namely Knudsen diffusion, surface diffusion, capillary condensation, and molecular sieving. Knudsen diffusion is characterized by a phenomenon in which gaseous molecules undergoing diffusion exhibit a higher frequency of collisions with the walls of the pore relative to other diffusing molecules. This differential interaction with the pore walls leads to distinct retention times. By surface diffusion, gaseous molecules attach themselves to the walls or surfaces of the pores and then migrate along a concentration gradient that gets smaller from one location to the next. Capillary condensation is a phenomenon that takes place in certain conditions where gas molecules undergoing diffusion condense within a particular pore, resulting in the generation of capillary forces that impede the rates of diffusion. The molecular sieving mechanism pertains to situations where the diffusing gas molecule's size and the pore size are nearby, necessitating an energy of activation that is directly proportional to the size of the molecule. Transport in non-porous or dense polymeric membranes is governed by solution diffusion (Baker, 2012). The process of solution diffusion involves the absorption of gaseous molecules onto the surface of a polymer membrane, followed by their dissolution into the bulk membrane. In the end, these molecules are carried through the membrane by a chemical potential gradient, going from the feed stream to the product stream (Stern, 1994).

3.4 POLYMERIC HYDROGEN SELECTIVE MEMBRANES

Polymeric membranes with hydrogen selectivity are engineered to enhance the concentration of hydrogen in the permeate while retaining the other constituents of the gas mixture in either the feed stream or a secondary waste stream, known as the retentate. Impurities in hydrogen feed streams vary depending on the process from which they are derived. The water electrolysis process is known to produce impurities such as CH_4, O_2, N_2, CO_2, and CO, while the impurities generated during the steam methane reforming process include CO, CO_2, and CH_4. The petroleum reforming process also produces contaminants such as C1–C6, benzene, toluene, and xylene (Shindo et al., 2014). The presence of impurities may cause interference with the separation membranes. But hydrogen gas exhibits a comparatively lower critical temperature (Tc) in comparison to the majority of other gases and possesses one of the most diminutive kinetic diameters among all gas molecules. The observed low Tc value implies a reduced capacity for hydrogen solubility, whereas the relatively small kinetic diameter is indicative of significantly greater diffusivities (Lin and Freeman, 2004). The present investigation concerning selective polymeric H_2 membranes is focused on maximizing diffusivities while minimizing the associated impacts of reduced solubilities. Enhancing the efficacy of polymeric membranes that selectively separate H_2 is primarily reliant on a focused separation process that excludes a greater quantity of condensable compounds (impurities) and facilitates the permeation of less condensable H_2.

The concept of an "upper bound" pertains to achieving a delicate equilibrium between permeability and selectivity. The aforementioned upper limit offers valuable understanding regarding the highest degree of selectivity that can be achieved when utilizing polymeric membranes for a specific gas composition in the H_2 feed stream, while taking into account the membrane's permeability. Various poly methyl methacrylates (PMMs) have been found to be effective in separating H_2 from a mixture of H_2/ N_2 at the lower end of H_2 permeability. Conversely, poly(1-trimethylsilyl-1-propyne) (PTMSP) has been utilized for H_2 permeability at the higher end. Therefore, membranes are presently being formulated utilizing polymers that conform to this upper limit. In cases where separations that surpass the aforementioned upper limit are necessary, it becomes imperative to resort to alternative membranes such as zeolitic or metallic, as well as alternative technologies (Ghalei et al., 2017; Hou et al., 2020).

The utilization of cross-linkable polymers is a contemporary approach that has been implemented to enhance the efficacy of polymeric membranes (Sazali et al., 2020). Notwithstanding, the intricacy of executing this methodology on extensive industrial-scale membranes remains unresolved and necessitates attention before achieving ubiquitous commercial application. The creation of novel membrane materials continues to get significant attention in the patent literature, although polymeric membrane study in the literature seems to be transitioning toward membrane processing membrane systems and new supports (Simmons, 2005).

Hydrogen rejective membranes are one particular use for polymer membranes. By utilizing the increased sorption of alternative gases, rejective membranes can offset any potential selective bias toward the minute size of hydrogen molecules

(Perry et al., 2006). Studies have demonstrated that the presence of rubbery polymers, such as polydimethylsiloxane (PDMS), can result in a decrease in the level of diffusion selectivity. The enhanced mobility exhibited by the chain structures present in rubbery polymers leads to an elevation in the diffusivity of gaseous species. The enhanced mobility of larger molecules resulting from chain mobility is more significant compared to that of smaller molecules, leading to a decrease in diffusion selectivity. Unfortunately, this phenomenon is not observed in glassy polymers due to the restricted mobility of the polymer chains, which hinders any substantial decrease in diffusion selectivities (Genduso et al., 2019).

As mentioned previously, enhancing the solubility selectivity is a viable approach to enhance the characteristics of H_2-selective polymeric membranes. Numerous techniques have been documented in the literature as a means to enhance solubility selectivities. The aforementioned options encompass the integration of polar entities, such as poly ether oxide, poly propylene oxide, poly ester-ether, or poly urethane-ether (Simmons, 2005). The polymers offer an enhanced medium for the dissolution of polarizable molecules present in the feed stream, including CO, CO_2, and NO_x. However, nonpolarizable components of the feed stream, such as H_2, do not exhibit a similar increase in solubility. The Research Institute of Innovative Technology for the Earth (RITE) reported a highly noteworthy advancement in the realm of polymeric membranes (Olajire, 2010). A novel gating membrane was developed by RITE, utilizing poly(amido-amine) dendrimers. The membrane exhibits a distinctive CO_2 gating characteristic and is anticipated to offer unparalleled CO_2/H_2 selectivity within the realm of polymeric substances. The selective characteristics of polymeric membranes have been attained by means of meticulous management and processing regulation of the fractional free volumes. The fractional free volumes of H_2-selective PTMSP membranes, for example, may be changed to make reverse-selective membranes that prefer hydrocarbons over hydrogen (Nagai, 2004).

Because the unique qualities of each gas (size, shape, and chemistry) influence the sorption selectivity and solubilities of a polymer, it is generally impossible to create a membrane that functions consistently for all gas compositions. Consequently, it is frequently imperative to devise membranes of this nature tailored to particular pairs of desired gases.

Currently, the separation and purification of hydrogen gas in SMR facilities is achieved by utilizing PSA and/or acid gas scrubbers based on amine. As indicated previously, polymeric membranes are highly suitable for the elimination of large quantities of CO_2 while simultaneously retaining H_2. Additionally, they represent a cost-effective alternative to the conventional methods currently in use. There exist multiple categories of membranes that could potentially serve the purpose of H_2 purification at SMR plants. However, the elevated temperature range of the product gas stream, which ranges between 450°C and 650°C, renders conventional polymeric membranes unsuitable for this application. Nevertheless, their utilization is feasible provided that the temperature of the gaseous flow is adequately reduced. If chilling the product gas stream is not a possibility, carbon molecular sieve membranes or proton conducting membranes based on palladium (Pd) are utilized. The minor component (CO_2) is separated from the H_2, and the product is recovered.

Nevertheless, given that these particular membranes typically exhibit permeability to both hydrogen and carbon dioxide, the hydrogen that is retrieved will inevitably contain carbon dioxide as well, as the latter also permeates to a certain degree (Bredesen et al., 2004).

3.5 SELECTIVE POLYMERIC MEMBRANES FOR H_2 VS CO_2

The permeation of CO_2 through a polymeric membrane that exhibits selectivity toward CO_2 is achieved through a reversible reaction mechanism. The reaction between an amine-based carrier gas species and CO_2 on the high-pressure feed side of the membrane results in the creation of an adduct, R_2N-CO_2. The R_2N-CO_2 adduct undergoes diffusion across the membrane toward the side of the product stream with lower pressure, leading to the regeneration of unbound carrier species and the liberation of CO_2. The process of Fickian diffusion facilitates the permeation of both CO_2 and H_2, wherein gas molecules dissolve at the feed side with high pressure, diffuse through the membrane, and subsequently desorb into the gas phase on the permeate side with lower pressure (Han and Ho, 2020).

An optimal polymeric membrane for H_2 separation is characterized by the simultaneous maximization of CO_2 permeability and CO_2/H_2 selectivity while maintaining the relative H_2 fluxes. The augmentation of CO_2 permeability results in a reduction of membrane surface area, thereby leading to decreased expenses. Attaining high permeability must not come at the cost of selectivity as it is deemed commercially and economically infeasible to incur losses of H_2. The permeation of H_2 occurs solely through a solution diffusion mechanism. However, the resolution of this matter lies in circumventing the utilization of a membrane with exceedingly low H_2 solubility, as it impedes the process of H_2 permeation. Numerous polymeric materials have been examined as potential candidates, such as ion exchange resins, hydrophilic polymers, blended polymers containing CO_2-reactive salts, polyelectrolytic membranes, and polyanilines. Membranes comprising ion exchange resins, polyelectrolytes, and polymer/salt blends possess mobile carrier species that exhibit a preference for reacting with CO_2 and subsequently undergo diffusion across the membranes (Barillas et al., 2011).

3.6 ION EXCHANGE AND IONIC POLYMER MEMBRANES

Ion exchange membranes were used for the first time in groundbreaking research in this field in the 1980s (LeBlanc Jr et al., 1980). Polyvinylpyridine membranes, when combined with basic anions such as carbonate (CO_3^-) or glycinate ($NH_2CH_2CO_2^-$), exhibit superior CO_2 permeability compared to polyvinylpyridine membranes containing chloride. The permselective properties of membranes composed of polystryenesulfonic acid (PSSA) were enhanced through neutralization with ethylenediamine (EDA). The monoprotonated form of ethylenediamine, $NH_2CH_2CH_2NH_3^+$ (EDAH+), can undergo a reversible reaction with CO_2 to form carbamates ($R_2NCO_2^-$). This reaction presents a reliable means of achieving reversible carbon dioxide sequestration.

Upon diffusion toward the low-pressure region of the membrane, the carbamate undergoes dissociation, leading to the formation of the initial constituents, namely

$EDAH^+$ and CO_2 in the gaseous state. However, carrier saturation (carbamate production) at higher CO_2 pressures caused the CO_2 permeabilities to fluctuate. The membrane's ionic properties create a significant hindrance to the penetration of non-polarizable gases, including H_2. $EDAH^+$-containing ion exchange membranes have undergone extensive study of membrane technology. Nafion, a distinct set of membranes, comprises poly(perfluorosulfonic acid) that encompasses EDAH+. According to reports, Nafion facilitates the transportation of CO_2, albeit with a limited CO_2/H_2 selectivity (Strathmann et al., 1992). Nafion offers just a modest H_2 penetration barrier due to its limited ion exchange capacity. Nevertheless, Nafion-EDAH membranes with high levels of hydration demonstrated elevated selectivity of CO_2/H_2. Noble et al. (1988) demonstrated that a membrane composed of sulfonated polybenzimidazole-EDAH (PBI-EDAH) exhibited a facilitated transport mechanism for CO_2 permeation, owing to its higher ionic site density in comparison to Nafion. It is reported that a sulfonated styrene-divinylbenzene material with a high ion exchange capacity of 2.2 mequiv/g and containing $EDAH^+$ demonstrated a notable CO_2 permeance of $5.0 \times 10^{-8} m^3/(m^2 \cdot s \cdot kPa)$ when exposed to a CO_2 feed pressure of 0.41 kPa. The material was swollen with water during the experiment. Despite the absence of H_2 data, the selectivity of CO_2/N_2 exhibited considerable potential. Sodium alginate, a polymer, was cross-linked and then exchanged for an $EDAH^+$-containing membrane to create a new ion exchange membrane. The augmentation of $EDAH^+$ concentration in the membrane led to exponential enhancements in both CO_2 presence and selectivity. However, the CO_2/N_2 selectivity of approximately 50 indicates that this particular membrane may not be appropriate for CO_2/H_2 separations (Benito et al., 2017).

The promising potential of ion exchange membranes has been demonstrated through the polymerization of monomers containing acid onto surfaces that are either highly permeable or microporous. The utilization of the plasma graft polymerization technique for acrylic acid resulted in the production of membranes that exhibit higher exchange capacities compared to conventional ion exchange membranes. The outcome yields an elevated density of CO_2 carriers and an increased ionic barrier against the permeation of H_2. Upon neutralization with EDA resulting in the formation of $EDAH^+$, a facilitated transport process exhibiting exceptional selectivity for CO_2 was observed. Specifically, the CO_2 permeance was measured to be $7.5 \times 10^{-6} m^3 /(m^2 \cdot s \cdot kPa)$, while the CO_2/N_2 selectivity was found to be 4,700 at a CO_2 feed pressure of 4.8 kPa. Given their significant selectivity, it is anticipated that these membranes will demonstrate efficacy in the separation of CO_2 from H_2.

Blends of polyvinyl alcohol and poly(acrylic acid) block copolymers that have been neutralized with EDA make up yet another innovative class of polymeric membranes (Sagitha et al., 2018). Ion exchange capacities varied from 1.3 to 4.5 mequiv per gram of polymer, depending on the kind of polymer and the ratio of polymers. The membrane with the greatest exchange capacity achieved a maximum CO_2/N_2 selectivity of 1,500 when subjected to a CO_2 supply pressure of 6.2 kPa. The significant selectivity of CO_2/N_2 implies potential applicability in the separation of CO_2/H_2.

Separations of CO_2 and H_2 have also been shown to be successful using hydrophilic polymers combined with basic or CO_2-reactive salts. The aforementioned membranes are comprised of carrier species that exhibit mobility and undergo reactions leading to the formation of bicarbonate or carbonates. Polyvinyl alcohol membranes blended with different amino acid salts displayed CO_2 permeability within the range of 0.72×10^{-11} to $1.6 \times 10^{-11} m^3/(m^2 \cdot s \cdot kPa)$ and CO_2/H_2 selectivity between 13 and 30, when subjected to a CO_2 feed pressure of 76 kPa. Polyethylenimine membranes have been used, either on their own or in combination with polyvinyl alcohol. The reported selectivity of CO_2/H_2 for polyvinyl alcohol membranes that contained 50 wt% of tetramethylammonium fluoride tetrahydrate was 19 when subjected to a CO_2 feed pressure of 76 kPa (Bai and Ho, 2011). Quinn (1998) found that membranes composed of a combination of cesium fluoride and either polyvinyl alcohol or cesium polyacrylate demonstrated a CO_2/H_2 selectivity of approximately 60 when subjected to a CO_2 feed pressure of roughly 4.4 kPa.

Polyelectrolytes are a type of polymer that possess a high concentration of ions, generally one ionic unit for every polymer repeat unit. Polyelectrolytes, in contrast to ion exchange polymers, dissolved in liquid. Polyelectrolytes exhibit a high density of ionic sites, which results in a significant concentration of CO_2-reactive sites and an ionic environment characterized by low solubility of H_2. Polyelectrolytes that possess practicality for hydrogen purification are characterized by the presence of cationic moieties on the polymer chain, which are associated with anions, specifically fluoride (F^-) and acetate ($CH_3CO_2^-$). Polyvinyl benzyl trimethylammonium fluoride (PVBTAF) is a frequently membrane cited instance. The polymer backbone of PVBTAF is characterized by the presence of quaternary nitrogen groups. The interaction between these functional groups and CO_2 is absent, and the reactivity of CO_2 is attributed to the counter anion. As elucidated in reaction 4, the aforementioned anions function as bases in order to facilitate the generation of HCO_3^- (Quinn and Laciak, 1997).

$$2F^- \cdot H_2O + CO_2 \rightarrow HCO_3^- + HF_2^- \cdot (2n-1)H_2O \tag{3.4}$$

The PVBTAF membranes and its counterparts demonstrated a notable capability to transport CO_2 through facilitated means, with selectivity for CO_2/H_2 surpassing 80 (Zhao et al., 2014). Permselective characteristics exhibited a notable reliance on the relative humidities of both the feed and sweep gases, as expected. The bilayer polyelectrolyte membranes demonstrated an unforeseen enhancement in selectivity up to 207, while maintaining CO_2 permeance. Similar to the hydrophilic polymers that were previously mentioned, the introduction of different salts containing fluoride and acetate into polyelectrolyte membranes resulted in enhanced permselective characteristics. The introduction of four moles of cesium fluoride per mole of repeat unit resulted in the production of membranes that exhibited a CO_2 permeance that was four to six times greater than that of the control. Additionally, at a CO_2 pressure of 4.1 kPa, these membranes demonstrated a CO_2/H_2 selectivity of 127 (Quinn et al., 1997).

3.7 CONCLUSION

While polymeric membranes have been utilized effectively for hydrogen recovery on a massive scale, there is still a lot of space for advancement in this area. It is usually desirable to have membranes with enhanced selectivity and productivity. It would also be highly helpful for these membranes to have greater resistance to hydrocarbons or other complicated and aggressive feed streams. The performance of cross-linking polymers in hydrogen separation has been deemed acceptable; however, further investigation is required to fully comprehend the properties of these materials. One of the primary obstacles in utilizing membranes for hydrogen retrieval is the fact that conventional membranes generate the refined hydrogen in the permeate at low pressures, thereby elevating the expenses associated with recompression. Additional research on hydrogen-repellent membranes may give a solution to this challenge in the future. In order to boost performance and maintain the polymer's processing capabilities, it may be required to combine different technologies into hybrid systems, such as the mixed matrix membranes, which include inorganic, highly selective particles into a polymeric membrane. Although some important technological obstacles must yet be addressed before any of these newer materials can be applied, the future of polymeric membranes for hydrogen separations is quite bright and has tremendous development potential.

REFERENCES

Bai, H. & Ho, W. W. 2011. Carbon dioxide-selective facilitated transport membranes for hydrogen purification. In: Yun Hang Hu, Xiaoliang Ma, Elise Fox, and Xinwen Guo (eds.) *Production and Purification of Ultraclean Transportation Fuels,* Washington, D.C. and Columbus, OH: American Chemical Society, pp. 115–141.

Baker, R. W. 2012. *Membrane Technology and Applications,* Hoboken, NJ: John Wiley & Sons.

Barillas, M. K., Enick, R. M., O' Brien, M., Perry, R., Luebke, D. R. & Morreale, B. D. 2011. The CO_2 permeability and mixed gas CO2/H_2 selectivity of membranes composed of CO_2-philic polymers. *Journal of Membrane Science,* 372, 29–39.

Benito, J., Sánchez- Laínez, J., Zornoza, B., Martín, S., Carta, M., Malpass- Evans, R., Téllez, C., Mckeown, N. B., Coronas, J. & Gascón, I. 2017. Ultrathin composite polymeric membranes for CO_2/N_2 separation with minimum thickness and high CO_2 permeance. *ChemSusChem,* 10, 4014–4017.

Bredesen, R., Jordal, K. & Bolland, O. 2004. High-temperature membranes in power generation with CO_2 capture. *Chemical Engineering and Processing: Process Intensification,* 43, 1129–1158.

Farrauto, R., Hwang, S., Shore, L., Ruettinger, W., Lampert, J., Giroux, T., LIU, Y. & Ilinich, O. 2003. New material needs for hydrocarbon fuel processing: generating hydrogen for the PEM fuel cell. *Annual Review of Materials Research,* 33, 1–27.

Genduso, G., Litwiller, E., MA, X., Zampini, S. & Pinnau, I. 2019. Mixed-gas sorption in polymers via a new barometric test system: Sorption and diffusion of CO_2-CH4 mixtures in polydimethylsiloxane (PDMS). *Journal of Membrane Science,* 577, 195–204.

Ghalei, B., Kinoshita, Y., Wakimoto, K., Sakurai, K., Mathew, S., YUE, Y., Kusuda, H., Imahori, H. & Sivaniah, E. 2017. Surface functionalization of high free-volume polymers as a route to efficient hydrogen separation membranes. *Journal of Materials Chemistry A,* 5, 4686–4694.

Gunardson, H. & Gunardson, H. 1998. Ch. 2 Synthesis gas manufacture. In: Harold H. Gunardson (ed.) *Industrial Gases in Petrochemical Processing,* Boca Raton: Marcel Dekker, Inc. Pub.

Han, Y. & Ho, W. W. 2020. Recent advances in polymeric facilitated transport membranes for carbon dioxide separation and hydrogen purification. *Journal of Polymer Science,* 58, 2435–2449.

Hemminger, J., Crabtree, G. & Kastner, M. 2008. A Report from the Basic Energy Sciences Advisory Committee. US Department of Energy, Washington DC.

Hou, R., O' Loughlin, R., Ackroyd, J., Liu, Q., Doherty, C. M., Wang, H., Hill, M. R. & Smith, S. J. 2020. Greatly enhanced gas selectivity in mixed-matrix membranes through size-controlled hyper-cross-linked polymer additives. *Industrial & Engineering Chemistry Research*, 59, 13773–13782.

Kluiters, S. 2004. *Status Review on Membrane Systems for Hydrogen Separation*. Energy Center of the Netherlands, Petten, The Netherlands.

Koros, W. J. & Mahajan, R. 2000. Pushing the limits on possibilities for large scale gas separation: which strategies? *Journal of Membrane Science*, 175, 181–196.

Leblanc JR, O. H., Ward, W. J., Matson, S. L. & Kimura, S. G. 1980. Facilitated transport in ion-exchange membranes. *Journal of Membrane Science*, 6, 339–343.

Lin, Y. 2001. Microporous and dense inorganic membranes: Current status and prospective. *Separation and Purification Technology*, 25, 39–55.

Lin, H. & Freeman, B. D. 2004. Gas solubility, diffusivity and permeability in poly (ethylene oxide). *Journal of Membrane Science*, 239, 105–117.

Meshksar, M., Salahi, F., Zarei- Jelyani, F., REZA Rahimpour, M. & Farsi, M. 2022. Synthesis, morphology control, and application of hollow Al_2O_3 spheres in the steam methane reforming process. *Journal of Industrial and Engineering Chemistry*, 111, 324–336.

Molburg, J. C. & Doctor, R. D. 2003. Hydrogen from steam-methane reforming with CO_2 capture. In: *20th Annual International Pittsburgh Coal Conference*, pp. 1–21.

Mosallanezhad, S., Haghighatjoo, F. & Rahimpour, H. R. 2023. *Fossil Fuels Storage Technologies and Challenges*. Netherlands: Elsevier

Nagai, K. 2004. Transport properties of large gas molecule permselective glassy polymer membranes. *Membrane*, 29, 42–49.

Noble, R., Pellegrino, J., Grosgogeat, E., Sperry, D. & Way, J. 1988. CO_2 separation using facilitated transport ion exchange membranes. *Separation Science and Technology*, 23, 1595–1609.

Ockwig, N. W. & Nenoff, T. M. 2007. Membranes for hydrogen separation. *Chemical Reviews*, 107, 4078–4110.

Olajire, A. A. 2010. CO_2 capture and separation technologies for end-of-pipe applications-A review. *Energy*, 35, 2610–2628.

Perry, J. D., Nagai, K. & Koros, W. J. 2006. Polymer membranes for hydrogen separations. *MRS Bulletin*, 31, 745–749.

Quinn, R. 1998. A repair technique for acid gas selective polyelectrolyte membranes. *Journal of Membrane Science*, 139, 97–102.

Quinn, R. & Laciak, D. 1997. Polyelectrolyte membranes for acid gas separations. *Journal of Membrane Science*, 131, 49–60.

Quinn, R., Laciak, D. & Pez, G. 1997. Polyelectrolyte-salt blend membranes for acid gas separations. *Journal of Membrane Science*, 131, 61–69.

Sagitha, P., Reshmi, C., Sundaran, S. P. & Sujith, A. 2018. Recent advances in post-modification strategies of polymeric electrospun membranes. *European Polymer Journal*, 105, 227–249.

Salahi, F., Zarei- Jelyani, F., Farsi, M. & Rahimpour, M. R. 2023a. Optimization of hydrogen production by steam methane reforming over Y-promoted Ni/Al2O3 catalyst using response surface methodology. *Journal of the Energy Institute*, 108, 101208.

Salahi, F., Zarei- Jelyani, F., Meshksar, M., Farsi, M. & Rahimpour, M. R. 2023b. Application of hollow promoted Ni-based catalysts in steam methane reforming. *Fuel*, 334, 126601.

Salahi, F., Zarei- Jelyani, F., Rahimpour, H. R. & Rahimpour, M. R. 2023a. Chapter 5- Dry reforming for syngas production. In: Rahimpour, M. R., Makarem, M. A. & Meshksar, M. (eds.) *Advances in Synthesis Gas: Methods, Technologies and Applications*. Netherlands: Elsevier.

Salahi, F., Zarei- Jelyani, F. & Rahimpour, M. R. 2023b. Hydrogen energy as a fuel for transportation. In: Mohammad Reza Rahimpour (ed.) *Reference Module in Earth Systems and Environmental Sciences*. Netherlands: Elsevier.

Sazali, N., Mohamed, M. A. & Salleh, W. N. W. 2020. Membranes for hydrogen separation: A significant review. *The International Journal of Advanced Manufacturing Technology*, 107, 1859–1881.

Shindo, R., Kishida, M., Sawa, H., Kidesaki, T., Sato, S., Kanehashi, S. & Nagai, K. 2014. Characterization and gas permeation properties of polyimide/ZSM-5 zeolite composite membranes containing ionic liquid. *Journal of Membrane Science*, 454, 330–338.

Simmons, J. W. 2005. Block polyester-ether gas separation membranes. Google Patents.

Sircar, S. & Golden, T. 2000. Purification of hydrogen by pressure swing adsorption. *Separation Science and Technology*, 35, 667–687.

Stern, S. A. 1994. Polymers for gas separations: The next decade. *Journal of Membrane Science*, 94, 1–65.

Stocker, J., Whysall, M. & Miller, G. 1998. *Years of PSA Technology for Hydrogen Purification*. United states: UOP LLC.

Strathmann, H., Winston, H. W. & Sirkar, K. 1992. *Membrane Handbook*. Vam Nostrand Reinhold, New York.

Vermaak, L., Neomagus, H. W. & Bessarabov, D. G. 2021. Recent advances in membrane-based electrochemical hydrogen separation: A review. *Membranes*, 11, 127.

Weller, S. 1950. Engineering aspects of separation of gases: Fractional permeation through membranes. *Chemical Engineering Progress*, 46, 585.

Zarei- Jelyani, F., Salahi, F., Farsi, M. & Reza Rahimpour, M. 2022. Synthesis and application of Ni-Co bimetallic catalysts supported on hollow sphere Al_2O_3 in steam methane reforming. *Fuel*, 324, 124785.

Zarei- Jelyani, F., Salahi, F., Meshksar, M., Farsi, M. & Rahimpour, M. R. 2023a. Response surface methodology for optimizing the activity of bimetallic Ni-Co-Ce/Al_2O_3 catalysts in the steam methane reforming. *Journal of the Energy Institute*, 110, 101363.

Zarei- Jelyani, F., Salahi, F. & Rahimpour, M. R. 2023b. Introduction to hydrogen as a clean source of energy. In: Mohammad Reza Rahimpour (ed.) *Reference Module in Earth Systems and Environmental Sciences*. Netherlands: Elsevier. doi:10.1016/B978-0-323-93940-9.00098-0

Zhao, Y., Jung, B. T., Ansaloni, L. & Ho, W. W. 2014. Multiwalled carbon nanotube mixed matrix membranes containing amines for high pressure CO_2/H_2 separation. *Journal of Membrane Science*, 459, 233–243.

Zolandz, R. R. & Fleming, G. K. 1992. Gas permeation. In: Ho, W. S. W., & Sirkar, K. K. (eds.) *Membrane Handbook*. Van Nostrand Reinhold, New York, pp. 365–385.

4 Zeolite Membranes for Hydrogen Separation

Samuel Eshorame Sanni,
Denen Ashiekaa Vershima,
Emeka Emmanuel Okoro, and Babalola Aisosa Oni

4.1 INTRODUCTION

For industrial-based processes, hydrogen (H_2), which is usually produced by processing methane (CH_4) accompanied by multiple purification stages, has been seen as a source of renewable energy that is environmentally friendly (Cebrucean et al., 2017). Hydrogen is one of the most critical molecules in industrial gas operations and one of the leading energy producers in places where it is used as a primary energy source (Acer and Dincer, 2018), hence, hydrogen separation technologies have drawn a lot of interest from both the industrial and academic worlds (Cebrucean et al., 2017; Chen et al., 2018a). It can be used in petroleum refining and upgrading, as a vital reagent in chemical production processes such as metallurgy and food processing, and as a clean, zero-emission fuel (Chen et al., 2018b). In the modern world, fossil fuels like coal and petroleum are the dominant fuel sources (Zaabout et al., 2019). Commercial fuel cell electric vehicles and stationary electricity producers use hydrogen as a fuel source. Nowadays, methane processing, coal processing, and stored ammonia are the primary sources of hydrogen production. The resulting gas frequently contains combinations of numerous tiny gases, including nitrogen, methane, carbon dioxide (CO_2), and oxygen (O_2) (Zaabout et al., 2019; Tso et al., 2018; Wassie et al., 2018a). It is impossible to use fuel cells with hydrogen that has a purity of only 60% as a direct energy source (Wassie et al., 2018b). Hydrogen is an important chemical produced primarily by steam (H_2O) processing (Wassie et al., 2018b). Hydrogen separation is the primary step in producing high-purity hydrogen (Wassie et al., 2018b; Spallina et al., 2017b).

To address issues related to global energy security and air pollution, new energy solutions are required as the availability of fossil fuels diminishes. One of the most exciting possibilities for renewable energy sources (for a sustainable future) appears to be hydrogen. Till date, 99% of the world's hydrogen come from fossil fuels (Stenberg et al., 2018). By fermenting materials rich in carbohydrates, hydrogen can be produced alternatively (Symonds et al., 2019). As a result of this process, carbon dioxide and hydrogen are made (Symonds et al., 2019; Song and Shen, 2018). For any suitable application, the hydrogen concentration in the mixture needs to be enhanced (Riva et al., 2018). Therefore, the resultant gaseous mixture must be improved and purified to increase the concentration of hydrogen (Riva et al., 2018; Nazir et al., 2018). Due to their comparable qualities, separating H_2 from carbon dioxide and

DOI: 10.1201/9781003382522-6

CH_4 using a traditional physisorption technique is difficult (Nadgouda et al., 2019). Membrane separation has attracted much interest in hydrogen extraction due to its various benefits (Arif et al., 2020; Cao et al., 2020; Luo et al., 2020). Over the last ten years, different membrane types have been synthesized for gas separation purposes (Villain-Gambier et al., 2020).

Most zeolite-based membranes have been tested for $CO_2/CH_4/N_2$ adsorption, whereas only a few studies have been dedicated to hydrogen adsorption. Beni and Shahrak (2020) synthesized zeolitic membranes (ZIF-8 and ZIF-90) in their pristine forms and compared them with their modified forms, which were functionalized with K, Na, and Li cations. The results of the CO_2 adsorption revealed that the Li-doped zeolite had the highest CO_2 uptake owing to its better electrostatic and dispersion interactions between the adsorbate and adsorbent molecules.

The presence of alkali/alkaline earth cations in zeolite matrices is also another notable area that needs to be exploited as this, in turn, alters its CO_2 adsorption capacity, which is also justified by the work of Balashankar and Rajendran (2019), where a zeolite screening for post-combustion CO_2 sequestration was conducted under vacuum swing adsorption. Despite the several strategies adopted to improve the gas trapping potential of zeolites, they still present some shortcomings, including their relatively low gas selectivity and high hydrophilicity, especially in feed mixtures containing both gases. Nonetheless, the CO_2 adsorptive capacities of zeolites may likely decrease, especially in situations where the gas mixtures are entrained with moisture; also, after adsorption, the regeneration of spent zeolites is only achievable at temperatures >300°C (Wang et al., 2011).

Compared to other separation techniques, zeolitic membranes have been utilized in processing gases at elevated temperatures and pressures (Villain-Gambier et al., 2020; Ji et al., 2018). Microporous inorganic membranes have also attracted much interest for gas separation and purification due to their superior characteristics relative to typical polymeric membranes (Chu et al., 2018; Cai et al., 2019; Ding et al., 2018). The economic viability of cryogenic and pressure swing adsorption (PSA) will likely face stiff competition with membrane systems because of the latter's potential for competitive hydrogen capture efficiency. Microporous ceramic membranes with excellent characteristics display remarkable selectivity and comparatively low gas permeance at higher temperatures (Acar and Dincer, 2019; Shen et al., 2018). Additionally, ceramic membrane processes may be applied in membrane technologies to improve conversion in CH_4 reforming and dehydrogenation processes which both produce H_2 (Cheng et al., 2019; Yang et al., 2018; Chen et al., 2019). Studies on inorganic membranes for hydrogen separation are still in progress. Nonetheless, numerous articles have been geared towards improving the characteristics of the inorganic membrane processes for hydrogen separation, specifically by developing new membrane materials or systems (Parra et al., 2019).

4.2 GAS SEPARATION MEMBRANES

Gas processing membranes have been utilized industrially since 1980, even though membranes have been studied for more than 150 years (Nuhnen et al., 2018; Caro and Noack, 2007). Many industrial processes, including the creation of an O_2-enriched atmosphere, the separation of carbon dioxide and H_2O from carbon-based materials, the purification of H_2, and the recovery of vapors from materials use gas separation membranes (Nuhnen et al., 2018; Caro and Noack, 2007; Liu et al., 2021). Reviews on gas processing membranes have been published in several research outlets (Liu et al., 2021). Membrane-mediated gas separation has several inherent advantages over more traditional processes like Pressure Swing Adsorption (PSA) and cryogenic distillation, including excellent energy efficiency, comparative cheap investment costs, easy maintenance, and small carbon foot prints (Liu et al., 2021; Gu et al., 2008). As a result, it has significant uses in energy-intensive industrial gas separation operations like upgrading natural gas, enriching nitrogen and oxygen, and recovering hydrogen and helium. Several researchers have proposed various methods for creating more effective membranes. For effective gas separation operations, many membrane materials have been developed. Conventional inorganic membranes frequently have low processability and workability but excellent permeability and selectivity (Gu et al., 2008). Robeson's upper limit, a trade-off limitation between permeability and selectivity, is a prevalent problem for numerous membranes. Figure 4.1 shows the structure of polymer membranes.

Almost half of the energy available in the chemical industry is used in separation processes. Compared to other separation methods, the membrane-based processing technique for gas processing has demonstrated a credible advantage. Mixed matrix membranes (MMMs), which combine a polymer-integrated membrane gas processing technique with micro- or nanoparticles, have drawn more and more interests among various gas separation membranes. They benefit from quickly processed polymers and nanofillers' benefits. The Robeson upper bound is likewise exceeded by MMMs, which have a strong separation capability for countering the usual trade-off between selectivity and permeability in traditional membranes (Gu et al., 2008; Joly et al., 1999).

The need for membranes has surged over the last 10 years, along with the demand from the energy-consuming industries, particularly for the creation of hydrogen. As a green method for producing hydrogen, photocatalytic separation of water has a significant potential use (Joly et al., 1999; Diawara et al., 2020). However, as hydrogen and oxygen are simultaneously created in this process, the oxygen must be extracted to obtain high-purity H_2. In a recent demonstration of the photocatalytic breakdown of water on a large scale, low H_2 extraction efficiency was found to be one of the primary causes of the low solar-to-H_2 ratio. The effectiveness of H_2 generation from water electrolysis supported by solar energy will be enhanced by using the membrane to split hydrogen and oxygen. However, there are only few reports on the use of membranes for hydrogen–oxygen separation. Inorganic membranes, such as dense membranes and microporous membranes, are one of the potential materials for hydrogen gas separation applications due to their excellent stability. Several research groups have effectively modified inorganic membranes over the past 30 years via various techniques as ways of enhancing membrane transport/their separation capabilities (Diawara et al., 2020).

Multiple methods have been employed in separating gasses. The nanoporous membrane whose pore size is 0.5 nm can serve as a molecular sieve in the membrane modules. The tiny pores allow for the passage of microscopic gas molecules but not larger ones. Therefore, a molecular sieve mechanism can create high and improved characteristics for the smaller component of a gas composition if the larger gas particles do not condense at the pores. In contrast, there is a significant correlation between the permeability of membranes and the length of the

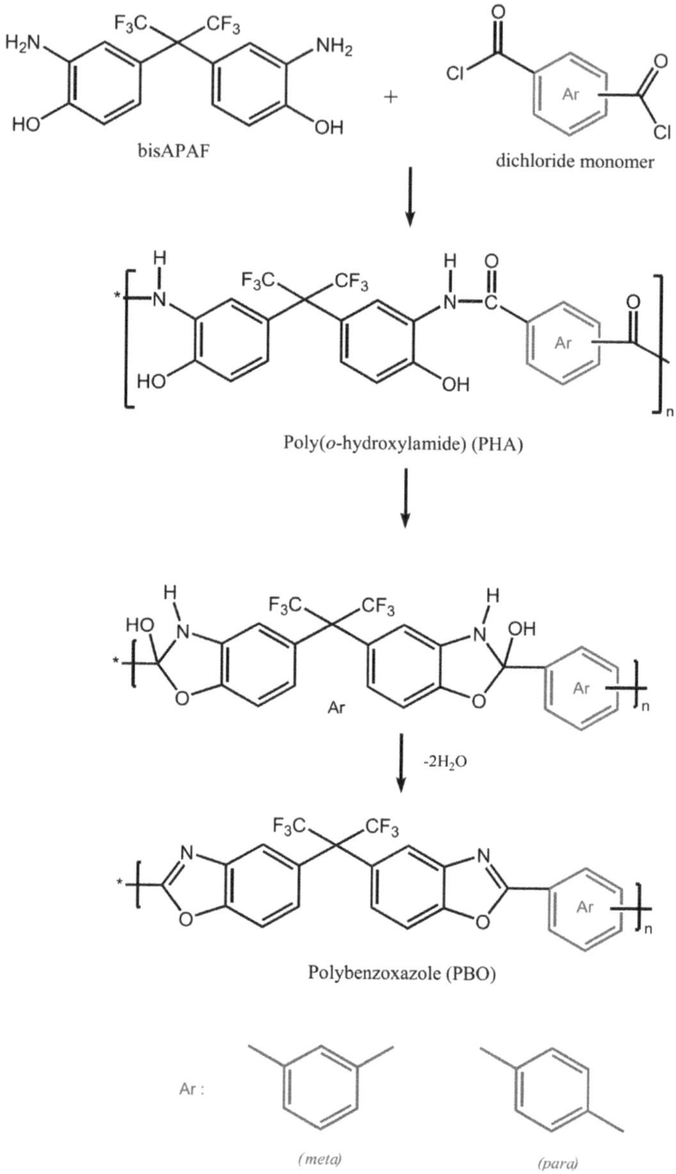

FIGURE 4.1 Structure of polymer membranes. (Adapted from Kim et al., 2015 with reprint permission and copyright license obtained from Elsevier.)

gas-transport channel. A longer path length will improve the gas transport characteristics because a thinner material should allow for greater gas permeance. It has been demonstrated that ultra-thin membranes with typical low-thickness characteristic, have outstanding gas separation capabilities. Ultra-thin membranes with nanopores smaller than 0.5 nm should have great benefits because of the effects of molecular sieving offered by them as well as their gas transport properties.

This can be explained by the presence of nanopores that allow smaller molecules pass through the membranes while preventing the larger ones from going through (Diawara et al., 2020; Métayer et al., 1999).

Creating ultra-thin inorganic membranes with evenly spaced pores to extract valuable gases like hydrogen, oxygen, and nitrogen is difficult. This is due to the kinetic diameters of light gases differing very slightly from one another. Though they have a high throughput, traditional techniques like physical vapor deposition (PVD) and typical chemical vapor deposition (CVD) sometimes fall short when creating conformal thin films. Moreover, these techniques are inappropriate for accurate pore size and gas-transport length regulation. Thus, a quick and efficient manufacturing process is required to create enhanced inorganic membranes with efficient particle size distribution. By using atomic layer deposition (ALD), it is possible to create ultra-thin inorganic membranes with precisely adjustable pore diameters that can even reach molecular dimensions. This method offers the chance to develop ultra-thin layers that can extract gases, towards achieving a considerable positive impact, especially in terms of gas separation performance (Bernardo et al., 2009).

Oxygen-permeable membrane reactors can be used to produce high-purity hydrogen. Both sides of the O_2-permeable materials of this catalytic membrane reactor are supplied with H_2O and low-purity H_2, respectively. In this catalytic membrane reactor, mixed ionic-electron conductor (MIEC) membranes move O_2 ions from one direction to another while catalysts quicken the two surface reactions. As opposed to traditional MIEC membrane processors, which have one side open to an oxidizing environment and the opposite end relaxed to a reducing environment (like CH_4) for hydrogen separation, MIEC membrane-based reactors have both sides exposed to reducing environments for H_2 extraction (Farnam et al., 2014). Even though reducing atmospheres are present on each part of the MIEC membrane, a gradient in the partial pressure of oxygen encourages direct ambipolar diffusion of O_2 ions and electrons. The H_2 separation process requires membranes and catalysts with excellent reducing atmospheric stability since the MIEC and the catalysts are open to strong reducing environments on both sides.

Due to their adaptable compositions and high levels of stability, dual-phase membranes are potential possibilities. In earlier researches, it was discovered that dual-phase materials for hydrogen separation are more stable than single-phase perovskite materials (Galizia et al., 2017). Hydrogen is adsorbed during the hydrogen-lean phase (Salaudeen et al., 2020), while oxygen is adsorbed during the hydrogen-rich phase (Wahab and Sunarti, 2015).

The results of a study on the impacts of surface reactions and bulk diffusion have not yet been published in previous studies. A noble metal catalyst known as Ruthenium/Samarium-doped ceria (Ru/SDC) was used in early works. It is essential to construct non-noble metal catalysts to reduce the material costs of the membrane reactor. Iron-, cobalt-, and nickel-based catalysts have drawn much interest because of their enormous abundance and low cost in real-world applications. A few of the reactions that exhibit considerable catalytic activity at high temperatures are water splitting in chemical looping processes (Chen et al., 2018a,b), the catalytic oxidation and reduction of fossil fuels, oxygen reduction, and hydrogen oxidation on the cathodes and anodes of solid oxide fuel cells (SOFCs). However, it is unknown how they catalyze the separation of hydrogen in oxygen-permeable membrane reactors. Good catalysts need both active centers and supports. This H_2 purification method uses membranes and catalysts and incorporates both oxygen ionic diffusion and electron transfer (Kanezashi et al., 2008). As a result, the supports need to have high ionic conductivity for oxygen and a robust electronic conductivity. Traditional supports like zeolites, SiO_2 (silicon dioxide), and Al_2O_3 are ineffective because they cannot carry oxygen ions and electrons. Doped ceria are desirable options due to their strong ionic conductivity and outstanding stability in decreasing conditions. The Samarium-doped ceria (SDC) has the highest ionic conductivity amongst all the doped ceria. Additionally, SDC becomes a mixed conductor with high ionic and electronic conductivity when Ce^{4+} to Ce^{3+} reduction occurs under favourable conditions. Figure 4.2 presents different morphology/images of ZIF-8+PBI membrane of varying ratios.

Glassy aromatic polyimides, particularly those based on (4,4′-hexafluoroisopropylidene) diphthalic anhydride (6FDA), have garnered much attention among the many reported gas extraction resources. This is due to their remarkable thermal and mechanical qualities, simplicity of processing, and structural versatility. However, conventional polyimide membranes frequently struggle with gas separation performance due to plasticization. In these conditions, as feed pressures increased, more

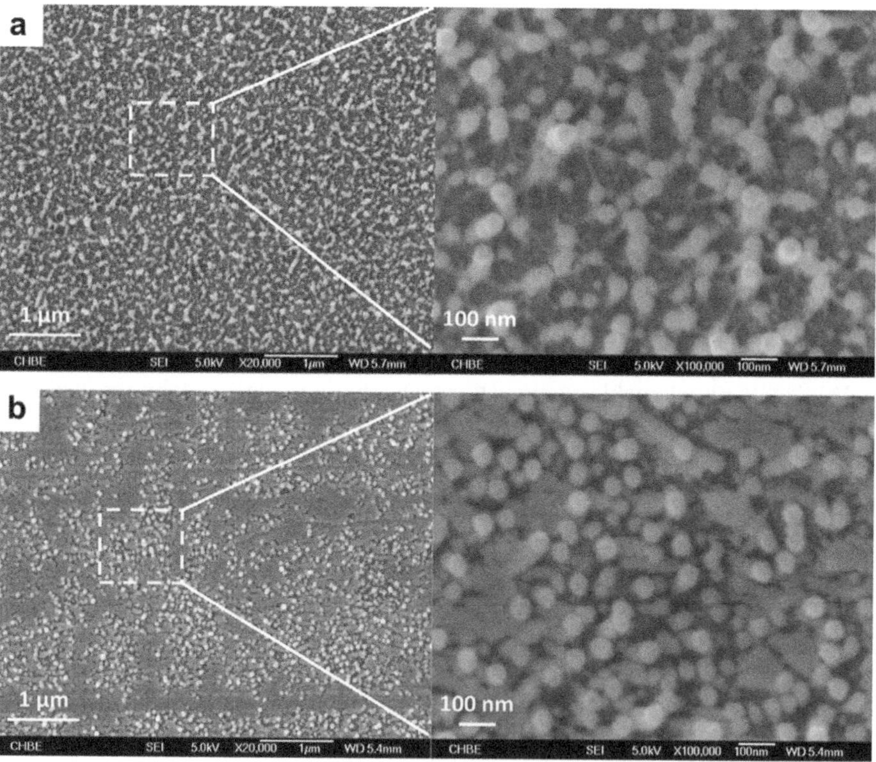

FIGURE 4.2 Field Emission Scanning Electron Microscopy (FESEM) images of (a) ZIF-8+PBI 30:70 %(wt:wt), (b) ZIF-8+PBI 60:40 %(wt/wt) membranes. (Adapted from Yang and Chung, 2013 with reprint permission and copyright license obtained from Elsevier.)

condensable gases (including carbon dioxide) would adsorb and swell the polyimide chains, leading to a more pronounced increase in methane over carbon dioxide characteristics and a sizable reduction in the carbon dioxide–methane selectivity (Wenwei et al., 1994). Furthermore, hollow fiber membranes used in the industry are more prone to plasticization, resulting in an increased selectivity drop; hybridized membranes were thoroughly explored in their previous study. Polyimide designs having minimal plasticization are greatly needed for membrane gas splitting. Several methods are being employed to prevent plasticization, such as increasing the stiffness of the polymer chains or increasing inter-chain connections through hydrogen bonding and electrostatic contact (Prager and Long, 1951; Xiao and Wei, 1992).

Incorporating polar pendant groups into polyimides has been suggested to limit polymer chain mobility through charge transfer complexes (CTCs), H_2 bonds, and electrostatic reactions. These polar pendant groups were shown to tighten the polyimide compositions and boost selectivity, but the materials still underwent significant plasticization at high pressures. These findings demonstrated that physical interactions, such as chain stiffness, hydrogen bonds, and electrostatic contact, which constrain chain mobility, are insufficient to prevent plasticization. Hence, limited segmental mobility is required under mixed-gas feed chemically crosslinked polyimide membranes. To increase plasticization resistance, thermally induced chemical crosslinking of polyimide membranes containing functional groups including -COOH, -Br, and -C $(CF3)^{2-}$, as well as diester linkages, have been used. The properties of polyimide membranes are thought to be significantly impacted by thermally induced chemical crosslinking. Decarboxylation-induced free radical crosslinking has garnered much attention among the many crosslinking methods because it produces stable interchain C–C bonds that increase plasticization resistance. Moreover, the evolution of carbon dioxide because of decarboxylation makes it easier to build open architecture as well as boost gas permeability (Marais et al., 2000; Follain et al., 2010).

Due to their inherent benefits over more traditional separation procedures, such as simplicity of usage and decreased environmental impact, polymeric membranes are a popular option for such separations. By using a solution (S)–diffusion (D) process, the gas can move across polymer membranes (Soto Puente et al., 2015; Chen et al., 2019; Hirata et al., 2005). Due to the substantially smaller diameter of hydrogen than that of nitrogen, methane, and carbon dioxide, hydrogen's diffusion coefficient (D) is much bigger than that of nitrogen, methane, and carbon dioxide. Instead, because the critical temperatures of carbon dioxide, nitrogen, and methane are much higher than the critical temperature of hydrogen, the solubility coefficients (S) of those gases are much higher than those of hydrogen (Yong et al., 2019; Hashino et al., 2011). This trade-off between diffusion and solubility significantly hampers the separation of hydrogen–N_2, hydrogen–CH_4, and hydrogen–CO_2. Thus, the only realistic way to create a membrane with high H_2 permeability and high hydrogen–nitrogen, hydrogen–methane, and hydrogen–carbon dioxide selectivity is to improve the diffusion coefficient and diffusion selectivity of hydrogen to nitrogen, methane, and carbon dioxide. By dramatically increasing the properties, adding micropores to create a polymer of intrinsic microporosity (PIM) is an effective technique to increase the benefits of hybrid membranes (Yong et al., 2019; Hashino et al., 2011). Due to their distinctive kinked topologies, which prevent the polymer main chains from packing tightly and create quick paths for gas transport, PIMs are distinguished by their significant fractional free volumes (Miao et al., 2015). The hydrogen permeability of PIMs can be extremely high; however, most porous PIMs have very poor hydrogen–nitrogen, hydrogen–methane, and hydrogen–carbon dioxide selectivity because their internal micropores are significantly larger than those gas pairs.

Several techniques, including functionalization, have been developed to enhance the micropore and improve the size-sieving impact of PIMs for hydrogen extraction. Unfortunately, the disadvantages limit the utilization of polymeric membranes. MMMs may significantly improve gas separation membrane performance, with molecular sieving components embedded in a polymer matrix. These membranes combine the polymers' simple processing capabilities with the molecular sieve material's size sieving capabilities (Yu et al., 2015). The qualities of the polymers and fillers, compatibility of the polymer fillers, and the process of membrane creation all impact the performance of MMMs. Due to poor compatibility between the polymer and the external zeolite surface, the fundamental issue with MMMs is the creation of voids at the polymer–filler interface. There have been numerous attempts to improve the compatibility of polymers and fillers used in membrane fabrication processes by adding functional groups (Mansor et al., 2020; Huang et al., 2014; Soto Puente et al., 2017).

With the emergence of nanoporous materials like zeolites and metal–organic frameworks (MOFs), MMMs, which incorporate flexible fillers into the polymeric matrix, have been shown to be a promising contender to overcome this restriction. However, several prerequisites, including interfacial compatibility, uniform dispersion (Stern, 1994), permeability matching, and plasticization resistance, are necessary to convert their intrinsic separation performance into MMMs efficiently. Since they combine the ease of processing polymers with the higher separating power of inorganic materials, MMMs, which comprise an organic continuous matrix and an inorganic filler, are promising membranes (Li et al., 2022). MMMs have faced significant difficulties with interfacial compatibility problems due to insufficient filler–polymer adhesion. The main causes of the creation of non-selective voids are often filler particle aggregation and interfacial gaps between the filler and polymer phases. Due to the unrestricted flow of gas molecules caused by the presence of interfacial defects, gas selectivity will be lowered (Stern, 1994; Vu et al., 2003).

Due to its strong chain stability and scalability, cellulose acetate (CA) is a polymer that has been thoroughly researched for industrial gas separation. The surface groups on polymer CA are sufficient, and it is simple to combine them with various inorganic fillers. Various fillers have been incorporated into the CA matrix to improve gas transport capabilities, including zeolites, carbon compounds, silicas (Kim et al., 2013), and MOFs. The production of CA-based MMMs, and MOFs with acceptable pore aperture and customizable surface characteristics holds the most promise among the organic fillers. However, the low compatibility of fillers and polymers poses a significant problem for MMMs. The presence of filler particle aggregation, non-selective vacancies, and interfacial defects severely hamper membrane performance. Numerous efforts have been made to create the perfect polymer–filler interface to combat this. According to earlier research, the reduction of interfacial voids can be achieved by NH_2-functionalozation the organic MOF moiety (Kaiser et al., 2017). Table 4.1 shows the synthetic membranes for hydrogen storage and their characteristics.

TABLE 4.1

Synthetic Membranes for Hydrogen Storage Alongside Their Strengths and Weaknesses

Membrane	Selectivity	Permeability (%/ mol·m^{-2}·s/ mL·min^{-1}·cm^{-2}/Barrer/ cm^3(STP)/GPU/ cm·s·cm·Hg)/ Permeability Flux (cm^3(STP)/min.cm^2)/ ppm.H$_2$S)	Temp. (°C)	Pressure (MPa/kPa)	Strengths	Weaknesses	References
Nickel hollow fiber membranes for hydrogen separation	100	100a	1,000	0.3	• Excellent H$_2$ capture selectivity • Excellent H$_2$ capture permeability • Excellent mechanical strength and stable morphology even after hydrogen separation • Increase permeability with an increase in temperature (400°C–1,000°C) • High hydrogen recovery with an increase in temperature • Excellent thermal stability • Excellent chemical stability	• High cost of materials	Wang et al. (2019)
High sulfur tolerance dual-functional cermet hydrogen separation membranes	–	220a	600	–	• Increase in permeability flux with an increase in temperature • Decrease in the cost of materials	• Permeation of O$_2$ which may have reacted with H$_2$ and hence reduced the actual H$_2$ partial pressure on the feed side	Jeona et al. (2011)

(Continued)

TABLE 4.1 (*Continued*)

Synthetic Membranes for Hydrogen Storage Alongside Their Strengths and Weaknesses

Membrane	Selectivity	Permeability (%/ mol·m^{-2}·s/ mL·min^{-1}·cm^{-2}/ Barrer/ cm^3(STP)/GPU/ cm·s·cm·Hg)/ Permeability Flux (cm^3(STP)/min.cm^2)/ ppm.H$_2$S)	Temp. (°C)	Pressure (MPa/kPa)	Strengths	Weaknesses	References
Hydrogen separation by dual functional cermet membranes	–	2.7b	1,500	200	• High thermal stability • High chemical stability	• SO$_2$ detected on the feed side	Jeon et al. (2013)
Asymmetric layered vanadium membranes for hydrogen separation	–	65a	350	0.10	• Increase in relative conversion with an increase in temperature (50°C–100°C), and it remains consistent with a further increase in temperature (200°C–400 °C)	• Low relative conversion at a temperature of 50°C • Reduction in Ni (nickel) thickness leads to low hydrogen permeance	Viano et al. (2015)

(*Continued*)

TABLE 4.1 (*Continued*)

Synthetic Membranes for Hydrogen Storage Alongside Their Strengths and Weaknesses

Membrane	Selectivity	Permeability (%/ mol·m^{-2}·s/ mL·min^{-1}·cm^{-2}/ Barrer/ cm^3(STP)/GPU/ cm·s·cm·Hg)/ Permeability Flux (cm^3(STP)/min.cm^2)/ ppm.H$_2$S)	Temp. (°C)	Pressure (MPa/kPa)	Strengths	Weaknesses	References
Metallic nickel hollow fiber membranes for hydrogen separation	100	7.66×10^{-3}b	1,000–1,400	0.15	• Excellent H$_2$ capture selectivity • Excellent H$_2$ capture permeability • Excellent mechanical strength and stable morphology even after hydrogen separation • Excellent thermal stability • Excellent chemical stability • Decrease in N$_2$ permeance at high temperature ranges (1150°C–1400°C)	• No noticeable rise in H$_2$ permeation flux at temperatures between 0°C and 400°C	Wang et al. (2016)
Graphene oxide gas separation membranes intercalated by UiO-66-NH$_2$	9.75	3.9×10^{-8}a	25	–	• Excellent H$_2$ capture selectivity • Excellent H$_2$ capture permeability • Excellent mechanical strength and stable morphology even after hydrogen separation • Excellent thermal stability • Excellent chemical stability		Jia et al. (2017)

(*Continued*)

TABLE 4.1 (*Continued*)

Synthetic Membranes for Hydrogen Storage Alongside Their Strengths and Weaknesses

Membrane	Selectivity	Permeability (%/ mol·m^{-2}·s/ mL·min^{-1}·cm^{-2}/ Barrer/ cm^3(STP)/GPU/ cm·s·cm·Hg)/ Permeability Flux (cm^3(STP)/min.cm^2)/ ppm.H$_2$S)	Temp. (°C)	Pressure (MPa/kPa)	Strengths	Weaknesses	References
Tailoring hydrogen separation performance through the ceramic lanthanum tungstate membranes by chlorine doping	–	0.15a	1,000	–	• Excellent H$_2$ capture selectivity • Excellent H$_2$ capture permeability • Excellent mechanical strength and stable morphology even after hydrogen separation • Excellent thermal stability • Excellent chemical stability	• Low hydrogen permeation flux at temperatures between 800°C and 950°C	Chen et al. (2019)
Non-noble metal catalysts coated on oxygen-permeable membrane reactors	100	7a	900	–	• Excellent H$_2$ capture selectivity • Excellent H$_2$ capture permeability • Excellent mechanical strength and stable morphology even after hydrogen separation • Excellent thermal stability • Excellent chemical stability	• Drop in hydrogen concentration with an increase in pressure gradient	Cai et al. (2020)

(Continued)

TABLE 4.1 (*Continued*)

Synthetic Membranes for Hydrogen Storage Alongside Their Strengths and Weaknesses

Membrane	Selectivity	Permeability (%/ mol·m^{-2}·s/ mL·min^{-1}·cm^{-2}/ Barrer/ cm^3(STP)/GPU/ cm·s·cm·Hg)/ Permeability Flux (cm^3(STP)/min.cm^2)/ ppm.H$_2$S)	Temp. (°C)	Pressure (MPa/kPa)	Strengths	Weaknesses	References
Cysteamine-crosslinked graphene oxide membrane	21.3	51.5a	25	0.15	• Excellent H$_2$ capture permeability • Excellent mechanical strength and stable morphology even after hydrogen separation • Excellent thermal stability • Excellent chemical stability • Increase in selectivity with an increase in time (1–5hr)	• Average H$_2$ capture selectivity • Drop in permeance with an increase in time (1–5hr)	Cheng et al. (2020)
Thermally rearranged polymer membranes containing highly rigid biphenyl ortho-hydroxyl diamine	50	325a	425	–	• Excellent H$_2$ capture permeability • Excellent mechanical strength and stable morphology even after hydrogen separation • Excellent thermal stability • Excellent chemical stability • Increase in permeability at temperatures between 350°C and 450°C	• Average H$_2$ capture selectivity • Drop in permeability at temperatures between 250°C and 350°C	Hu et al. (2020)

(*Continued*)

TABLE 4.1 (*Continued*)

Synthetic Membranes for Hydrogen Storage Alongside Their Strengths and Weaknesses

Membrane	Selectivity	Permeability (%/ mol·m^{-2}·s/ mL·min^{-1}·cm^{-2}/ Barrer/ cm^3(STP)/GPU/ cm·s·cm·Hg)/ Permeability Flux (cm^3(STP)/min.cm^2)/ ppm.H$_2$S)	Temp. (°C)	Pressure (MPa/kPa)	Strengths	Weaknesses	References
Alicyclic segments/Troger's Base (TB)-based polyimide backbones	259	10^{-10}a	35	1	• Excellent H$_2$ capture selectivity • Excellent mechanical strength and stable morphology even after hydrogen separation • Excellent thermal stability	• Average H$_2$ capture permeability • Mechanical strength drops with an increase in temperature	Zhang et al. (2020)
Few-layered transition metal dichalcogenide Nanosheets membrane	7.0 ± 0.2	47,100 ± 174a	–	–	• Excellent H$_2$ capture permeability • Excellent mechanical strength and stable morphology even after hydrogen separation • Excellent thermal stability • Environmentally friendly for use • Relatively easy to synthesize	• Average H$_2$ capture selectivity	Keshebo et al. (2021)
Asymmetric carbon molecular sieve hollow fiber membranes	511	430a	600	-	• Excellent H$_2$ capture permeability • Excellent H$_2$ capture selectivity • Excellent mechanical strength and stable morphology even after hydrogen separation • Excellent thermal stability • Excellent chemical stability		Liu et al. (2022)

(*Continued*)

TABLE 4.1 (*Continued*)

Synthetic Membranes for Hydrogen Storage Alongside Their Strengths and Weaknesses

Membrane	Selectivity	Permeability (%/ mol·m^{-2}·s/ mL·min^{-1}·cm^{-2}/ Barrer/ cm³(STP)/GPU/ cm·s·cm·Hg)/ Permeability Flux (cm³(STP)/min.cm²)/ ppm.H$_2$S)	Temp. (°C)	Pressure (MPa/kPa)	Strengths	Weaknesses	References
Fabricated flexible hydrogen-bonded organic framework-based mixed matrix membrane	61.7	428.1a	25	1	• No reduction of permeability and separation factor with an increase in temperature • Excellent H$_2$ capture permeability • Excellent mechanical strength and stable morphology even after hydrogen separation • Excellent thermal stability • Excellent chemical stability	• Average H$_2$ capture selectivity	Li et al. (2022)
Externally self-supported metallic nickel hollow fiber membranes	–	73.17a	1,000	–	• Increase in permeability flux between 500°C and 1,000°C • Excellent H$_2$ capture permeability • Excellent mechanical strength and stable morphology even after hydrogen separation • Excellent thermal stability	• Low permeability flux with an increase in temperature between 0°C and 400°C	Wang et al. (2022)

(*Continued*)

TABLE 4.1 (*Continued*)

Synthetic Membranes for Hydrogen Storage Alongside Their Strengths and Weaknesses

Membrane	Selectivity	Permeability (%/ mol·m^{-2}·s/ mL·min^{-1}·cm^{-2}/ Barrer/ cm^3(STP)/GPU/ cm·s·cm·Hg)/ Permeability Flux (cm^3(STP)/min.cm^2)/ ppm.H$_2$S)	Temp. (°C)	Pressure (MPa/kPa)	Strengths	Weaknesses	References
Facile tailoring molecular sieving effect of PIM-1 by in situ O^3 treatment	121	1,294a	35	2	• Excellent H$_2$ capture permeability • Excellent mechanical strength and stable morphology even after hydrogen separation • Excellent thermal stability • Excellent H$_2$ capture selectivity	• Drop in weight at temperatures above 500°C	Ji et al. (2022)
Atomic layer deposited aluminum oxide membranes	5,148.2	3.03×10^{-7}a	21 ± 2	102	• Excellent recycling potential • Excellent H$_2$ capture permeability • Excellent mechanical strength and stable morphology even after hydrogen separation • Excellent thermal stability • Excellent H$_2$ capture selectivity	• Drop in separation factor with an increase in kinetic diameter	Liu and He (2022)

Note: [a] = permeability; [b] = permeability flux.

4.3 ZEOLITE-BASED MEMBRANES FOR TRAPPING HYDROGEN GAS

In the chemical industry, zeolites, which are hybrid materials with regular dimensions and angstrom-scale openings made of TO_4 tetrahedra, are frequently used as adsorbents, catalysts for heterogeneous reactions, and ion exchangers. Additionally, because of the unmatched benefits, such as molecule recognition at the sub-nanometer scale and improved characteristics, their generated films/membranes have also been extensively researched and explored in extraction and catalytic reactions. More than twenty different types of zeolite materials and membranes have been manufactured up to this point, including NaX (compounds of sodium) (Jiang et al., 2019).

Zeolite can be created when volcanic rock or ash reacts with alkaline fluids. The mineral zeolite can create a wide variety of aluminosilicates that form because of combinations of elements such as O_2, SiO_2, and Al (Nenoff et al., 2006; Bernardo et al., 2020). Due to their structure and composition, they are highly prone to cation exchange abilities and are small. The efficiency of these exchange capacities varies between types due to potential impurities, and it is significantly less in naturally formed zeolites (Bernardo et al., 2020; Japip et al., 2020; Hillman et al., 2018; Zhou et al., 2020; Keizer et al., 1998; Avila et al., 2009; Rey et al., 2017; Olson et al., 2004; Wang et al., 2017). Nevertheless, zeolites are also produced artificially by boiling a solution of NaOH (sodium hydroxide), Al, and SiO_2. Using voids between the grains, traps materials too big to pass through. Everything is caught up at the top of the system described above, and the lowest phases offer support and space for processing. By adding more pores to the treatment media, the zeolite filters' filtering efficiency may be increased (Wang et al., 2017; Kim et al., 2016; Chisholm et al., 2015; Nian et al., 2018; Luo et al., 2018; Fasolin et al., 2019). Zeolite media have numerous pores, which allow them to take in particles before trapping them. As a result, they collect and absorb particles that are trapped between grains. The mineral zeolite's capacity for cation exchange, which entails soaking up positive ions from H_2O and substituting them for other ions, makes this possible. Zeolite may be able to capture a lot of impurities without the requirement for back-washing because of its characteristics (Mei et al., 2018; Yu et al., 2019). The medium uses the adsorption process to both absorb and remove particles. In contrast to passively becoming stuck between grains, materials adhere to the medium's body during this process (Yu et al., 2019).

Zeolite does not clog up quickly. Therefore, it also experiences less pressure loss during use. Since it may reduce or eliminate some hard chemicals and is more stable than other materials, this medium can function as a water softener. The characteristics of the H_2O to be processed, the supply of natural zeolite, the extent of alteration to create novel zeolites, and other factors will significantly affect how effective zeolite-based filters are. Zeolite can be classified as either natural or synthetic types, with the only difference being that the latter is produced by processing natural ore materials. At the same time, the former is synthesized using energy-intensive chemicals. Synthetic zeolites have a balanced ratio, whereas clinoptilolite zeolites have a ratio of 5:1 (Tang et al., 2019). Some important factors are to be considered when evaluating zeolites. This results in complicated and tough methods of synthesis, frequently made even more difficult by producing grain boundary defects during activation. One of the main obstacles preventing the use of zeolite frameworks with pore structures suitable for H_2-sieving, such as sodalite, into gas separation membranes is their complexity (Khan et al., 2010). Table 4.2 consists of different zeolite membranes alongside their strengths and weaknesses.

The ability of zeolite membranes to separate gases mostly depends on the gas components' tendency to bind to the membrane's interface and their diffusion rates via their inherent nanopores. It is generally known that these two processes interact, as the adsorption of gases in the zeolite's channels affects how easily molecules move through the crystal. The competitive adsorption of gas molecules significantly impacts on the diffusion of the target penetrating gas in a multi-component gas mixture. Therefore, knowledge of multicomponent adsorption and diffusion algorithms in microporous materials is necessary to predict the separation performance of molecular sieve membranes. The permeability of gas mixtures with more than one element is highly limited through the nanopores of the zeolite membrane despite substantial studies on the prediction of multicomponent gas permeation through such membranes based on single gas results (Farjoo et al., 2017).

Under challenging circumstances, zeolite is known to be stable. Thanks to its unique features; zeolite is suitable for various applications, including membrane separation, integrated membrane reactors, electrodes, chemical sensors, and insulators. The manner of preparation affects the characteristics of zeolite membranes. Additionally, natural zeolites are ineffective for separating hydrogen. As a result, adjustments are needed to personalize and enhance the features of zeolite into necessary properties. The method of synthesis may need to be adjusted, or some composite materials may be created because of inherent alterations. Hydrothermal synthesis is the most employed technique for creating dense zeolite membranes. Hydrothermal synthesis can be carried out with seeding, and a microwave can be used for the heating process. Zeolite membranes require a porous support to be attached; this support could be a tube, a hollow fiber, or another membrane module design. The procedures used to create the zeolite membrane will influence how dense it is, thus impacting how well it separates (Makertihartha et al., 2017). Figure 4.3 shows carbon-templated zeolite membranes with visible adsorption sites.

A robust economic incentive has driven the development of techniques to recover hydrogen from steam reformer off-gas, catalytic reformer off-gas, and ethylene plant effluent gas due to the rising need for hydrogen in petroleum refineries and petrochemical processes. One of the usual multibed PSA processes that incorporate several PSA phases in various ways is the H_2 recovery PSA process which is widely utilized in the chemical and petroleum refining industries. The established hydrogen PSA process used for recovering hydrogen from coke oven gas (COG) is a promising new application. About 60% of COG is made up of hydrogen, 25% is methane, 8% is carbon monoxide, and the rest is carbon dioxide, nitrogen, oxygen, etc. Because most of the time some impurities must be pretreated before any large impurities are absorbed in the central adsorbent portion, an adsorption bed is constructed using various adsorbents, including alumina, activated carbon, and zeolite. Researching a PSA process that uses zeolite 5A is crucial because it is commonly used as an adsorbent at the main section of the adsorption bed (Chang, 1997). Some adsorbents suitable for PSA are illustrated in Figure 4.4.

TABLE 4.2

Zeolite Membranes for Hydrogen Storage

Zeolite	Selectivity	Permeance (GPU/ kg·m⁻²·h⁻²)	Temp. (°C)	Pressure (bar/kPa/ atm)	Strengths	Weaknesses	References
PSf-acrylate-zeolite membrane	3.57	15.10	25	12	• With the addition of zeolite concentration, permeance increases • Low selectivity • High permeance • Excellent mechanical strength • Excellent chemical stability • Excellent mechanical strength	• Low selectivity • With increased CO_2 concentration in the feed gas, the selectivity decreases • An increase in temperature leads to a decrease in permeance	Khan et al. (2010)
Adsorbent/ membrane hybrid (AMH) system	199.0	48.82×10^{11}	50	100	• Excellent regeneration capacity • High selectivity • High permeance	• Drop in permeation flux with an increase in the time of the experiment	An et al. (2011)
Fabrication of high-stability W-MFI zeolite membranes for ethanol/water mixture separation	–	2.81	–	–	• Excellent permeance • Increase in permeance with an increase in temperature • Excellent chemical stability • Excellent mechanical strength	• Drop in selectivity with an increase in pressure	Moon et al. (2010)
Dense natural zeolite membranes	57	5.2×10^{-7}	500	–	• High permeance • Excellent chemical stability • Excellent mechanical strength	• Average selectivity	Moon et al. (2010)
Highly stable bilayer MFI zeolite membranes	180	1.2×10^{-7}	500	6	• High permeance • Excellent chemical stability • Excellent mechanical strength	• Average selectivity	Wang et al. (2014)

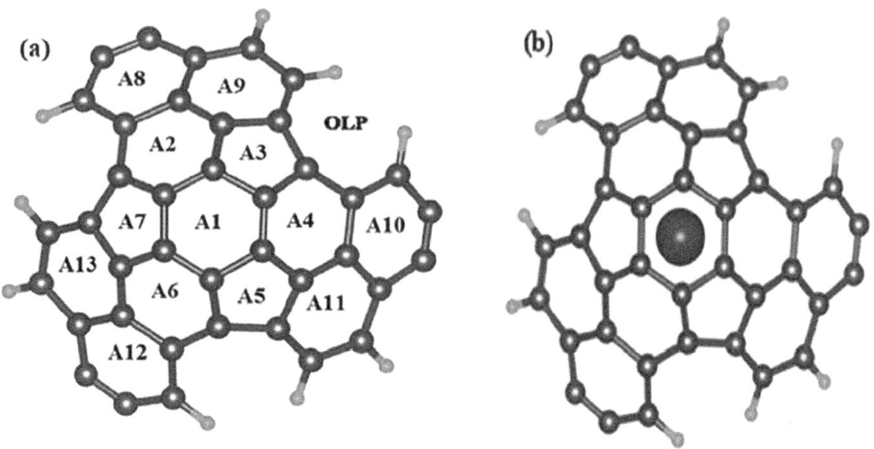

FIGURE 4.3 Structure of (a) zeolite templated carbon showing all possible adsorption sites and (b) doped zeolite templated carbon. The sites marked A1–A12 are possible doping sites. (Adapted from Kundu et al., 2022 with reprint permission and copyright license obtained from Elsevier.)

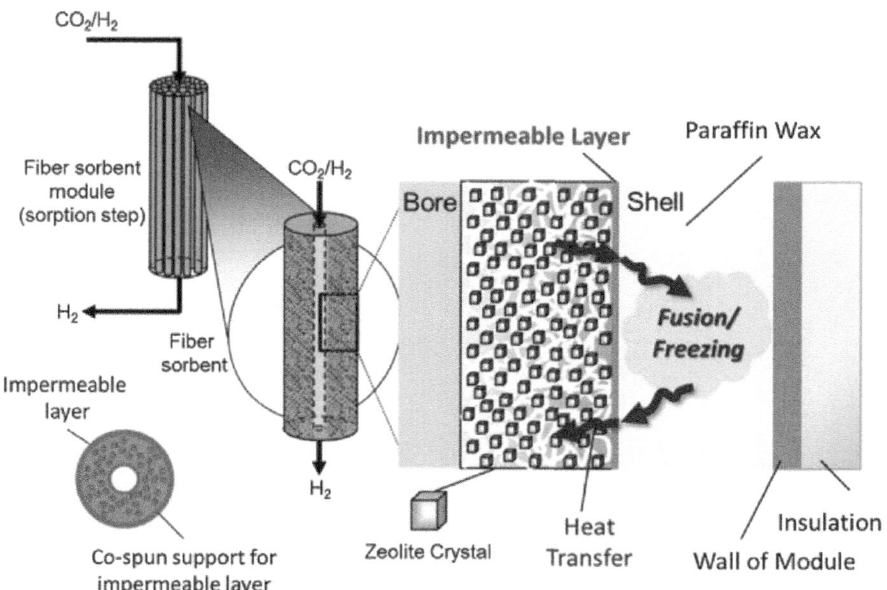

FIGURE 4.4 Adsorbents for pressure swing adsorption. (Adapted from Lider et al., 2023 with reprint permission and copyright license obtained from Elsevier.)

Due to their significance in gas sorption, the solid adsorbent's internal porosity and surface characteristics are crucial for separation efficiency. Instead, the adsorbent's molecular characteristics, such as chemical affinity or the size of the separated components' molecules, are crucial to the separation process. Meanwhile, the species' selective adsorption from gas mixtures separates and purifies these materials. Additionally, because porous materials have a limited volumetric capacity and the separated gas must be delivered at or above ambient pressure, gas storage requires increased pressures. Using molecular sieving and preferential adsorption mechanisms, crystalline zeolite and MOF membranes can separate molecules with high flux and selectivity. The effects of impure gases, such as higher hydrocarbons, are complex and heavily reliant on separation pairs and membrane microstructures (Al-Naddaf et al., 2018; Das and Das, 2020). The characteristics of 13X zeolite, which makes it suitable for hydrogen adsorption, are presented in Figure 4.5.

Although hydrogen storage in zeolites has been well studied experimentally, its modeling features have not yet been investigated. To the author's knowledge, no methods

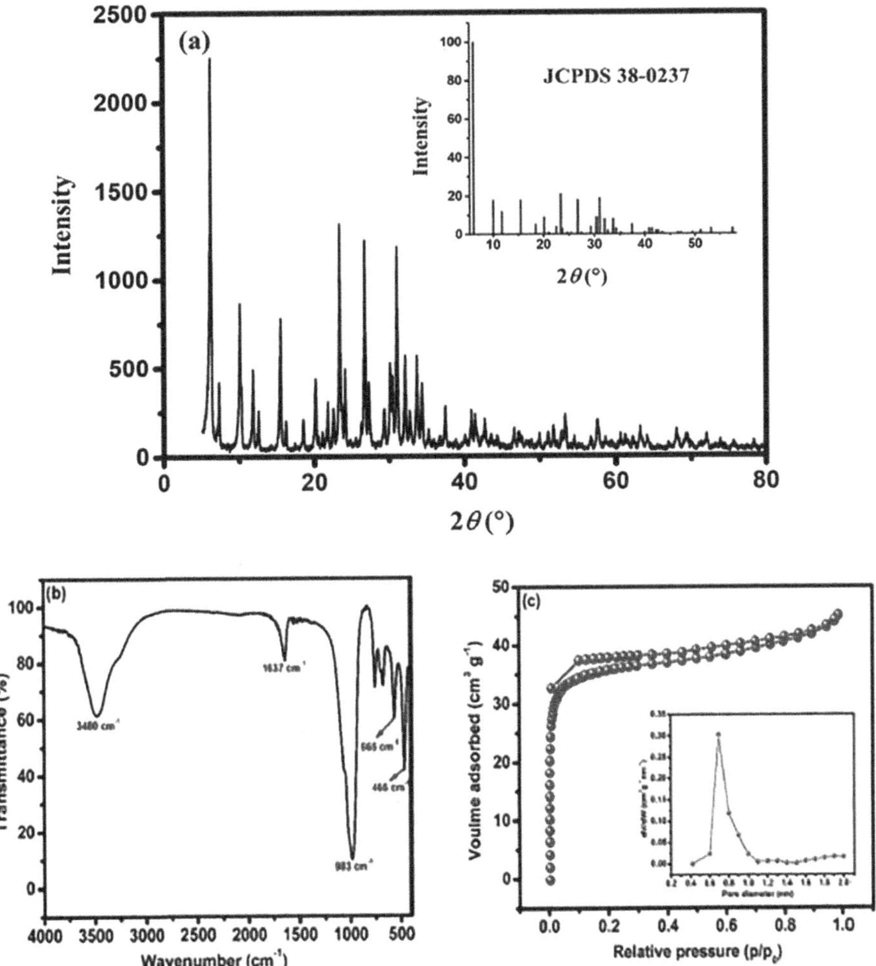

FIGURE 4.5 13X Zeolite's (a) XRD, (b) FTIR, (c) adsorption isotherm, and (c) size distribution. (Adapted from Vinodh et al., 2020 with reprint permission and copyright license obtained from Elsevier.)

for calculating the hydrogen adsorption capability of zeolites have been proposed. A technique like this makes it possible to compare the ability of various zeolites' hydrogen storage capacity and indicates the impact of critical factors on hydrogen adsorption. Four ANN types have been used to forecast or compare the hydrogen adsorption capacity of 28 zeolites because artificial neural networks (ANNs) have lately been involved in accurately mimicking a wide range of phenomena (Wu et al., 2020).

4.4 CONCLUSION

Membranes are special materials that separate distinct substances from their parent mixtures. There are different types of membranes, and their properties are geared toward specificity in the substance they are synthesized to extract. Other membranes require different pore sizes to remove distinct gases selectively. The diameters of the distinct gases are critical. Hydrogen is significant to the chemical and power plant industries. It serves different purposes and, as such, is required to be able to carry out essential processes. Zeolite membranes are unique materials synthesized from silica and alumina varying ratios for designated operations. This chapter highlighted several membranes and zeolites to determine their strengths, weaknesses, and potential upgrades to enhance their operations. Hybridizing zeolites with other membranes helps to increase their mechanical strength and chemical and thermal stability, which allows them to operate at the required conditions.

REFERENCES

Acar, C., Dincer, I. (2018). Hydrogen production. *Comprehensive Energy Systems*, 3–5, 1–40. https://doi.org/10.1016/B978-0-12-809597-3.00304-7.

Acar, C., Dincer, I. (2019). Review and evaluation of hydrogen production options for better environment. *Journal of Cleaner Production*, 218, 835–849.

Al-Naddaf, Q., Thakkar, H., Rezaei, F. (2018). Novel zeolite-5A@MOF-74 composite adsorbents with core-shell structure for H_2 purification. *ACS Applied Materials & Interfaces*, 29. https://doi.org/10.1021/acsami.8b10494.

An, W., Swenson, P., Wu, L., Waller, T., Ku, A., Kuznicki, S.M. (2011). Selective separation of hydrogen from C1/C2 hydrocarbons and CO_2 through dense natural zeolite membranes, *Journal of Membrane Science*, 369, 414–419. https://doi.org/10.1016/j.memsci.2010.12.025.

Arif, A., Rizwan, A., Elkamel, A., Hakeem, L., Zaman, M. (2020). Optimal selection of integrated electricity generation systems for the power sector with low greenhouse gas (GHG) emissions. *Energies*, 13 (17), 4571. https://doi.org/10.3390/en13174571.

Avila, A.M., Funke, H.H., Zhang, Y., Falconer, J.L., Noble, R.D. (2009). Concentration polarization in SAPO-34 membranes at high pressures. *Journal of Membrane Science*, 335 (1), 32–36.

Balashankar, V.S., Rajendran, A. (2019). Process optimization-based screening of zeolites for post-combustion CO_2 capture by vacuum swing adsorption. *ACS Sustainability and Chemical Engineering*, 7 (21), 17747–17755.

Beni, F.A., Shahrak, M.N. (2020). Alkali metals-promoted capacity of ZIF-8 and ZIF-90 for carbon capturing: A molecular simulation study. *Polyhedron*, 178, 114338.

Bernardo, G., Araujo, T., da Silva Lopes, T., Sousa, J., Mendes, A. (2020). Recent advances in membrane technologies for hydrogen purification. *International Journal of Hydrogen Energy*, 45 (12), 7313–7338.

Bernardo, P., Drioli, E., Golemme, G. (2009). Membrane gas separation: A review/state of the art. *Industrial & Engineering Chemistry Research*, 48, 4638–4663. https://doi.org/10.1021/ie8019032.

Caro, J., Noack, M. (2007). Zeolite membranes – from Barrer's vision to technical applications: New concepts in zeolite membrane R&D. *Studies in Surface Science and Catalysis*, 170, 96–109.

Cai, L.L., Hu, S.Q., Cao, Z.W., Li, H.B., Zhu, X.F., Yang, W.S. (2019). Dual-phase membrane reactor for hydrogen separation with high tolerance to CO_2 and H_2S impurities. *AIChE Journal*, 65, 1088–1096.

Cao, L., Yu, I.K.M., Xiong, X., Tsang, D.C.W., Zhang, S., Clark, J.H., Hu, C., Ng, Y.H., Shang, J., Ok, Y.S. (2020). Biorenewable hydrogen production through biomass gasification: A review and future prospects. *Environmental Research*, 186, 109547. https://doi.org/10.1016/j.envres.2020.109547.

Cai, L., Zhu, Z., Cao, Z., Li, W., Li, H., Zhu, Z., Yang, W. (2020). Non-noble metal catalysts coated on oxygen-permeable membrane reactors for hydrogen separation. *Journal of Membrane Science*, 594, 117463.

Cebrucean, D., Cebrucean, B., Ionel, I., Spliethoff, H. (2017). Performance of two iron-based syngas-fueled chemical looping systems for hydrogen and/or electricity generation combined with carbon capture. *Clean Technologies and Environmental Policy*, 19, 451–470. https://doi.org/10.1007/s10098-016-1231-y.

Chang, J.-W. (1997). Separation of hydrogen mixtures by a two-bed pressure swing adsorption process using zeolite 5A. *Industrial & Engineering Chemistry Research*, 36, 2789–2798,

Chen, C., Chen, C.H., Chang, M.H., Chang, Y.C., Shen, C.H., Wan, H.P. (2018a). 30-kWth moving bed chemical looping system progress for hydrogen production. In: *The Proceedings of 5 International Conference on Chemical Looping 2018*. Park City, Utah, USA. Sept 24–27, 3C-1. https://doi.org/10.1016/j.ijggc.2019.102954.

Chen, C., Lee, H.H., Chen, W., Chang, Y.C., Wang, E., Shen, C.H., Huang, K.E. (2018b). Study of an iron-based oxygen carrier on the moving bed chemical looping system. *Energy Fuels*, 32, 3660–3667. https://doi.org/10.1021/acs.energyfuels.7b03721.

Chen, L., Liu, L., Xue, J., Zhuang, L., Wang, H. (2019). Tailoring hydrogen separation performance through the ceramic lanthanum tungstate membranes by chlorine doping. *Journal of Membrane Science*, 573, 117–125. https://doi.org/10.1016/j.memsci.2018.11.073.

Cheng, L., Guan, K., Liu, G., Jin, W. (2020). Cysteamine-crosslinked graphene oxide membrane with enhanced hydrogen separation property. *Journal of Membrane Science*, 595, 11756. https://doi.org/10.1016/j.memsci.2019.117568.

Cheng, L., Liu, G., Jin, W. (2019). Recent progress in two-dimensional-material membranes for gas separation. *Acta Physico - Chimica Sinica*, 35, 1090–1098.

Chisholm, N.O., Anderson, G.C., McNally, J.F., Funke, H.H., Noble, R.D., Falconer, J.L. (2015) Increasing H_2/N_2 separation selectivity in CHA zeolite membranes by adding a third gas. *Journal of Membrane Science*, 496, 118–124.

Chu, W.L., Liu, Y.C., Wang, H.K., Cai, R., Yang, W.S. (2018). Highly efficient removal of CO in effluent streams from real-life propane oxidation process over CuO-CeO_2-based catalysts. *ChemCatChem*, 10, 4292–4299.

Das, N., Das, J. K. (2020). Zeolites: An emerging material for gas storage and separation applications. In: Margeta K. and Farkaš A. (eds.) *Zeolites-New Challenges*, pp. 1–27. IntechOpen. https://dx.doi.org/10.5772/intechopen.91035

Diawara, B., Fatyeyeva, K., Chappey, C., Colasse, L., Ortiz, J., Marais, S. (2020). Evolution of mechanical and barrier properties of thermally aged polycarbonate films. *Journal of Membrane Science*, 607, 117630. https://doi.org/10.1016/j.memsci.2019.117630.

Ding, L., Wei, Y., Li, L., Zhang, T., Wang, H., Xue, J., Ding, L-X., Wang, S., Caro, J., Gogotsi, Y. (2018). MXene molecular sieving membranes for highly efficient gas separation. *Nature Communications*, 9, 155. https://doi.org/10.1038/s41467-017-02529-6.

Farnam, M., Mukhtar, H., Shariff, A. (2014). A review on glassy polymeric membranes for gas separation. *Applied Mechanics and Materials*, 625, 701–703. https://doi.org/10.4028/www.scientific.net/AMM.625.701.

Farjoo, A., Kuznicki, S.M, Sadrzadeh, M. (2017). Hydrogen separation by natural zeolite composite membranes: Single and multicomponent gas transport. *Materials*, 10, 1159. https://doi.org/10.3390/ma10101159.

Fasolin, S., Romano, M., Boldrini, S., Ferrario, A., Fabrizio, M., Armelao, L., Barison, S. (2019) Single-step process to produce alumina supported hydroxy-sodalite zeolite membranes. *Journal of Materials Science*, 54 (3), 2049–2058.

Follain, N., Valleton, J.-M., Lebrun, L., Alexandre, B., Schaetzel, P., Metayer, M., Marais, S. (2010). Simulation of kinetic curves in mass transfer phenomena for a concentration-dependent diffusion coefficient in polymer membranes. *Journal of Membrane Science*, 349 195–207. https://doi.org/10.1016/j.memsci.2009.11.044.

Galizia, M., Chi, W.S., Smith, Z.P., Merkel, T.C., Baker, R.W., Freeman, B.D. (2017). 50th Anniversary perspective: polymers and mixed matrix membranes for gas and vapor separation: a review and prospective opportunities. *Macromolecules*, 50 (20), 7809–7843. https://doi.org/10.1021/acs.macromol.7b01718.

Gu, X., Tang, Z., Dong, J. (2008). On-stream modification of MFI zeolite membranes for enhancing hydrogen, separation at high temperature. *Microporous Mesoporous Materials*, 111, 441–448.

Hashino, M., Hirami, K., Ishigami, T., Ohmukai, Y., Maruyama, T., Kubota, N., Matsuyama, H. (2011). Effect of kinds of membrane materials on membrane fouling with BSA. *Journal of Membrane Science*, 384, 157–165. https://doi.org/10.1016/j.memsci.2011.09.015.

Hillman, F., Brito, J., Jeong, H.-K. (2018). Rapid one-pot microwave synthesis of mixed-linker hybrid zeolitic-imidazolate framework membranes for tunable gas separations. *ACS Applied Materials & Interfaces*, 10 (6), 5586–5593.

Hirata, Y., Marais, S., Nguyen, Q.T., Cabot, C., Sauvage, J.-P. (2005). Relationship between the gas and liquid water permeabilities and membrane structure in homogeneous and pseudo-bilayer membranes based on partially hydrolyzed poly(ethylene-co-vinyl acetate). *Journal of Membrane Science*, 256, 7–17. https://doi.org/10.1016/j.memsci.2005.01.002.

Hu, X., Lee, W.H., Zhao, J.J., Kim, J.S., Wang, Z., Yan, J., Zhuang, Y., Lee, Y.M. (2020). Thermally rearranged polymer membranes containing highly rigid biphenyl ortho-hydroxyl diamine for hydrogen separation. *Journal of Membrane Science*, 604, 118053. https://doi.org/10.1016/j.memsci.2020.118053.

Huang, Y., Merkel, T.C., Baker, R.W. (2014). Pressure ratio and its impact on membrane gas separation processes. *Journal of Membrane Science*, 463, 33–40. https://doi.org/10.1016/j.memsci.2014.03.016 906.

Japip, S., Xiao, Y., Chung, T.-S. (2020). Particle-size effects on gas transport properties of 6FDA-Durene/ZIF-71 mixed matrix membranes. *Industrial & Engineering Chemistry Research*, 55 (35), 9507–9517.

Jeon, S.-Y., Choi, M.-B., Singh, B., Song, S.-J. (2013). Hydrogen separation by dual functional cermet membranes with self-repairing capability against the damage by H2S. *Journal of Membrane Science*, 428, 46–51. https://dx.doi.org/10.1016/j.memsci.2012.11.009.

Jeona, S.-Y., Choi, M.-B., Parka, C.-N., Wachsmanb, E.D., Songa, S.-J. (2011). High sulfur tolerance dual-functional cermet hydrogen separation membranes. *Journal of Membrane Science*, 382, 323–327. https://doi.org/10.1016/j.memsci.2011.08.024.

Ji, W., Geng, H., Chen, Z., Dong, H., Matsuyama, H., Wang, H., Wang, H., Li, J., Shi, W., Ma, X. (2022). Facile tailoring molecular sieving effect of PIM-1 by in-situ O_3 treatment for high performance hydrogen separation. *Journal of Membrane Science*, 662, 120971. https://doi.org/10.1016/j.memsci.2022.120971.

Ji, F.Z., Yao, J.G., Clough, P.T., Diniz da Costa, J.C., Anthony, E.J., Fennell, P.S., Wang, W., Zhao, M. (2018). Enhanced hydrogen production from thermochemical processes. *Angewandte Chemie International Edition*, 11, 2647–2672.

Jia, M., Feng, Y., Liu, S., Qiu, J., Yao, J. (2017). Graphene oxide gas separation membranes intercalated by UiO-66-NH2 with enhanced hydrogen separation performance. *Journal of Membrane Science*, 539, 172–177. https://dx.doi.org/10.1016/j.memsci.2017.06.005.

Jiang, X., Li, S., Bai, Y., Shao, L. (2019). Ultra-facile aqueous synthesis of nanoporous zeolitic imidazolate framework membranes for hydrogen purification and olefin/paraffin separation. *Journal of Materials Chemistry A*, 7 (18), 10898–10904.

Joly, C., Le Cerf, G., Chappey, C., Langevin, D., Muller, G. (1999). Residual solvent effect on the permeation properties of fluorinated polyimide films. *Separation and Purification Technology*, 16, 47–54. https://doi.org/10.1016/S1383-5866(98)00118-X. 797.

Kaiser, A., Stark, W.J., Grass, R.N. (2017). Rapid production of a porous cellulose acetate membrane for water filtration using readily available chemicals. *Journal of Chemical Education*, 94, 483–487. https://doi.org/10.1021/acs.jchemed.6b00776.

Kanezashi, M., O'Brien-Abraham, J., Lin, Y.S., Suzuki, K. (2008). Gas permeation through DDR-type zeolite membranes at high temperatures. *AIChE Journal*, 54, 1478–1486.

Keizer, K., Burggraaf, A.J., Vroon, A.A.E.P., Verweij, H. (1998). Two component permeation through thin zeolite MFI membranes. *Journal of Membrane Science*, 147 (2), 159–172.

Keshebo, D.L., Hu, C.-P., Hu, C.-C., Hung, W.-S., Wang, C.-F., Tsai, H.-C., Lee, K.-R., Lai, J.-Y. (2021). Effect of composition of few-layered transition metal dichalcogenide nanosheets on separation mechanism of hydrogen selective membranes. *Journal of Membrane Science*, 634, 119419. https://doi.org/10.1016/j.memsci.2021.119419.

Khan, A.L., Cano-Odena, A., Gutiérrez, B., Minguillón, C., Vankelecom, I.F.J. (2010). Hydrogen separation and purification using polysulfone acrylate-zeolite mixed matrix membranes. *Journal of Membrane Science*, 350, 340–346. https://doi.org/10.1016/j.memsci.2010.01.009.

Kim, E., Cai, W., Baik, H., Choi, J. (2013). Uniform Si-CHA zeolite layers formed by a selective sonication-assisted deposition method. *Angewandte Chemie International Edition*, 52 (20), 5280–5284.

Kim, S., Seong, J.G., Do, Y.S., Lee, Y.M. (2015). Gas sorption and transport in thermally rearranged polybenzoxazole membranes derived from polyhydroxylamides. *Journal of Membrane Science*, 474, 122–131. https://dx.doi.org/10.1016/j.memsci.2014.09.051.

Kim, T., Shamsaei, E., Lin, X.C., Hu, Y.X., Simon, G.P., Seong, J.G., Kim, J.S., Lee, W.H., Lee, Y.M., Wang, H.T. (2016). The enhanced hydrogen separation performance of mixed matrix membranes by incorporation of two-dimensional ZIF-L into polyimide containing hydroxyl group. *Journal of Membrane Science*, 549, 260–266. https://dx.doi.org/10.1016/j.memsci.2017.12.022.

Kundu, A., Trivedi, R., Garg, N., Chakraborty, B. (2022). Novel permeable material "yttrium decorated zeolite templated carbon" for hydrogen storage: Perspectives from density functional theory. *International Journal of Hydrogen Energy*, 47, 28573–28584. https://doi.org/10.1016/j.ijhydene.2022.06.159.

Li, W.P., Cao, Z.W., Zhu, X.F., Yang, W.S. (2019). Effects of membrane thickness and structural type on the hydrogen separation performance of oxygen-permeable membrane reactors. *Journal of Membrane Science*, 573, 370–376.

Li, W., Li, Y., Caro, J., Huang, A. (2022). Fabrication of a flexible hydrogen-bonded organic framework based mixed matrix membrane for hydrogen separation. *Journal of Membrane Science*, 643, 120021. https://doi.org/10.1016/j.memsci.2021.120021.

Lider, A., Kudiiarov, V., Kurdyumov, N., Lyu, J., Koptsev, M., Travitzky, N., Hotza, D. (2023). Materials and techniques for hydrogen separation from methane-containing gas mixtures. *International Journal of Hydrogen Energy*. https://doi.org/10.1016/j.ijhydene.2023.03.345

Liu, L., Doherty, C.M., Ricci, E., Chen, G.Q., De Angelis, V., Kentish, V. (2021). The influence of propane and n-butane on the structure and separation performance of cellulose acetate membranes. *Journal of Membrane Science*, 638, 119677. https://doi.org/10.1016/j.memsci.2021.119677.

Liu, X., He, D. (2022). Atomic layer deposited aluminium oxide membranes for selective hydrogen separation through molecular sieving. *Journal of Membrane Science*, 662, 121011. https://doi.org/10.1016/j.memsci.2022.121011.

Liu, L., Liu, D., Zhang, C. (2022). High-temperature hydrogen/propane separations in asymmetric carbon molecular sieve hollow fiber membranes. *Journal of Membrane Science*, 642, 119978. https://doi.org/10.1016/j.memsci.2021.119978.

Luo, Z., Hu, Y., Xu, H., Gao, D., Li, W. (2020). Cost-economic analysis of hydrogen for China's fuel cell transportation field. *Energies*, 13 (24), 6522. https://doi.org/10.3390/en13246522.

Luo, S., Zhang, Q., Zhu, L., Lin, H., Kazanowska, B., Doherty, C., Hill, A., Gao, P., Guo, R. (2018). Highly selective and permeable microporous polymer membranes for hydrogen purification and CO_2 removal from natural gas. *Chemistry of Materials*, 30 (15), 5322–5332.

Makertihartha, I.G.B.N., Zunita, M., Rizki, Z., Dharmawijaya, P.T. (2017). Advances of zeolite-based membrane for hydrogen production via water gas shift reaction. *IOP Conference Series: Journal of Physics*, 877, 012076. doi :10.1088/1742-6596/877/1/012076.

Mansor, E.S., Abdallah, H., Shaban, A.M. (2020). Fabrication of high selectivity blend membranes based on poly vinyl alcohol for crystal violet dye removal. *Journal of Environmental Chemical Engineering*, 8, 103706. https://doi.org/10.1016/j.jece.2020.103706.

Marais, S., Métayer, M., Nguyen, Q.T., Labbé, M., Langevin, D. (2000). New methods for the determination of the parameters of a concentration-dependent diffusion law for molecular penetrants from transient permeation or sorption data. *Macromolecular Theory and Simulations*, 9, 207–214. https://doi.org/10.1002/(SICI)1521-3919(20000401)9:4 3.0.CO;2-Q.

Mei, W., Du, Y., Wu, T., Gao, F., Wang, B., Duan, J., Zhou, V., Zhou, R. (2018). High-flux CHA zeolite membranes for H_2 separations. *Journal of Membrane Science*, 565, 358–369.

Métayer, M., Labbé, M., Marais, S., Langevin, D., Chappey, C., Dreux, F., Brainville, M., Belliard, P. (1999). Diffusion of water through various polymer films: A new high-performance method of characterization. *Polymer Testing*, 18, 533–549. https://doi.org/10.1016/S0142-9418(98)00052-803 X.

Miao, R., Wang, L., Gao, Z., Mi, N., Liu, T., Lv, Y., Wang, X. (2015). Polyvinylidene fluoride/poly(ethylene co-vinyl alcohol) blended membranes and a systematic insight into their antifouling properties. *RSC Advances*, 5, 36325–36333. https://doi.org/10.1039/C5RA03875H.

Moon, J.-H., Bae, J.-H., Han, Y.-J., Lee, C.-H. (2010). Adsorbent/membrane hybrid (AMH) system for hydrogen separation: Synergy effect between zeolite 5A and silica membrane. *Journal of Membrane Science*, 356, 58–69. https://doi.org/10.1016/j.memsci.2010.03.027.

Nadgouda, S.G., Guo, M., Tong, A., Fan, L.S. (2019). High purity syngas and hydrogen coproduction using copper-iron oxygen carriers in chemical looping reforming process. *Applied Energy*, 235, 1415–1426. https://doi.org/10.1016/j.apenergy.2018.11.051.

Nazir, S.M., Morgado, J.F., Bolland, O., Quinta-Ferreira, R., Amini, S. (2018). Technoeconomic assessment of chemical looping reforming of natural gas for hydrogen production and power generation with integrated CO_2 capture. *International Journal of Greenhouse Gas Control*, 78, 7–20. https://doi.org/10.1016/j.ijggc.2018.07.022.

Nenoff, T., Spontak, R., Aberg, C. (2006). Membranes for hydrogen purification: an important step toward a hydrogen-based economy. *MRS Bulletin*, 31 (10), 735–741.

Nian, P., Li, Y., Zhang, X., Cao, Y., Liu, V., Zhang, X. (2018). ZnO nanorod-induced heteroepitaxial growth of SOD type Co-based zeolitic imidazolate framework membranes for H_2 separation. *ACS Applied Materials & Interfaces*, 10 (4), 4151−4160.

Nuhnen, A., Dietrich, D., Millan, S., Janiak, C. (2018). Role of filler porosity and filler/polymer interface volume in metalorganic framework/polymer mixed-matrix membranes for gas separation. *ACS Applied Materials and Interfaces*, 10, 33589–33600. https://doi.org/10.1021/acsami.8b12938.

Olson, D.H., Camblor, M.A., Villaescusa, L.A., Kuehl, G.H. (2004). Light hydrocarbon sorption properties of pure silica Si-CHA and ITQ-3 and high silica ZSM-58. *Microporous Mesoporous Materials*, 67 (1), 27−33.

Parra, V., Valverde, L., Pino, V., Patel, M.K. (2019). A review on the role, cost and value of hydrogen energy systems for deep decarbonization. *Renewable and Sustainable Energy Reviews*, 101, 279–294.

Prager, S., Long, F.A. (1951). Diffusion of hydrocarbons in polyisobutylene. *Journal of American Chemical Society*, 73, 4072–4075. https://doi.org/10.1021/ja01153a004.

Rey, M.F., Valencia, S., Corcoran, E.W., Kortunov, P., Ravikovitch, P.I, Burton, A., Yoon, C., Wang, Y., Paur, C., Guzman, J., Bishop, A.R., Casty, G.L. (2017). Control of zeolite framework flexibility and pore topology for separation of ethane and ethylene. *Science*, 358, 1068−1071.

Riva, L., Martínez, I., Gallucci, F., Annaland, M.V.S., Romano, M.C. (2018). Technoeconomic analysis of the Ca-Cu process integrated in hydrogen plants with CO_2 capture. *International Journal of Hydrogen Energy*, 43, 15720–15738. https://doi.org/10.1016/j.ijhydene.2018.07.002.

Salaudeen, S.A, Acharya, B., Heidari, M., Al-Salem, S.M., Dutta, A. (2020). Hydrogen-rich gas stream from steam gasification of biomass: Eggshell as a CO_2 sorbent. *Energy Fuels*, 34 (4), 4828− 4836.

Shen, J., Liu, G., Ji, Y., Liu, Q., Cheng, L., Guan, K., Zhang, M., Liu, G., Xiong, J., Yang, J., Jin, W. (2018). 2D MXene nanofilms with tunable gas transport channels. *Advanced Functional Materials*, 28, 1801511.

Song, T., Shen, L. (2018). Review of reactor for chemical looping combustion of solid fuels. *International Journal of Greenhouse Gas Control*, 76, 92–110. https://doi.org/10.1016/j.ijggc.2018.06.004.

Soto Puente, J.A., Fatyeyeva, K., Marais, S., Dargent, E. (2015). Multifunctional hydrolyzed EVA membranes with tunable microstructure and water barrier properties. *Journal of Membrane Science*, 480, 93–103. https://doi.org/10.1016/j.memsci.2015.01.008.

Soto Puente, J.A., Fatyeyeva, K., Chappey, C., Marais, S., Dargent, E. (2017). Layered poly(ethylene- co -vinyl acetate)/poly(ethylene- co -vinyl alcohol) membranes with enhanced water separation selectivity and performance. *ACS Applied Materials & Interfaces*, 9, 6411-6423. https://doi.org/10.1021/acsami.6b14909.

Spallina, F., Shams, A., Battistella, A., Gallucci, F., Annaland, M.V.S. (2017b). Chemical looping technologies for H_2 production with Co_2 capture: Thermodynamic assessment and economic comparison. *Energy Procedia*, 114, 419–428. https://doi.org/10.1016/j.egypro.2017.03.1184.

Stenberg, V., Rydén, M., Mattisson, T., Lyngfelt, A. (2018). Exploring novel hydrogen production processes by integration of steam methane reforming with chemical looping combustion (CLC-SMR) and oxygen carrier aided combustion (OCAC-SMR). *International Journal of Greenhouse Gas Control*, 74, 28–39. https://doi.org/10.1016/j.ijggc.2018.01.008.

Stern, S.A. (1994). Polymers for gas separation: The next decade. *Journal of Membrane Science*, 94, 1–65. https://doi.org/10.1016/0376-7388(94)00141-3.

Symonds, R.T., Sun, Z., Ashrafi, O., Navarri, P., Lu, D.Y., Hughes, R.W. (2019). Ilmenite ore as an oxygen carrier for pressurized chemical looping reforming: characterization and process simulation. *International Journal of Greenhouse Gas Control*, 81, 240–258. https://doi.org/10.1016/j.ijggc.2018.12.006.

Tang, H., Bai, L., Wang, M., Zhang, Y., Li, M., Wang, M., Kong, L., Xu, N., Zhang, Y., Rao, P. (2019). Fast synthesis of thin high silica SSZ-13 zeolite membrane using oil-bath heating. *International Journal of Hydrogen Energy*, 44 (41), 23107–23119.

Tso, W.W., Niziolek, A.M., Onel, O., Demirhan, C.D., Floudas, C.A., Pistikopoulos, E.N. (2018). Enhancing natural gas-to-liquids (GTL) processes through chemical looping for syngas production: Process synthesis and global optimization. *Computers & Chemical Engineering*, 113, 222–239. https://doi.org/10.1016/j.compchemeng.2018.10.017.

Villain-Gambier, M., Courbalay, M., Klem, A., Dumarcay, S., Trebouet, D. (2020). Recovery of lignin and lignans enriched fractions from thermomechanical pulp mill process water through membrane separation technology: Pilot-plant study and techno-economic assessment. *Journal of Cleaner Production*, 249, 119345 https://doi.org/10.1016/j.jclepro.2019.119345.

Viano, D.M., Dolan, M.D., Weiss, F., Adibhatla, A. (2015). Asymmetric layered vanadium membranes for hydrogen separation. *Journal of Membrane Science*, 487, 83–89. https://dx.doi.org/10.1016/j.memsci.2015.03.048.

Vinodh, R., Deviprasath, C., Gopi, C.V.V.M, Kummara, V.G.R, Atchudan, R., Ahamad, T., Kim, H.-J., Yi, M. (2020). Novel 13X Zeolite/PANI electrocatalyst for hydrogen and oxygen evolution reaction. *International Journal of Hydrogen Energy*. https://doi.org/10.1016/j.ijhydene.2020.07.194.

Vu, D.Q., Koros, W.J., Miller, S.J. (2003). Mixed matrix membranes using carbon molecular sieves. I. Preparation and experimental results. *Journal of Membrane Science*, 211, 311–334. https://doi.org/10.1016/S0376-7388(02)00429-5.

Wang, H., Dong, X., Lin, Y.S. (2014). Highly stable bilayer MFI zeolite membranes for high temperature hydrogen separation. *Journal of Membrane Science*, 450, 425–432. https://dx.doi.org/10.1016/j.memsci.2013.08.030.

Wang, D., Sentorun-Shalaby, C., Ma, X., Song, C. (2011). High-capacity and low-cost carbon-based molecular basket sorbent for CO_2 capture from flue gas. *Energy & Fuels*, 25 (1), 456–458.

Wang, M., Song, J., Wu, X., Tan, X., Meng, B., Liu, S. (2016). Metallic nickel hollow fiber membranes for hydrogen separation at high temperatures. *Journal of Membrane Science*, 509, 156–163. https://dx.doi.org/10.1016/j.memsci.2016.02.025.

Wang, V., Wang, Z., Tan, X., Liu, S. (2022). Externally self-supported metallic nickel hollow fiber membranes for hydrogen separation. *Journal of Membrane Science*, 653, 120513. https://doi.org/10.1016/j.memsci.2022.120513.

Wang, Q., Wu, A., Zhong, S., Wang, B., Zhou, R. (2017). Highly (h0h)-oriented silicalite-1 membranes for butane isomer separation. *Journal of Membrane Science*, 540, 50–59.

Wang, V., Zhou, Y., Tan, V., Gao, J., Liu, S. (2019). Nickel hollow fiber membranes for hydrogen separation from reformate gases and water gas shift reactions operated at high temperatures. *Journal of Membrane Science*, 575, 89–97. https://doi.org/10.1016/j.memsci.2019.01.009.

Wahab, M.S.A., Sunarti, A.R. (2015). Development of PEBAX based membrane for gas separation: A review. *International Journal of Membrane Science*, 2, 78–84. https://doi.org/10.15379/2410-1869.2015.02.02.08.

Wassie, T.A., Cloete, S., Spallina, V., Gallucci, F., Amini, S., Annaland, M.V.S. (2018a). Techno-economic assessment of membrane-assisted gas switching reforming for pure H2 production with CO_2 capture. *International Journal of Greenhouse Gas Control*, 72, 163–174. https://doi.org/10.1016/j.ijggc.2018.03.021.

Wassie, S.A., Medrano, J.A., Zaabout, A., Cloete, S., Melendez, J., Tanaka, D.A.P., Amini, S., van Sint Annaland, M., Gallucci, F. (2018b). Hydrogen production with integrated CO_2 capture in a membrane assisted gas switching reforming reactor: Proof-of concept. *International Journal of Hydrogen Energy*, 43, 6177–6190. https://doi.org/10.1016/j.ijhydene.2018.02.040.

Wenwei, Z., Xiaoguang, Z., Li, Y., Yuefang, Z., Jiazhen, S. (1994). Determination of the vinyl acetate content in ethylene-vinyl acetate copolymers by thermogravimetric analysis. *Polymer*, 35, 3348–3350. https://doi.org/10.1016/0032-3861(94)90148-1.

Wu, T., Shu, C., Liu, S., Xu, B., Zhong, S., Zhou, R. (2020). Separation performance of Si-CHA zeolite membrane for a binary H2/CH4 mixture and ternary and quaternary mixtures containing impurities. *Energy & Fuels*. https://dx.doi.org/10.1021/acs.energyfuels.0c01720.

Xiao, J.R., Wei, J. (1992). Diffusion mechanism of hydrocarbons in zeolites. II. Analysis of experimental observations. *Chemical Engineering Science*, 47, 1143–1159.

Yang, T., Chung, T.-S. (2013). High performance ZIF-8/PBI nano-composite membranes for high temperature hydrogen separation consisting of carbon monoxide and water vapor. *International Journal of Hydrogen Energy*, 38, 229–239. https://dx.doi.org/10.1016/j.ijhydene.2012.10.045.

Yang, J., Gong, D., Li, G., Zeng, G., Wang, Q., Zhang, Y., Liu, G., Wu, P., Vovk, E., Peng, Z. (2018). Self-assembly of thiourea-crosslinked graphene oxide framework membranes toward separation of small molecules. *Advanced Materials*, 30, 1705775.

Yong, M., Zhang, Y., Sun, S., Liu, W. (2019). Properties of polyvinyl chloride (PVC) ultra-filtration membrane improved by lignin: Hydrophilicity and antifouling. *Journal of Membrane Science*, 575, 50–59. https://doi.org/10.1016/j.memsci.2019.01.005.

Yu, Z., Liu, X., Zhao, F., Liang, X., Tian, Y. (2015). Fabrication of a low-cost nano-SiO_2/PVC composite ultrafiltration membrane and its antifouling performance. *Journal of Applied Polymer Science*, 132. https://doi.org/10.1002/app.41267.

Yu, L., Nobandegani, M.S., Holmgren, A., Hedlund, J. (2019). Highly permeable and selective tubular zeolite CHA membranes. *Journal of Membrane Science*, 588, 117224.

Zaabout, A., Dahl, P.I., Ugwu, A., Tolchard, J.R., Cloete, S., Amini, S. (2019). Gas switching reforming (GSR) for syngas production with integrated CO_2 capture using iron-based oxygen carriers. *International Journal of Greenhouse Gas Control*, 81, 170–180. https://doi.org/10.1016/j.ijggc.2018.12.027.

Zhang, Y., Lee, W.H., Seong, J.G., Bae, J.Y., Zhuang, Y., Feng, S., Wan, Y., Lee, Y.M. (2020). Alicyclic segments upgrade hydrogen separation performance of intrinsically microporous polyimide membranes. *Journal of Membrane Science*, 611, 118363. https://doi.org/10.1016/j.memsci.2020.118363.

Zhou, Z., Wu, C., Zhang, B. (2020). ZIF-67 membranes synthesized on α-Al2O3-plate-supported cobalt nanosheets with amine modification for enhanced H2/CO2 permselectivity. *Industrial & Engineering Chemistry Research*, 59 (7), 3182–3188.

5 Hydrogen Separation Using Ceramic Membranes

Fatemeh Kavousi, Anna O'Sullivan,
Archishman Bose, and Sudipta De

5.1 INTRODUCTION

The demand for membrane systems for low cost and sustainable production and purification of hydrogen is on the rise. Process and manufacturing flexibilities, small footprints and low-cost separation with moderate operating conditions make membrane systems favourable over conventional costly and low-throughput processes. The flexibility and simplicity of operation of membrane systems increase with their ability to absorb unconditioned feeds without the need for pretreatment (Lu et al., 2007). Hence, the chemical stability and/or thermal stability of any form of membrane are crucial for their long-term function and efficacy. Selective membranes such as polymers, ceramics, silica-based glasses and dense metals have the capability of purifying hydrogen from gas mixtures. Recently, metallic and polymer membranes have been the main focus of research for hydrogen separations due to their scalability and effectiveness (Yang et al., 2021). However, despite their traditionally fragile structure and costly production, ceramic membranes offer good thermal and chemical stability, more robust performance and anti-fouling properties, which are very beneficial in separating small gas molecules such as He or H_2 from gas mixtures at unfavourable (harsh) operating conditions (Amanipour et al., 2012a; Serhiienko, et al., 2020). In addition, they offer superior resistance to contaminants due to their chemical stability. As a result, ceramic membranes are favourably utilised as the support structure for hybrid and mixed matrix membranes.

One of the earliest industrial uses of ceramic membranes was to separate uranium isotopes (Lin, 2001). Since then, they have been adopted in the food and pharmaceutical sectors for polish filtration to remove bacteria due to their ease of sterilisation and microfiltration capabilities (Abdullayev et al., 2019; Michaels, 1989a). While the materials for ceramic membranes are generally more expensive than polymeric membranes, the material's high stability can lead to favourable industrial use (Michaels, 1989a). Furthermore, ceramic membranes are easy to maintain as they have a high cleaning tolerance without renewing materials (Michaels, 1989a). Ceramic microfilters are proven to have high rigidity levels, allowing them to handle fluxes that are two to three times higher than symmetric polymer and symmetric sintered-metal media, as well as five to ten times higher than asymmetric polymerics (Kamarudin et al., 2003;

DOI: 10.1201/9781003382522-7

Winston Ho et al., 1992). These inorganic membranes have a significantly longer service life than other membrane materials due to their high resistance to contamination and degradation (Winston Ho et al., 1992). Ceramic materials are resistant to almost all substances apart from those with a pH over thirteen or under three and hydrogen fluoride solutions (Michaels, 1989a). Although much progress has been made to improve the product purity and production capacity, membrane-based purification systems are still deemed impractical to compete against pressure swing adsorption (PSA) and cryogenic distillation in large scale due to performance limitations (Yang et al., 2021). To this point, there are currently no ceramic membranes in commercial use for gas separation. Given these shortcomings, researchers are looking for ways to improve ceramic membranes' properties while also reducing their fabrication cost (Koutsonikolas et al., 2021a). Indeed, ceramic membranes have recently come back into research focus for three gas separation processes, namely, CO_2 capture membranes (Smart et al., 2010), O_2 purification membranes and H_2 purification membranes (Deibert et al., 2015; Korelskiy et al., 2015; Koutsonikolas et al., 2021b; Rosensteel et al., 2016). Much study has been focused on identifying improved hydrogen selective membranes (Balachandran et al., 2005; David and Kopac, 2011; Llosa Tanco et al., 2021; Rosensteel et al., 2016). This study focuses on the use of ceramic membranes as a potential membrane for hydrogen separation from hydrogen-rich mixtures.

5.2 MEMBRANE MATERIAL

Based on the membrane material, synthetic membranes can be classified into organic, inorganic and mixed matrix or hybrid membranes. Currently, industrial gas production mainly utilises organic (polymer) membranes due to their scalability, low cost and moderate operating conditions. However, organic membranes suffer from thermal instability. Inorganic membranes, while generally more expensive, offer high thermal, chemical and mechanical stability with minimal plasticisation, as well as controllable pore size, which leads to tunability of selectivity and permeability. Ceramic membranes are made from inorganic materials, with the more notable materials being silica, alumina, oxides, ceria, zirconia and titania and silicon carbide or their combinations (Amin et al., 2016). Figure 5.1 depicts the classification of hydrogen selective membranes based on their properties.

Morphologically, membranes generally can be categorised into porous and dense (non-porous) membranes (Amanipour et al., 2022; Ghouil et al., 2015; Hao et al., 2022). Ceramic proton-conducting membranes are dense ceramic membranes, and silica and zeolite membranes are examples of porous ceramic membranes. There are significant differences between their separation mechanisms which will further be discussed in Section 5.5.

5.2.1 POROUS MEMBRANES

5.2.1.1 Silica

Silica is the least stable amongst common membrane materials; so they are often utilised as a composite material with the other oxides, forming ceramic–ceramic matrixes. Nonetheless, these inorganic amorphous microporous membranes are

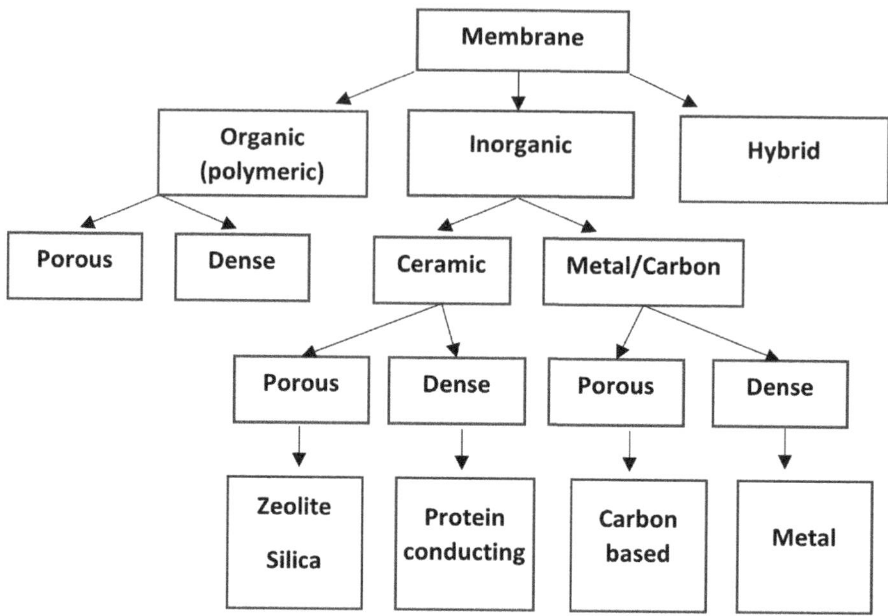

FIGURE 5.1 Classification of membranes used for hydrogen separation.

formed through covalent bonds between SiO_4 tetrahedrons, forming unique porous structures with tuneable pore size distribution. This makes them capable of separating small molecules such as H_2, He and O_2 (Table 5.1). Silica membranes are often fabricated via the solution–gelation (sol–gel) or chemical vapour deposition (CVD) methods. These methods will be covered in Section 5.3.

These membranes show high selectivity for hydrogen over other feed gases as well as reasonable permeabilities even at elevated temperatures (Koutsonikolas et al., 2021a). Despite the overall superior stability of microporous silicon membranes, an estimated 50% loss in their permeability was reported when operated in a continuously humid environment (Bui et al., 2023). When the membrane surface is in contact with steam, the microporous structure collapses, resulting in failure of separation performance and loss of gas permeation flux (Hao et al., 2022). To overcome the hydrothermal instability of silica membranes at high temperatures in contact with steam, studies have explored options that would increase the hydrophobicity of the silica layer. Adding hydrophobic groups such as methyl groups to the silica film was found to be successful in limiting the impact of condensation and the resulting decomposition reaction. Another attractive method that has not only shown significant increase in the humidity tolerance of silica-based membranes, but also has enhanced their structural stability is the introduction of metals or metal oxides such as Pd, Nb, Co and Nb (Agoudjil et al., 2005).

5.2.1.2 Alumina

Alumina oxide (Al_2O_3) is mostly used to prepare commercial ceramic membranes (Chathurappan and Anandkumar, 2023). Available in flat sheet or hollow fibre, they are often found as support layers (α-Al_2O_3), intermediate layers (α-Al_2O_3 or γ-Al_2O_3)

TABLE 5.1

Hydrogen Separation Capability of Representative Silica Membranes

Precursor	Support	Pore Size (nm)	Temperature (K)	H_2 Permeance $(10^{-8}\ mol \cdot m^{-2} \cdot s^{-1} \cdot Pa^{-1})$	H_2/Selectivity			Source
					N_2	CH_4	CO_2	
TEOS	α-Alumina	0.31	373–673	51.0–70.0	5.7		36	Gopalakrishnan and da Costa (2008)
TEOS	α-Alumina	0.38–0.55	473	200		>500		De Vos and Verweij (1998)
TMOS	α-Alumina		873	4.4	2,900	2,400	3,200	Gopalakrishnan et al. (2006)
TPMS	α-Alumina	0.5	573	120	7.5			Zhang et al. (2016)
TEOS	α-Alumina	0.6	373	1,000		27		De Vos et al. (1999)
TEOS	α-Alumina		873	50		5900	1,500	Gu and Oyama (2007)
TEOS+ ATSB	α-Alumina		623	6.8			600	Lim et al. (2012)
TEOS+ MOTMS	α-Alumina	0.6–0.7	573	70.0		132	36	Kim et al. (2001)

TEOS, tetraethylorthosilicate; ATSB, aluminium tri-sec butoxide; MOTMS, methacryoxpropyltrimethoxysilane.

and active layers (α-Al$_2$O$_3$ or γ-Al$_2$O$_3$) in ceramic membranes as well as multilayer membranes such as Pd/PD-Cu, metal–organic framework (MOF) and carbon membranes.

5.2.1.3 Zeolites

Zeolites are conventionally crystalline porous structures composed of aluminosilicate building blocks with oxygen atoms connecting adjacent TO$_4$ tetrahedrons. They avail well-defined pores in the range of 0.2–20 nm, making them attractive for gas separation. Zeolite membranes are often fabricated as thin films on macro/mesoporous supports such as alumina and silica. The most common zeolite membranes include Linda Type A (LTA), Mobile Type Five (MFI) and Faujasite type (FAU), which have been the focus of many studies trying to improve their selectivities for a variety of applications (Dong et al., 2008). A variety of synthesis methods have led to the creation of zeolite membranes with different morphology, composition and characteristics of the separation process as shown in Table 5.2. Morphological defects such as intracrystalline pores, pinholes and cracks remain one of the main failure factors in zeolite membrane fabrication. Mitigating techniques such as optimisation of the hydrothermal fabrication method, support pretreatment and post-synthesis treatment of the membrane have been applied to variable levels of success.

Although many attempts in fine tuning the membrane material and properties have led to enhanced performance of porous ceramic membranes for H$_2$ separation (Amanipour et al., 2012b), the compromise between high selectivity and high permeation rates still remains. On the other hand, hydrogen separation using dense membranes has attracted a considerable interest.

5.2.2 Dense Ceramic Membranes

Dense ceramic-based membranes are mixed ionic electronic conductors (MIECs) as the transport through these occurs via ion rather than molecules (Figure 5.2). Dense membranes can separate molecules according to their solubility and diffusivity. Some of the earlier dense hydrogen permeable membranes were oxides that were mixed with proton–electron conductors (Norby and Haugsrud, 2006). These proton–electron conducting oxides separate hydrogen through a bipolar co-current diffusion of protons and electrons. Following a surface dissociation reaction of H$_2$, ionised H$^+$ molecules diffuse to the permeate side as a result of H$_2$ partial pressure gradient, where they form H$_2$ molecules again at the membrane interface. Originally developed for oxygen transport, perovskite materials are at the forefront of dense ceramic membranes (Amanipour et al., 2022; Athayde et al., 2016; Heidari et al., 2015). Perovskites are made of a combination of metal and oxygen atoms arranged in a specific pattern, which gives them unique properties. As such, they consist of a cubic crystal structure in which cations A and B are surrounded by 12 and 6 oxygen atoms, respectively, rendering its general formula to be ABO$_3$ (Figure 5.3).

Dense membranes for gas separation have shown success for oxygen purification (Athayde et al., 2016; Zhang et al., 2011). However, hydrogen purification using dense membranes is far more challenging (Norby and Haugsrud, 2006). This is because dense ceramic membranes only function at high temperatures, requiring high energy

TABLE 5.2

Zeolite Membranes for Hydrogen Separation (Hao et al., 2022)

Membrane	Name	T (°C)	ΔP (bar)	H_2 Permeance (mol·m^{-2}·s^{-1}·Pa^{-1})	H_2/Selectivity N_2	CO_2	CH_4
Zeolite36	FAU	100	1	4×10^{-7}	6	6.5	4
Zeolite43	SSZ-13	200	2	4.2×10^{-8}	28.4		263
Zeolite45	SAPO-34	250	10	3.7×10^{-7}	32		–
Zeolite34	Linda Type A (LTA)	30	1	2.2×10^{-6}		16.2	–
Zeolite41	Mobile Type Five (MFI)	450		1.1×10^{-7}		8.2	–
Zeolite32	Si-CHA	25	2	1.4×10^{-6}			85
Zeolite31	STT	25	2	2.8×10^{-8}			49.6
Zeolite35	DD_3R	30	1	1.8×10^{-8}			12.4
Zeolite40	FAU	50	1	1.9×10^{-7}			9.9

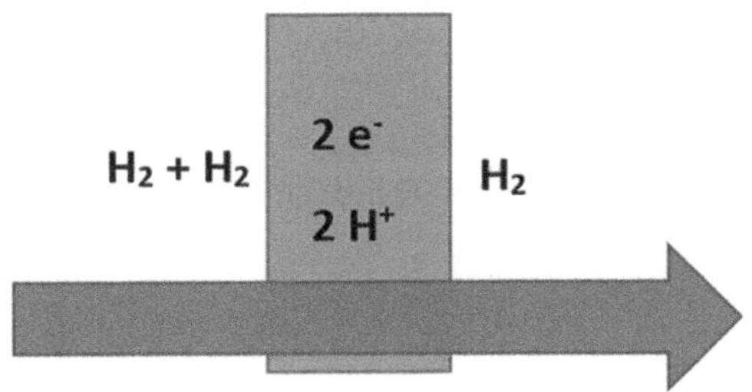

FIGURE 5.2 Simplified mechanism of proton–electron transfer through dense ceramic membranes.

expenditures (Yin and Yip, 2017). At these high temperatures, proton transfer is fast; however, laws of thermodynamics prevent a high concentration of protons in the material at high temperatures (Norby and Haugsrud, 2006). Hydrogen complicates the membranes' defect chemistry, causing rapid degradation over time (Rosensteel et al., 2016). Much research has been dedicated to develop highly hydrogen selective perovskite-type membranes (Amanipour et al., 2022; Athayde et al., 2016; Heidari et al., 2015, 2013). A new type of $BaCe_{0.9}Y_{0.1}O_{3-\delta}$ perovskite membrane developed by Heidari et al. (2013) showed great selectivities of around 99%, yet suffered from permeation fluxes lower than porous ceramic membranes. A study by Zhu et al. involved a new method for improving hydrogen permeation performance of BZCY membranes by developing a BZCY $(Ba(Zr_{0.1}Ce_{0.7}Y_{0.2})O_{3-\delta})$ asymmetric ceramic membranes with external short circuit (ESC). Their ESC asymmetrical membrane showed significantly superior (almost 10 times) hydrogen permeation flux for 20% H_2/N_2 feed and dry Ar as sweep, in comparison to Ni-BZCY symmetrical membrane and Ni-BZCY asymmetrical membranes. Its activation energy was however found to be higher than

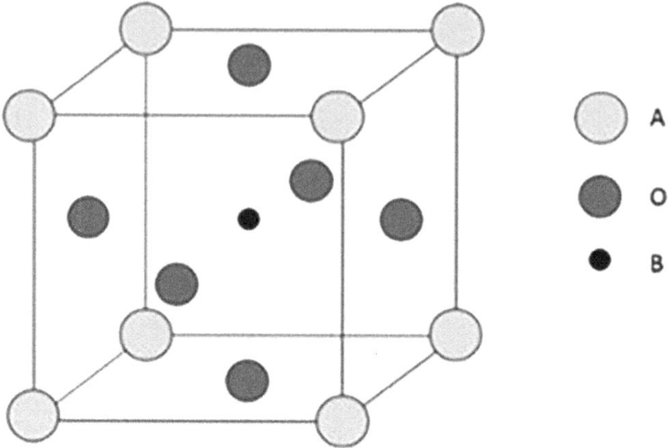

FIGURE 5.3   Crystal structure of ABO_3 perovskite (Leo et al., 2009).

the activation energies of Ni-BZCY symmetrical and asymmetrical membranes. This study also evaluated the impact of temperature (ranging 700°C–900°C) on hydrogen permeation flux, leading to the conclusion that typically hydrogen fluxes increase with rise in temperature, with ESC asymmetrical membranes exhibiting a pronounced increase (Zhu et al., 2016).

While dense ceramic membranes are deemed practical for hydrogen purification, more research is required into the complex transportation equations and defect models (Norby and Haugsrud, 2006).

5.3   MEMBRANE FABRICATION

Aside from the development of new material for hydrogen separation membranes, more robust membrane fabrication methods that enable maximised diffusivity, increased permeability and selectivity with high mechanical strength are essential (Shahbaz et al., 2020). Some of the most used methods for porous ceramic membranes fabrication are extrusion (Chathurappan and Anandkumar, 2023), solution–gelation, hydrothermal technique and CVD from the gas phase (Athayde et al., 2016). These methods allow for the fabrication of both selective layer as well as the intermediate and support layers with specified characteristics by controlling various parameters during synthesis (Bouwmeester, 2003).

5.3.1   SOLUTION–GELATION (SOL–GEL)

Solution–gelation is the most common fabrication method, which involves the formation of fine particles (from 1 to 0.1 mm) in a liquid medium (sol), followed by the formation of a three-dimensional gel via consolidation, agglomeration and dehydration. This is then followed by application to a support surface and subsequent heat treatment (Amanipour et al., 2012b). The sol particle size along with the degrees of hydrolysis and condensation impact the performance of the fabricated membranes and determines if they fall under the micro-, ultra- or nano-categories. Some of the most common precursors are inorganic metal salts and alkoxides (Me(OR)n, where Me is Zr, Ti, Al, Si, etc., and R is an alkyl group) (Amanipour et al., 2022). Variations in precursor types and their ratios with respect to water and solvents in the synthesis solution significantly impact the membrane pore size and subsequently its transfer mechanism and capabilities. In addition to precursors, addition of modifiers as well as controlling process parameters such as temperature and pH can be implemented in sol–gel fabrication. For instance, increasing the water:precursor ratio resulted in increased pore sizes, while further increases in this ratio lead to tighter structures (Bui et al., 2023). In addition, increasing calcination temperature was linked to increased pore size (Bockute et al., 2015).

5.3.2 Chemical Vapour Deposition (CVD)

Through the deposition of vapour phase of precursors and reactants (water, oxygen and/or air) at high temperatures in the CVD method, a thin silica coating is formed on the porous support layers. This process involves thermal decomposition, hydrolysis or oxidation of the precursors. Depending on the flow configuration (one-sided or counter diffusion), precursor residence time, reaction temperature and deposition pathway, the membrane performance changes considerably (Khatib and Oyama, 2013a). As the building blocks produced in this method are much smaller than the particles in sol–gel method, silica membranes produced via CVD exhibit tight structures offering lower gas permeance and higher gas selectivities. For instance, Nanosil membranes fabricated from TEOS precursors via CVD offered 10% lower H_2 permeance and over 30 times higher H_2/CH_4 selectivities in comparison to the same membrane prepared via sol–gel (Bui et al., 2023). Silica membranes manufactured via CVD are found to provide excellent H_2/CO_2 selectivities. Liu et al. (2023) summarised H_2/CO_2 selectivities of a variety of silica membranes as shown in Table 5.3.

Despite the success of CVD methods in membrane fabrication with great size-sieving abilities, the gas permeances achieved are often low. In addition, due to high temperature requirements of the original technique, the precursor selection is limited. Plasma-enhanced CVD (PE-CVD) and plasma oxidation methods allow for membrane fabrication at low temperatures showing promising prospects for scalable rapid production of highly hydrogen-selective silica membranes.

A similar technique to CVD in fabrication of multilayer ceramic membranes is chemical vapour infiltration (CVI), which showed substantial improvements in membrane separation efficiency rendering them promising for H_2 purification applications based on single gas and binary mixture evaluations. The modified membranes had high H_2 permeance (1.5×10^7 mol·m²s¹ Pa¹, at 250°C) and selectivities towards larger molecules ($H_2/CO_2 = 61.3$, $H_2/CH_4 = 460.5$) (Koutsonikolas et al., 2021a).

5.3.3 Sol-Gel vs CVD

While sol–gel and CVD are the most common methods for fabrication of ceramic membranes, there are benefits and drawbacks to each. Typically, membranes prepared via sol–gel show higher permeance of hydrogen in comparison to those

TABLE 5.3

Representative H_2/CO_2 Selective Silica Membranes Fabricated via Chemical Vapour Deposition (CVD) Method (Liu et al., 2023)

Membrane Material	Temperature (°C)	H_2 Permeance (Gas Permeation Units (GPU))	CO_2 Permeance (GPU)	H_2/CO_2 Selectivity
Dimethoxydimethylsilane (DMDMS)	500	960	0.75	1,280
Methyltrimethoxysilane (MTMOS)	600	720	0.24	3,000
Tetramethyl orthosilicate (TMOS)	500	290	0.13	2,200
Vinyltriethoxysilane (VTES)	600	1600	17	95

prepared via CVD (Amanipour et al., 2012a). Moreover, pore size control is easier in membranes prepared by sol–gel in comparison to those fabricated via high temperature CVD (Amanipour et al., 2022). However, CVD is typically simpler to exploit, takes less processing time and leads to the production of membranes with much higher selectivity when compared to sol–gel fabricated membranes (Amanipour et al., 2012b; Heidari et al., 2015). It has also been shown that membranes prepared via CVD may suffer from low thermal stability, in particular, in contact with hot vapour (Amanipour et al., 2022). This occurs because of restructuring of silica (due to Si–O–Si bonds forming in the presence of water), ultimately leading to the compression of the porous structure. Many studies have been dedicated to resolving this issue, ranging from increasing the pore hydrophobic properties to calcination in the presence of steam to avoid structural failure.

5.3.4 HYDROTHERMAL/SOLVOTHERMAL FABRICATION METHODS

While hydrothermal methods of fabrication are significantly popular in MOF membrane synthesis, the ability to produce nanostructures of varied morphology is a plausible advantage of solvothermal method. Following their success in MOF fabrication, greater prospects are envisaged for using this method for the fabrication of ceramic membrane active layers (Potapov et al., 2020).

5.4 MEMBRANE UNIT CONFIGURATION

The general principles of gas separation in a membrane can be based on differences in molecular mass, shape or size and variances in the gas molecules' affinity to the membrane material (Meinema et al., 2005). The driving force for separation of gaseous mixtures is the partial pressure gradient across either side of the membrane (Michaels, 1989). An applied electrical potential difference is another possible driving force (Bernardo et al., 2020). Another factor to consider when designing a membrane capsule is the membrane configuration and whether the module operates in

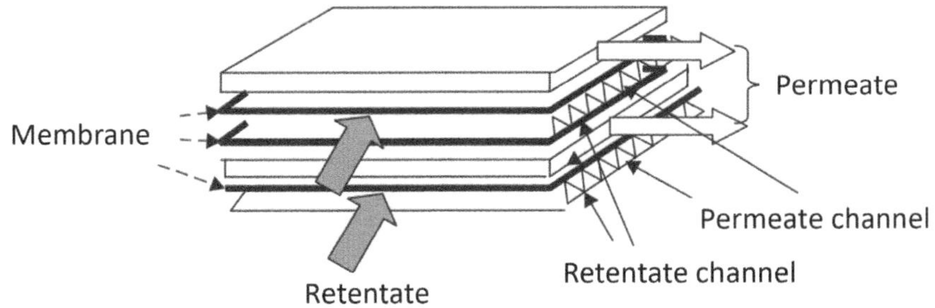

FIGURE 5.4 Diagram of plate and frame configuration (Berk and Berk, 2009).

co-current or counter-current flow (Bernardo et al., 2009). The configuration of a membrane module can affect factors such as the pressure drop across the membrane, the amount of fouling that occurs on a membrane surface and the ease of cleaning the membrane.

5.4.1 Sheet/Plate and Frame

Sheet membranes are composed of multiple membrane sheets separated by spacers. Although this membrane configuration allows easy access to clean and replace, they cannot handle large pressure drops and are not an efficient use of membrane area. These are used for laboratory investigations. Figure 5.4 demonstrates the concept of a sheet membrane.

5.4.2 Spiral Wound

Spiral wound membranes are flat sheet membranes that have been wound around a perforated central collector tube to increase the packing density of the module, as shown in Figure 5.5. The typical packing density of a spiral wound configuration is up to $600 \, m^2/m^3$. Spiral wound configurations are difficult to clean and require a high pressure drop across the membrane (Balster, 2013).

5.4.3 Hollow Fibre/Capillary

Hollow fibre and capillary membrane configurations comprise multiple bundles inside an outer casing or shell. It is the equivalent of mass transfer to the shell and tube heat exchanger (Mat et al., 2014), as shown in Figure 5.6. Hollow fibre and capillary membrane configurations have the highest packing density but are susceptible to fouling. Hollow fibre membranes are the most widely adopted membrane configuration in industry due to their high packing density of $1,200 \, m^2/m^3$ (Doran, 1995).

Each of these configurations can be run in co-current flow where the feed and permeate stream run in the same direction, or in counter-current flow, where the permeate stream runs in the opposite direction to the feed stream.

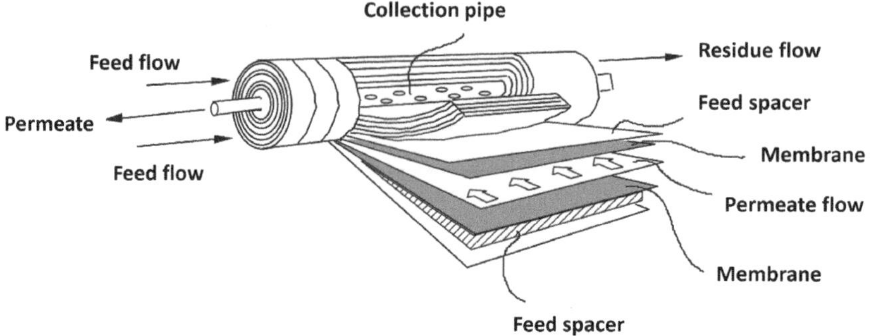

FIGURE 5.5 Diagram of spiral wound configuration (Balster, 2013).

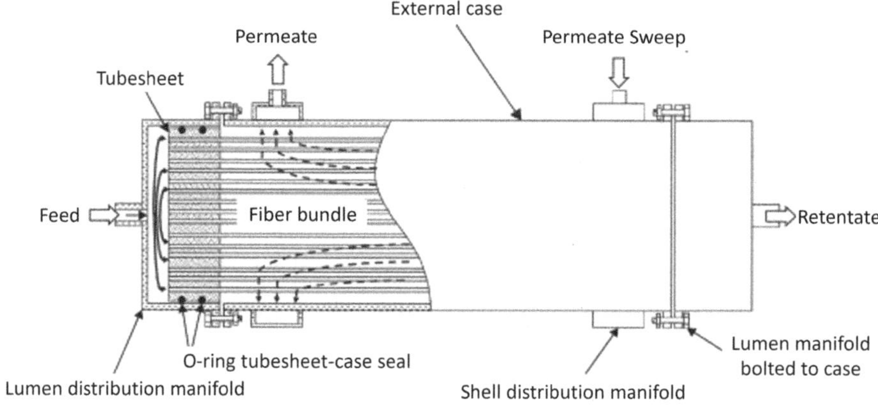

FIGURE 5.6 Diagram of hollow fibre configuration (Mat et al., 2014).

5.5 HYDROGEN SEPARATION MECHANISMS

Ceramic membranes separate molecules depending on different sizes, shapes and affinity between permeable molecules and membranes. Key separation mechanisms have been recognised: (i) Knudsen diffusion, (ii) surface diffusion, (iii) capillary condensation, (iv) molecular screening and (v) solution diffusion. The contribution of these mechanisms to specific materials ultimately determines their overall performance and separation efficiency. As mentioned previously, ceramic membranes are often asymmetrical structures consisting of support, intermediate and active layers. The support layer often acts as the structural backbone and is usually made of porous ceramic materials to provide high mechanical and permeation performance, as well as sufficient chemical resistance. Intermediate layer prevents the penetration of particles from active layer into support during the preparation process. More importantly, active layers are thin top selective layers with rationally designed pore structure and functions. The membrane separation mechanism depends on the morphology and functionality of these layers, which may be porous or dense in nature.

TABLE 5.4

Membrane Classification according to Pore Size (Danewalia and Singh 2021)

Classification	Pore Diameter
Microporous	<2 nm
Mesoporous	>2 nm and <50 nm
Macroporous	>50 nm

5.5.1 POROUS MEMBRANES

The porosity and pore size have a significant impact on the separation mechanism of these membranes. Porous ceramic membranes can further be broken up into three subcategories based on their pore size, namely, macroporous, mesoporous and microporous as per the International Union of Pure and Applied Chemistry (IUPAC) convention, as shown in Table 5.4.

The predominant forms of transport through porous ceramic membranes are Knudsen flow, surface diffusion and microporous sieving (Kentish et al., 2010; Vaezi et al., 2019). A rough approximation of pore size and the mechanism present is visualised in Figure 5.7.

5.5.1.1 Macroporous Ceramic Membranes

Viscous flow is the prominent flow regime in macroporous ceramic membranes, which can be modelled using the Hagen–Poiseuille mechanism (De Meis, 2017). This model is valid when the pore diameter is large compared to the mean free path (λ) of the gas molecule and the bulk of the fluid transports through the large pores

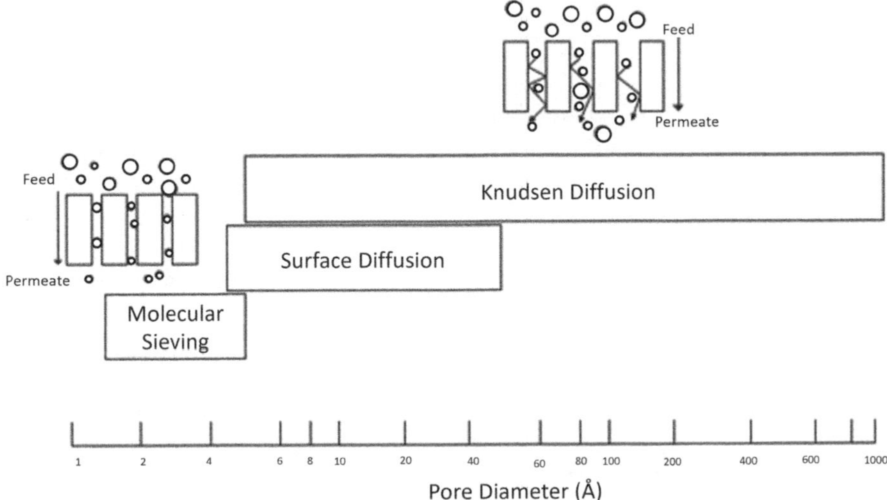

FIGURE 5.7 Diagram of the mass transfer mechanisms of porous membranes at different pore sizes.

(De Meis, 2017). The model correlates gas permeability (Pe) to gas viscosity of the Poiseuille regime according to the following equation:

$$Pe = \frac{\varepsilon \eta r^2}{8 \mu RT} p_{av} \tag{5.1}$$

where ε is the porosity, μ is the viscosity, η is the shape factor (1/tortuosity), r is the pore radius and P_{av} is the mean pressure.

Macroporous membranes have a deficient gas separation factor as these membranes allow the flow of all molecules and are not selective to any gases, hence not critical to the purity of separated gases.

5.5.1.2   Mesoporous Ceramic Membranes

Mesoporous membranes are membranes where the primary gas transport mechanism is the Knudsen diffusion mechanism, which occurs when the pore size is bigger than the molecule's size but smaller than the free path (λ) (Kamarudin et al., 2003; Pandey and Chauhan, 2001). Knudsen diffusivity is present when the number of collisions with the pore wall far outnumbers the number of collisions with other molecules in the system (Vaezi et al., 2019), as shown in Figure 5.8.

The ideal Knudsen separation factor (α_{Kn}) for a single gas is defined as per the following equation (de Lange et al., 1995):

$$\alpha_{Kn} = \left(\frac{M_B}{M_A} \right)^{\frac{1}{2}} \tag{5.2}$$

The Knudsen permeance (P_{MA}) is defined using the following equation (de Lange et al., 1995; Wang et al., 2013):

$$P_{MA} = \frac{\varepsilon_P d_p}{3t\tau} \sqrt{\frac{8RT}{\pi M_A}} \tag{5.3}$$

From Equation 5.3, the selectivity of Knudsen diffusion is low ($\alpha_{Kn,H2/CO2} \approx 4.5$ and $\alpha_{Kn,H2/CH4} \approx 2.7$), meaning these mesoporous membranes are not the most desirable for membrane separation. However, these membranes do allow for high permeance values. For this reason, mesoporous membranes are often used as support layers for microporous membranes.

5.5.1.3   Microporous Ceramic Membranes

For microporous membranes, there are two main models for mass transfer: the surface diffusion model and the activated Knudsen model.

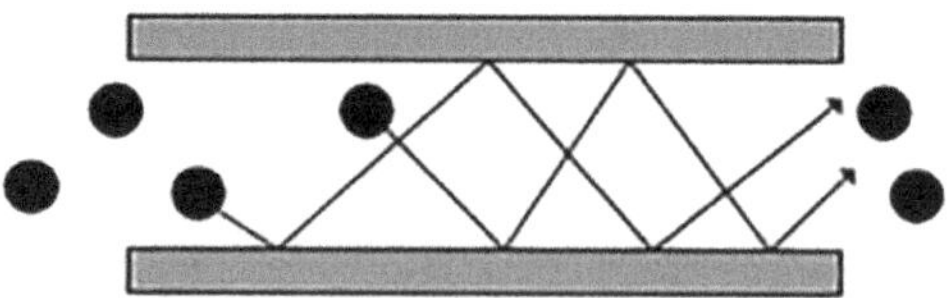

FIGURE 5.8   Diagram of Knudsen diffusion (Vaezi et al., 2019).

Microporous membranes support activated Knudsen gas transport across the membrane (Thornton et al., 2009) and have a pore diameter of less than 2 nanometres (nm) (Shelekhin et al., 1995; Vaezi et al., 2019). This activated transport is due to the need for activation energy caused by the molecule's attraction to the pore walls. Activated transport results in higher separation factors than Knudsen diffusion, which is desirable for product purity (de Lange et al., 1995). The interaction between pore walls and the gas molecule permeating through the membrane must be analysed to understand activated transport involved in microporous membranes. Every molecule entering the membrane will interact with the membrane's wall through van der Waals forces with the atoms that make up the pore wall (Thornton et al., 2009). These van der Waals forces increase significantly as the pore size decreases. In the case of microporous membranes, an activation of transport is required to overcome these forces as the particle size is significant compared to the pore diameter (Thornton et al., 2009).

Figure 5.9 shows how the activation energy of a particle diffusing through a microporous membrane relates to the activation energy required to allow permeation through the membrane (Roberts et al., 1992). McElfresh and Howitt (1986) studied the activation energy (E_{Act}) in micropores and found the following equation to be true (McElfresh and Howitt, 1986):

$$E_{\mathrm{Act}} = \delta W G \pi \left(r_p - r_{\mathrm{pd}} \right)^2 \tag{5.4}$$

where δW is the pore length, G is the shear modulus, r_p is the pore radius and r_{pd} is the dilated pore radius. Values for the dilated pore radius are not standard measurements, so it was decided in this model that the user would input the activation energy

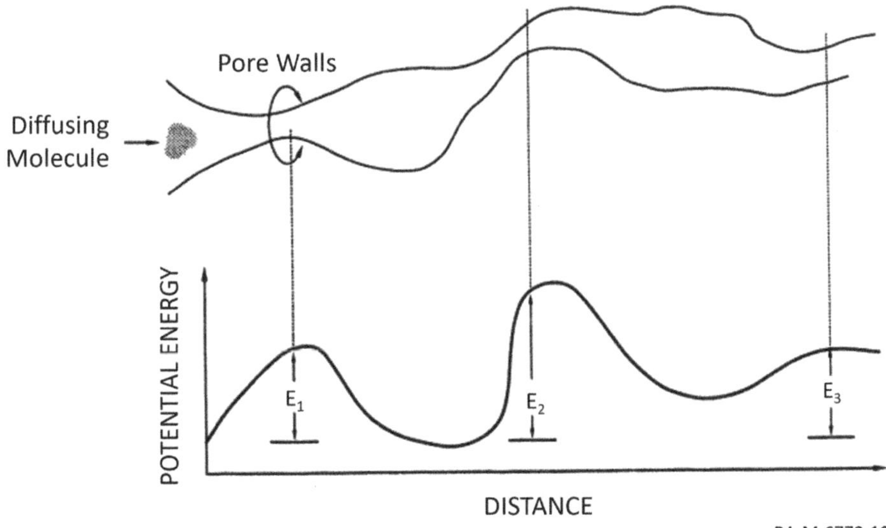

FIGURE 5.9　Visualisation of activation energy vs pore size changes. (Roberts et al., 1992).

TABLE 5.5
Table Detailing the Minimum Pore Size for Knudsen Diffusion (Thornton et al., 2009)

	$d_{p,Kn}$ (Å)
Silica tube	13.64
Silica slit	12.96

for the specified model as it is more likely that this value has been calculated from experiments (de Lange et al., 1995; Shelekhin et al., 1995). This would also lead to a more accurate model as it does not rely on assumptions that would broadly impact the value of permeance.

The concept of suction energy has been used to determine that Knudsen diffusion is the dominant mass transfer mechanism with pore sizes above 13 Angstrom (Å) (Thornton et al., 2009). Pore size definitions for the prediction of when Knudsen flow occurs are shown in Table 5.5.

Knudsen flow is the dominant gas transfer mechanism in pore sizes of > 13.64 Å for silica tubes and 12.96 Å for silica slits (Thornton et al., 2009). Consequently, a molecular sieving activated transport mechanism must be used for this model for membranes with pore sizes less than 13 Å (1.3 nm). However, membranes displaying microporous properties are noted up to 2 nm (Pakizeh et al., 2007).

As a microporous membrane is classified as less than 2 nm, this model will employ the microporous mass transfer mechanism for pore diameters less than 2 nm. Membranes above this size are classified as mesoporous membranes and will be modelled using the Knudsen diffusion mechanism.

5.5.2 SURFACE DIFFUSION

Surface diffusion occurs when surface interactions with gas molecules outweigh their kinetic energy, hence the gas molecules are adsorbed preferentially to the membrane surface at low temperatures (De Meis, 2017). It involves adsorption of molecules from fluid bulk to the vicinity of the membrane surface, sorption to the surface, pore transport by surface diffusion and ultimately desorption from the membrane surface to permeate bulk (Figure 5.10).

The point at which the amount of gas on the surface and the amount of gas not absorbed are equal is known as the isoconcentration point of the gas. The isoconcentration point is dependent on temperature. According to Shelekhin et al., the iso-concentration point of H_2 is at approximately $-250°C$ (Shelekhin et al., 1993). The Langmuir adsorption model predicts gas transfer for surface diffusion (Oyama et al., 2011). As most separation conditions are carried out above room temperature, it can be assumed that the surface diffusion mechanism will not occur in this model (Shelekhin et al., 1995). The surface diffusion model is not used in this model as the temperature range in which surface diffusion transport would take place is outside functional membrane operating conditions (Oyama et al., 2011).

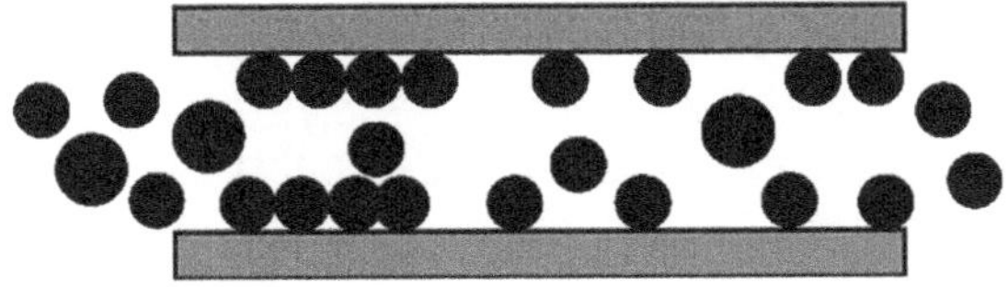

FIGURE 5.10   Schematic of surface diffusion mechanism.

5.5.3   ACTIVATED KNUDSEN/TRANSLATION

The equation for the permeability of species through a microporous membrane under activated Knudsen, also called the gas translation model [58], [59], is given by Equation 5.5. This equation thus combines the Knudsen gas transport model and the surface diffusion model. This model applies to gases with enough activation energy to overcome the diffusion barrier above the isoconcentration point (Oyama et al., 2011; Pakizeh et al., 2007).

$$P_{\mathrm{MA}} = \frac{\rho_g d_p}{t} \times \left(\frac{8}{\pi M_A \mathrm{RT}} \right)^{0.5} \frac{\varepsilon}{\tau} \mathrm{e}^{\frac{\Delta E_{\mathrm{Act}}}{\mathrm{RT}}} \tag{5.5}$$

Table 5.6 summarises the meaning of the inputs to Equation 5.5. ρ_g is the geometrical probability factor, which can be assigned the value of 0.33 if the diffusing gas can move through the pore without spatial obstruction unless otherwise specified (Shelekhin et al., 1995). This model does not account for physical blockages of the pore (Oyama et al., 2011).

5.5.4   MULTILAYER MEMBRANES

Typically, ceramic membrane structures consist of a matrix of a base (support) layer to maintain the mechanical stability, an intermediate layer allowing the transition to the final and most important layer – the selective layer (Figure 5.11). The selective (aka active) layer, which can have a thickness of tens to hundreds of nanometres, determines the membrane functional properties and degree of separation.

TABLE 5.6

Summary of Typical Parameter Ranges for Equation 5.5 (Shelekhin et al., 1995, Pakizeh et al., 2007, Oyama et al., 2011)

Parameter Symbol	Parameter Description	Value Ranges
ρ_g	Geometry factor	0.1–0.33
d_p	Pore diameter	2–20 Å
ϵ	Porosity	0–1
τ	Tortuosity	0–1
ΔE_{Act}	Activation energy	J

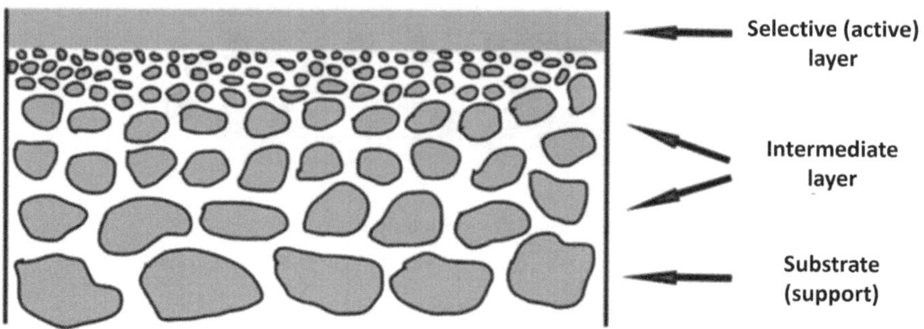

FIGURE 5.11 Multilayer ceramic membrane structure.

Using mesoporous membranes as supports offers structural integrity to membrane modules and therefore allows the microporous layer to be designed thinner, which allows a higher permeance value while gaining the benefit of the higher selectivity values of microporous membranes.

In order to calculate the permeance through multiple layers of different membrane materials, the overall permeance can be calculated using the sum of the resistances (Equation 5.6) (de Lange et al., 1995):

$$\frac{1}{P_{M,Total}} = \frac{1}{P_{M,\,Toplayer}} + \frac{1}{P_{M,Support,1}} + \cdots + \frac{1}{P_{M,Support,n}} \tag{5.6}$$

where there are n-amount of support layers. As stated in Equation 5.6, the selectivity of a membrane can be calculated as the permeability of species A divided by the permeance of species B.

5.6 CURRENT AND FUTURE PROGRESS IN CERAMIC MEMBRANE APPLICATION FOR HYDROGEN PRODUCTION

The majority of research efforts in the field of ceramic membrane purification of hydrogen have been focused on the development of highly selective membranes that exhibit high chemical and mechanical stability. This involves optimisation of existing fabrication methods as well as development of novel manufacturing techniques, modification and functionalisation of ceramic membranes and expanding hybrid and nanocomposite ceramic membranes. The inverse relationship between selectivity and permeability observed in many studies remains a challenge in producing suitable membrane structures, which therefore limits the feasibility of large-scale application of such membranes for hydrogen separation and purification. Although common ceramic membranes did not historically show promising industrial use due to their high fragility, their chemical and thermal stability continually prompts further research and development in the synthesis of stronger, more selective membranes. In addition, this has prompted the formation of a variety of ceramic–ceramic composites, composites utilising metal nanoparticles, nano carbons or metal oxides, ceramic–metal and ceramic–polymer composites further enhancing the selective, mechanical and chemical properties of membranes in comparison to individual ceramic phases. Ceramic membranes dispersed on polymer membrane surfaces or immobilised ceramic material in organic matrixes via grafting (formation of covalent bonds) are some of the methods utilised to form mixed matrix membranes, particularly favourable for hydrogen separation. Another area of significant importance is manufacturing the membranes into configurations suitable for their target application. Scale up of viable lab and pilot separation units is a key priority in the successful implementation of ceramic membrane technologies on an industrial scale.

5.7 CONCLUSION

Membrane separation of hydrogen using ceramic membranes has gained significant attention due to their high stability and good hydrogen selectivity. Amongst the porous variety, silica and zeolite membranes have shown significant viability for large scale applications. Many silica membrane varieties have been studied, where the tuneable pore structures of these membranes led to high hydrogen selectivities over other gases. High temperature environments were shown to improve their hydrogen permeability, while high humidities were found to be detrimental to their performance. Zeolite membranes have achieved variable levels of success when it comes to their selectivity for hydrogen and their structural longevity. Much effort is put into mitigating their morphological defects as well as overcoming the significant compromise between permeance and selectivities demonstrated by zeolite membrane varieties. Dense ceramic membranes such as the perovskite type membrane operate based on proton transfer, and therefore, their hydrogen separation performance is not impacted by issues common with porous structures. Studies have demonstrated selectivities up to 99% using these membranes; however, they suffer from low permeation fluxes in comparison to porous ceramic membranes. Current and future research is focused on optimisation of membrane fabrication methods to enhance the functionality of pure ceramic membranes, as well as utilising ceramic membranes' superior chemical and structural properties in hybrid/mixed matrix membrane varieties.

NOMENCLATURE

P: Permeability ($\mathrm{mol \cdot m \cdot m^{-2} \cdot s^{-1} \cdot Pa^{-1}}$)
P_M: Permeance ($\mathrm{mol \cdot m^{-2} \cdot s^{-1} \cdot Pa^{-1}}$)
α: Membrane selectivity
C: Concentration
CVD: Chemical Vapour Deposition
D: Diffusivity coefficient
GPU: Gas permeation units
M: Molecular weight
ε: Porosity
τ: Tortuosity
d_p: Pore diameter
t: Thickness

R: Universal gas constant
T: Temperature (K)
E_{Act}: Activation energy
r_p: Pore radius
r_{pd}: Dilated pore radius
δW: Pore length
ρ_g: Geometry factor
r: Pressure ratio

REFERENCES

Abdullayev, A., Bekheet, M.F., Hanaor, D.A.H., Gurlo, A., 2019. Materials and applications for low-cost ceramic membranes. *Membranes* 9, 105. https://doi.org/10.3390/MEMBRANES9090105

Agoudjil, N., Benmouhoub, N., Larbot, A., 2005. Synthesis and characterization of inorganic membranes and applications. *Desalination* 184, 65–69. https://doi.org/10.1016/j.desal.2005.04.034

Amanipour, M., Ganji Babakhani, E., Safekordi, A., Zamaniyan, A., Heidari, M., 2012a. Effect of CVD parameters on hydrogen permeation properties in a nano-composite SiO2-Al2O3 membrane. *Journal of Membrane Science* 423–424, 530–535. https://doi.org/10.1016/j.memsci.2012.09.007

Amanipour, M., Heidari, M., Walberg, M., 2022. Comparison of hydrogen permeation and structural properties of a microporous silica membrane and a dense BaCe0.9Y0.1O3-δ (BCY) perovskite membrane. *Results in Materials* 15, 100314. https://doi.org/10.1016/J.RINMA.2022.100314

Amanipour, M., Safekordi, A., Ganji Babakhani, E., Zamaniyan, A., Heidari, M., 2012b. Effect of synthesis conditions on performance of a hydrogen selective nano-composite ceramic membrane. *International Journal of Hydrogen Energy* 37, 15359–15366. https://doi.org/10.1016/J.IJHYDENE.2012.07.085

Amin, S.K., Hassan, M., Abdallah, H., 2016. An overview of production and development of ceramic membranes. *International Journal of Applied Engineering Research* 11(12), 7708–7721.

Athayde, D.D., Souza, D.F., Silva, A.M.A., Vasconcelos, D., Nunes, E.H.M., Da Costa, J.C.D., Vasconcelos, W.L., 2016. Review of perovskite ceramic synthesis and membrane preparation methods. *Ceramics International*. https://doi.org/10.1016/j.ceramint.2016.01.130

Balachandran, U., Lee, T.H., Chen, L., Song, S.J., Picciolo, J.J., Dorris, S.E., 2005. Current status of dense cermet membranes for hydrogen separation. In: *22nd Annual International Pittsburgh Coal Conference 2005, PCC 2005*. Elsevier, Pittsburgh, PA, pp. 771–777. https://doi.org/10.1016/s0140-6701(03)82114-8

Balster, J., 2013. Spiral wound membrane module. In: Drioli, E., Giorno, L. (eds) *Encyclopedia of Membranes*, Springer, Berlin, Heidelberg, pp. 1–3. https://doi.org/10.1007/978-3-642-40872-4_1586-1

Berk, Z., 2009. Chapter 10 - Membrane processes. In: Berk, Z. (ed) *Food Process Engineering and Technology*, Academic Press, San Diego, pp. 233–257. ISBN 978-0-12-812018-7. https://doi.org/10.1016/B978-0-12-373660-4.00010-7

Bernardo, G., Araújo, T., da Silva Lopes, T., Sousa, J., Mendes, A., 2020. Recent advances in membrane technologies for hydrogen purification. *International Journal of Hydrogen Energy* 45, 7313–7338. https://doi.org/10.1016/j.ijhydene.2019.06.162

Bernardo, P., Drioli, E., Golemme, G., 2009. Membrane gas separation: A review/state of the art. *Industrial & Engineering Chemistry Research* 48, 4638–4663. https://doi.org/10.1021/IE8019032/ASSET/IMAGES/LARGE/IE-2008-019032_0001.JPEG

Bockute, K., Prosyčevas, I., Lazauskas, A., Laukaitis, G., 2015. Preparation and characterisation of mixed matrix Al2O3- and TiO2-based ceramic membranes prepared using polymeric synthesis route. *Ceramics International* 41, 8981–8987. https://doi.org/10.1016/j.ceramint.2015.03.172

Bouwmeester, H.J.M., 2003. Dense ceramic membranes for methane conversion. *Catalysis Today*, 82, 141–150. https://doi.org/10.1016/S0920-5861(03)00222-0

Bui, V., Tandel, A.M., Satti, V.R., Haddad, E., Lin, H., 2023. Engineering silica membranes for separation performance, hydrothermal stability, and production scalability. *Advanced Membranes* 3. https://doi.org/10.1016/j.advmem.2023.100064

Chathurappan, R., Anandkumar, J., 2023. Synthesis and characterization of tubular ceramic ultrafiltration membrane from coal fly ash. *Environmental Quality Management*. https://doi.org/10.1002/tqem.21957

Danewalia, S.S., Singh, K., 2021. Bioactive glasses and glass-ceramics for hyperthermia treatment of cancer: State-of-art, challenges, and future perspectives. *Materials Today Bio* 10, https://doi.org/10.1016/j.mtbio.2021.100100.

David, E., Kopac, J., 2011. Development of palladium/ceramic membranes for hydrogen separation. *International Journal of Hydrogen Energy* 36, 4498–4506. https://doi.org/10.1016/J.IJHYDENE.2010.12.032

Deibert, W., Ivanova, M.E., Meulenberg, W.A., Vaßen, R., Guillon, O., 2015. Preparation and sintering behaviour of La5.4WO12−δ asymmetric membranes with optimised microstructure for hydrogen separation. *Journal of Membrane Science* 492, 439–451. https://doi.org/10.1016/J.MEMSCI.2015.05.065

De Lange, R.S.A., Keizer, K., Burggraaf, A.J., 1995. Analysis and theory of gas transport in microporous sol-gel derived ceramic membranes. *Journal of Membrane Science* 104, 81–100. https://doi.org/10.1016/0376-7388(95)00014-4

De Meis, Domenico. (2017). *Gas Transport through Porous Membranes*. Italian National Agency for New Technologies, Energy and Sustainable Economic Development (ENEA), Frascati. Report Number RT/2017/7/ENEA. https://doi.org/10.13140/RG.2.2.25450.72641.

De Vos, R.M., Maier, W.F., Verweij, H., 1999. Hydrophobic silica membranes for gas separation. *Journal of Membrane Science*, 158, 277–288. https://doi.org/10.1016/S0376-7388(99)00035-6

De Vos, R.M., Verweij, H., 1998. Improved performance of silica membranes for gas separation. *Journal of Membrane Science*, 143, 37–51. https://doi.org/10.1016/S0376-7388(97)00334-7

Dong, J., Lin, Y.S., Kanezashi, M., Tang, Z., 2008. Microporous inorganic membranes for high temperature hydrogen purification. *Journal of Applied Physics* 104. https://doi.org/10.1063/1.3041061

Doran, P.M., 1995. *Bioprocess Engineering Principles*. 1st Edition, Academic Press. United Kingdom. ISBN 978-0122208560

Gopalakrishnan, S., da Costa, J.C.D., 2008. Hydrogen gas mixture separation by CVD silica membrane. *Journal of Membrane Science*, 323, 144–147. https://doi.org/10.1016/j.memsci.2008.06.016.

Gopalakrishnan, S., Nair, B.N., and Nakao, S., 2006. High performance hydrogen selective membranes prepared using rapid processing method. *International Conference on Nanoscience and Nanotechnology*, Brisbane, QLD, Australia. https://doi.org/10.1109/ICONN.2006.340665.

Gopalakrishnan, S., Yamaguchi, T., Nakao, S., 2006. Permeation properties of templated and template-free ZSM-5 membranes. *Journal of Membrane Science* 274, 102–107.

Ghouil, B., Harabi, A., Bouzerara, F., Boudaira, B., Guechi, A., Demir, M.M., Figoli, A., 2015. Development and characterization of tubular composite ceramic membranes using natural alumino-silicates for microfiltration applications. *Materials Characterization* 103, 18–27. https://doi.org/10.1016/j.matchar.2015.03.009

Gu, Y., Oyama, S.T., 2007. Ultrathin, hydrogen-selective silica membranes deposited on alumina-graded structures prepared from size-controlled boehmite sols. *Journal of Membrane Science*, 306, 216–227., https://doi.org/10.1016/j.memsci.2007.08.045

Hao, A., Wan, X., Liu, X., Yu, R., Shui, J., 2022. Inorganic microporous membranes for hydrogen separation: Challenges and solutions. *Nano Research Energy*. https://doi.org/10.26599/nre.2022.9120013

Heidari, M., Safekordi, A., Zamaniyan, A., Ganji Babakhani, E., Amanipour, M., 2015. Comparison of microstructure and hydrogen permeability of perovskite type $ACe0.9Y0.1O3-\delta$ (A is Sr, Ba, La, and BaSr) membranes. *International Journal of Hydrogen Energy* 40, 6559–6565. https://doi.org/10.1016/J.IJHYDENE.2015.03.113

Heidari, M., Zamaniyan, A., SafeKordi, A., Ganji Babakhani, E., Amanipour, M., 2013. Effect of sintering temperature on microstructure and hydrogen permeation properties of perovskite membrane. *Journal of Materials Science & Technology* 29, 137–141. https://doi.org/10.1016/J.JMST.2012.12.003

Kamarudin, S.K., Daud, W.R.W., Mohammad, A.W., Som, A.M., Takriff, M.S., 2003. Design of a tubular ceramic membrane for gas separation in a PEMFC system. *Fuel Cells* 3, 189–198. https://doi.org/10.1002/FUCE.200330119

Kentish, S.E., Scholes, C.A., Stevens, G.W., 2010. Carbon dioxide separation through polymeric membrane systems for flue gas applications. *Recent Patents on Chemical Engineering* 1, 52–66. https://doi.org/10.2174/1874478810801010052

Khatib, S.J., Oyama, S.T., 2013a. Silica membranes for hydrogen separation prepared by chemical vapor deposition (CVD). *Separation and Purification Technology*. https://doi.org/10.1016/j.seppur.2013.03.032

Kim, Y.-S., Kusakabe, K., Morooka, S., Yang, S.M., 2001. Preparation of microporous silica membranes for gas separation. *Korean Journal of Chemical Engineering*, 18:106–112. https://doi.org/10.1007/BF02707206

Korelskiy, D., Ye, P., Fouladvand, S., Karimi, S., Sjöberg, E., Hedlund, J., 2015. Efficient ceramic zeolite membranes for CO2/H2 separation. *Journal of Materials Chemistry A* 3, 12500–12506. https://doi.org/10.1039/C5TA02152A

Koutsonikolas, D.E., Pantoleontos, G., Karagiannakis, G., Konstandopoulos, A.G., 2021a. Development of H2 selective silica membranes: Performance evaluation through single gas permeation and gas separation tests. *Separation and Purification Technology* 264, 118432. https://doi.org/10.1016/J.SEPPUR.2021.118432

Koutsonikolas, D., Kaldis, S., & Lappas, A., 2021b. Semi-pilot tests of ethanol dehydration using commercial ceramic pervaporation membranes. *Journal of Membrane Science and Research* 7(4), 268–272. https://doi.org/10.22079/jmsr.2021.130702.1401

Leo, A., Liu, S., da Costa, J.C.D., 2009. Development of mixed conducting membranes for clean coal energy delivery. *International Journal of Greenhouse Gas Control*. https://doi.org/10.1016/j.ijggc.2008.11.003

Lim, H., Gu, Y., Oyama, S.T., 2012. Studies of the effect of pressure and hydrogen permeance on the ethanol steam reforming reaction with palladium- and silica-based membranes. *Journal of Membrane Science*, 396, 119–127. https://doi.org/10.1016/j.memsci.2012.01.004

Lin, Y.S., 2001. Microporous and dense inorganic membranes: Current status and prospective. *Separation and Purification Technology* 25, 39–55. https://doi.org/10.1016/S1383-5866(01)00089-2

Liu, C., Zhang, X., Zhai, J., Li, X., Guo, X., He, G., 2023. Research progress and prospects on hydrogen separation membranes. *Clean Energy* 7, 217–241. https://doi.org/10.1093/ce/zkad014

Llosa Tanco, M.A., Medrano, J.A., Cechetto, V., Gallucci, F., Pacheco Tanaka, D.A., 2021. Hydrogen permeation studies of composite supported alumina-carbon molecular sieves membranes: Separation of diluted hydrogen from mixtures with methane. *International Journal of Hydrogen Energy* 46, 19758–19767. https://doi.org/10.1016/J.IJHYDENE.2020.05.088

Lu, G.Q., Diniz Da Costa, J.C., Duke, M., Giessler, S., Socolow, R., Williams, R.H., Kreutz, T., 2007. Inorganic membranes for hydrogen production and purification: A critical review and perspective. *Journal of Colloid and Interface Science* 314, 589–603. https://doi.org/10.1016/j.jcis.2007.05.067

Mat, N.C., Lou, Y., Lipscomb, G.G., 2014. Hollow fiber membrane modules. *Current Opinion in Chemical Engineering* 4, 18–24. https://doi.org/10.1016/j.coche.2014.01.002

McElfresh, D.K., Howitt, D.G., 1986. Activation enthalpy for diffusion in glass. *Journal of the American Ceramic Society* 69, C-237–C-238. https://doi.org/10.1111/j.1151-2916.1986.tb07347.x

Meinema, H.A., Dirrix, R.W.J., Brinkman, H.W., Terpstra, R.A., Jekerle, J., Kösters, P.H., 2005. Ceramic membranes for gas separation-recent developments and state of the art. *International Ceramic Review (INTERCERAM)*, 54, 86–91

Michaels, S.L., 1989a. Crossflow microfilters: The ins and outs. *Chemical Engineering* 96, 84.

Norby, T., Haugsrud, R., 2006. Dense ceramic membranes for hydrogen separation. In: Anthony F. Sammells and Michael V. Mundschau (eds) *Nonporous Inorganic Membranes: For Chemical Processing*, Wiley-VCH, Germany, pp. 1–48. https://doi.org/10.1002/3527608796.ch1

Oyama, S.T., Yamada, M., Sugawara, T., Takagaki, A., Kikuchi, R., 2011. Review on mechanisms of gas permeation through inorganic membranes. *Journal of the Japan Petroleum Institute*. https://doi.org/10.1627/jpi.54.298

Pakizeh, M., Omidkhah, M.R., Zarringhalam, A., 2007. Study of mass transfer through new templated silica membranes prepared by sol-gel method. *International Journal of Hydrogen Energy* 32, 2032–2042. https://doi.org/10.1016/j.ijhydene.2006.10.004

Pandey, P., Chauhan, R.S., 2001. Membranes for gas separation. *Progress in Polymer Science* 26, 853–893. https://doi.org/10.1016/S0079-6700(01)00009-0

Potapov, V., Fediuk, R., Gorev, D., 2020. Membrane concentration of hydrothermal SiO2 nanoparticles. *Separation and Purification Technology*. https://doi.org/10.1016/j.seppur.2020.117290

Roberts, D.L., Abraham, I.C., Blum, Y., Way, J.D., 1992. Gas separation with glass membranes. Report DOE/MC/25204-3133.

Rosensteel, W.A., Ricote, S., Sullivan, N.P., 2016a. Hydrogen permeation through dense BaCe0.8Y0.2O3−δ - Ce0.8Y0.2O2−δ composite-ceramic hydrogen separation membranes. *International Journal of Hydrogen Energy* 41, 2598–2606. https://doi.org/10.1016/J.IJHYDENE.2015.11.053

Serhiienko, A., Dontsova, T.A., Yanusevska, O.I., Nahirniak, S., Hoesseini-Bandegharaei, A., 2020. Ceramic membranes. New trends and prospects (short review). *Water and Water purification Technologies. Scientific and Technical News* 27, 4–31. https://doi.org/10.20535/2218-93002722020208817

Shahbaz, M., Al-Ansari, T., Aslam, M., Khan, Z., Inayat, A., Athar, M., Naqvi, S.R., Ahmed, M.A., McKay, G., 2020. A state of the art review on biomass processing and conversion technologies to produce hydrogen and its recovery via membrane separation. *International Journal of Hydrogen Energy*. https://doi.org/10.1016/j.ijhydene.2020.04.009

Shelekhin, A.B., Dixon, A.G., Ma Y.H., 1993. Adsorption, diffusion and permeation of gases in microporous membranes. III. Application of percolation theory to interpretation of porosity, tortuosity, and surface area in microporous glass membranes. *Journal Membrane Science*, 83 (2), 181–198.

Shelekhin, A.B., Dixon, A.G., Ma, Y.H., 1995. Theory of gas diffusion and permeation in inorganic molecular-sieve membranes. *AIChE Journal* 41, 58–67. https://doi.org/10.1002/aic.690410107

Smart, S., Lin, C.X.C., Ding, L., Thambimutha, K., Diniz da costa, J., 2010. Ceramic membranes for gas processing in coal gasification. *Energy & Environmental Science* 3, 268–278. https://doi.org/10.1039/B924327E

Thornton, A.W., Hilder, T., Hill, A.J., Hill, J.M., 2009. Predicting gas diffusion regime within pores of different size, shape and composition. *Journal of Membrane Science* 336, 101–108. https://doi.org/10.1016/j.memsci.2009.03.019

Vaezi, M.J., Kojabad, M.E., Beiragh, M.M., Babaluo, A.A., 2019. Transport mechanism and modeling of microporous zeolite membranes, In: Basile, A., Ghasemzadeh, K. (eds) *Current Trends and Future Developments on (Bio-) Membranes: Microporous Membranes and Membrane Reactors*. Elsevier, pp. 185–203. https://doi.org/10.1016/B978-0-12-816350-4.00008-8

Wang, Z., Achenie, L.E.K., Khatib, S.J., Oyama, T., 2013. Mixed mechanism model for permeation of gases in hybrid inorganic–organic membranes. *Industrial & Engineering Chemistry Research* 52 (9), 3258–3265. https://doi.org/10.1021/ie300912b

Winston Ho, W.S., Sirkar, K. K., Reinhold, V. N., New York, 1992. *Membrane Handbook*. https://doi.org/10.1007/978-1-4615-3548-5

Yang, E., Alayande, A.B., Goh, K., Kim, C.M., Chu, K.H., Hwang, M.H., Ahn, J.H., Chae, K.J., 2021. 2D materials-based membranes for hydrogen purification: Current status and future prospects. *International Journal of Hydrogen Energy* 46, 11389–11410. https://doi.org/10.1016/J.IJHYDENE.2020.04.053

Yin, H., Yip, A.C.K., 2017. A review on the production and purification of biomass-derived hydrogen using emerging membrane technologies. *Catalysts*. https://doi.org/10.3390/catal7100297

Zhang, K., Sunarso, J., Shao, Z., Zhou, W., Sun, C., Wang, S., Liu, S., 2011. Research progress and materials selection guidelines on mixed conducting perovskite-type ceramic membranes for oxygen production. *RSC Advances* 1, 1661–1676. https://doi.org/10.1039/C1RA00419K

Zhang, X.-L., Yamada, H., Saito, T., Kai, T., Murakami, K., Nakashima, M., Ohshita, J., Akamatsu, K., Nakao, S.I., 2016. Development of hydrogen- selective triphenylmethoxysilane-derived silica membranes with tailored pore size by chemical vapor deposition. *Journal of Membrane Science* 499, 28–35. https://doi.org/10.1016/j.memsci.2015.09.025

Zhu, Z., Hou, J., He, W., Liu, W., 2016. High-performance Ba(Zr0.1Ce0.7Y0.2)O3−δ asymmetrical ceramic membrane with external short circuit for hydrogen separation. *Journal of Alloys and Compounds* 660, 231–234. https://doi.org/10.1016/j.jallcom.2015.11.065

6 Hybrid Membranes for Hydrogen Separation

Henry Bryan Trujillo Ruales,
Enrico Drioli, and Adolfo Iulianelli

6.1 INTRODUCTION

Currently, the world energy production systems are not sustainable as the primary resources used come from fossil hydrocarbons that are not renewable and responsible for collateral huge emissions of greenhouse gases in the environment. The Committee on the Environment, Public Health and Consumer Protection considers the Commission's proposal as a possible solution to the existing problem. This proposal is based on the use of renewable sources and alternative systems for sustainable energy production (Liu et al. 2009; Dunbar, 2015), which is performed in conventional reformers operated at hard and harsh conditions ($T > 800°C$, $p > 20$ bar), producing a gas mixture rich in hydrogen, carbon dioxide, carbon monoxide, and water, along with additional impurities that vary based on the quality of the raw material. To purify hydrogen, it is essential to extract it from the reformer gas mixture, which is generally performed through the multistage conventional transformation/separation processes: Water gas shift (low and high temperature process) converts carbon monoxide into hydrogen and CO_2, followed by other processes such as pressure swing adsorption (PSA) and cryogenic distillation. Unfortunately, these processes involve the use of a large amount of energy; therefore, over the past thirty years, there has been significant interest in membrane technology for the separation of gases and then hydro red (Vadrucci et al., 2013) is used as a key factor for the sustainable energy transition. Nowadays, hydrogen plays a crucial role in the chemical, oil and energy sectors, particularly in oil refining, it is used to produce fuels for cleaner transport and its use will increase in the coming years. Hydrogen is a fundamental product in several chemicals such as the combined cycle of integrated gasification (IGCC) (Gernot et al., 2018).

Currently, the majority of the global hydrogen supply is generated industrially through the steam reforming process of natural gas (Dunbar, 2015), which is performed in conventional reformers operated at hard and harsh conditions ($T > 800°C$, $p > 20$ bar), producing a gas mixture rich in hydrogen, carbon dioxide, carbon monoxide, and water, along with additional impurities that vary based on the quality of the raw material. To purify hydrogen, it is essential to extract it from the reformer gas mixture, and this is generally performed through the multistage conventional transformation/separation processes: Water gas shift (low and high temperature process) converts carbon monoxide into hydrogen and CO_2, followed by other processes such as PSA and cryogenic distillation. Unfortunately, these processes involve the use of

DOI: 10.1201/9781003382522-8

"

a large amount of energy; therefore, over the past thirty years, there has been significant interest in membrane technology for the separation of gases and, then, hydrogen (Ghasemzadeh et al., 2013).

Membrane technology represents an economical and efficient solution for hydrogen separation and purification; additionally, it provides benefits in terms of minimal energy usage, straightforward installation, operational ease, and is environmentally friendly. Membranes are roughly classified into polymeric and inorganic, and also they can be dense or porous.

Hybrid membranes or mixed matrix membranes (MMMs) represent a kind of mixed membrane solution in which their organic polymer structures and inorganic particles (zeolite, metal structures, etc.) are inserted to enhance or highlight the membrane's performance regarding increased permeability and selectivity, thermal stability, and mechanical durability. Nevertheless, there is considerable room for the research on hybrid membranes, as this kind of membrane solution shows some relevant disadvantages, such as little affinity between the materials, which is responsible for lower permselectivity, and the complex procedure during the manufacturing process, limiting the possible scale-up at larger scale production.

The aim of this chapter is to showcase the capabilities of membrane gas separation specially hybrid membranes in separating and purifying hydrogen-rich streams. This involves presenting the latest advancements in this field and discussing the potential of these membrane solutions in developing an advanced technology for hydrogen separation and purification.

6.2   HYDROGEN

The current world demand for energy coming from fossil fuels is about 80% (Adhikari and Fernando, 2006). Unlike the utilization of fossil fuels, employing hydrogen as an energy source results solely in water as the by-product when it undergoes combustion, without creating any impact to the environment in terms of climate change and air pollution (Adhikari and Fernando, 2006).

Over the last few years, hydrogen production has frequently emerged as a feasible substitute for the use of fossil fuels. However, transitioning from the current carbon-dependent economy to a more sustainable hydrogen-based economy will require intricate scientific and technological resolutions, a process that is expected to take several years for implementation (Armaroli and Balzani, 2011).

Among the main importance in the use of hydrogen, we have:

- Hydrogen ranks among the Earth's most plentiful elements.
- Hydrogen has the capability to be directly utilized in an internal combustion engine and supplied to fuel cells.
- When molecular hydrogen combines with oxygen, combustion generates heat, while their fusion within a fuel cell produces electricity and heat.
- Water is the sole by-product, contrasting the combustion of fossil fuels, which yields CO_2 and an array of pollutants (Armaroli and Balzani, 2011).

In addition, we can mention that it possesses the highest energy content with respect to any other fuel, and these factors have led to increased demand for energy and hydrogen production in recent years. Due to this overturning, a quite important role is assuming the research into the development of new and more sustainable hydrogen production, separation, and purification methods. New processes such as membrane separation have become an interesting alternative to traditional technologies such as oscillating pressure adsorption and cryogenic distillation. Membrane separation processes are accountable for reduced energy usage, offering the potential for continuous operation (Adhikari and Fernando, 2006).

6.2.1 Insights, Characteristics, Data, and Industrial Applications of Hydrogen

The hydrogen element is the most prevalent and simplest in the universe, constituting 75% of normal matter by mass. It represents the third most abundant element in the planet. It has the following characteristics: it is colorless, odorless, and tasteless; we find it mainly in the form of the diatomic molecule H_2 (Dunbar, 2015) in gaseous form under standard conditions. It is a lightweight, non-toxic, and invisible gas. Unlike traditional petroleum-based fuels and derivatives of natural gas, hydrogen possesses an extremely small and light molecular structure. Its presence in the Earth's atmosphere is approximately 1 part per million (ppm) by volume due to its low weight, enabling it to escape the Earth's gravitational pull (Dunbar, 2015). Some of the characteristics of a hydrogen molecule are outlined in Table 6.1.

Primarily employed as a foundational component in the chemical and petrochemical sectors, hydrogen serves as a key raw material in the production of substances like ammonia, methanol, refined fossil fuels through hydrocracking, and various chemicals. Additionally, hydrogen is currently utilized as a vehicle fuel (Armaroli and Balzani, 2011).

6.2.2 Hydrogen Energy Systems

To take advantage of the potential of hydrogen as one of the environmentally friendly energy carriers, the so-called "clean energy system" based on the use of hydrogen as a promising solution to the depletion of fossil fuels and especially in relation to

TABLE 6.1

Properties of Hydrogen (Mazloomi and Gomes, 2012)

Name, Symbol and Number	Hydrogen, H, 1
Phase	Gas
Density	0.089 gL^{-1}
Energy per kilogram	143 MJ kg^{-1}
Octave number	>130
Flammability range in air	4%–75%
Flash point	−253°C

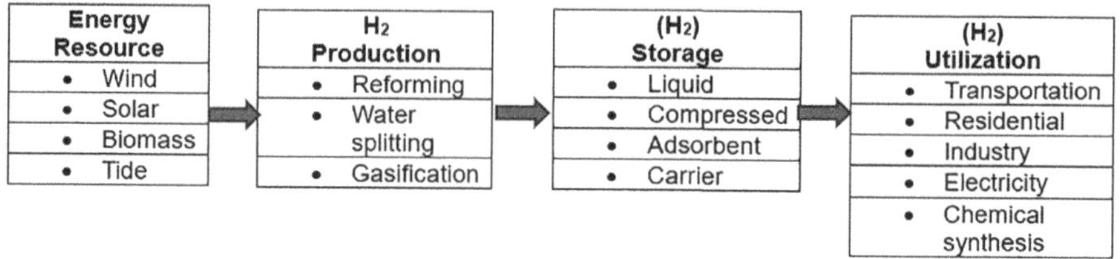

FIGURE 6.1 Hydrogen energy system.

environmental problems is introduced. This concept provided will become one of the energy systems in terms of being alternative and renewable, as is illustrated in Figure 6.1; this system can be composed of energies obtained from different primary sources (solar, wind, biomass, etc.). These energies undergo different reactions—such as gasification, water splitting, and reforming—to convert into a hydrogen carrier. Subsequently, hydrogen is transported, stored, and ultimately utilized by end-users (Yin and Yip, 2017).

6.2.3 Processes to Obtain Hydrogen

6.2.3.1 Biomass

Biomass is as one of the most plentiful renewable energy sources and is an integral phase in the energy and carbon life cycle on Earth. Abundant in every region globally, biomass resources encompass a diverse array of materials derived from agriculture, forestry, energy crops, and industrial waste. Biomass holds the potential for conversion into hydrogen through gasification or pyrolysis, combined with steam reforming. Typically, both processes are executed at elevated temperatures, surpassing 800 K, to attain a greater yield of hydrogen (Yin and Yip, 2017).

$$\text{Pyrolysis} = \text{Biomass} + \text{heat} \rightarrow \text{Syngas} \tag{6.1}$$

$$\text{Gasification} = \text{Biomass} + \text{heat} + \text{Air (Steam)}$$
$$\rightarrow \text{Syngas} + \text{Heavy and light HC} + H_2O \tag{6.2}$$

Following this, additional hydrogen can be generated by transforming other gases produced, such as CH_4, hydrocarbons, and CO, through subsequent reactions:

$$CH_4 + H_2O \rightarrow CO + 3H_2 \text{ Steam Reforming Reaction} \tag{6.3}$$

$$CO + H_2O \rightarrow CO_2 + H_2 \text{ Water Gas Shift Reaction} \tag{6.4}$$

6.2.3.2 Fossil Fuels

Fossil fuels possess a molecular structure that is considerable and weighty, predominantly composed of hydrocarbons. Extracting hydrogen involves severing the bonds between hydrogen and carbon elements within these fuels. Presently, steam methane reforming (SMR) is the most extensively utilized and cost-effective method for this extraction, contributing to approximately 48% of the world's hydrogen production (Mazloomi and Gomes, 2012). The reaction for this process is highly endothermic, as described below:

$$\text{Steam Reforming } C_nH_m + nH_2O \rightarrow nCO + \left(\frac{m}{2} + n\right)H_2 \qquad (6.5)$$

6.2.3.3 Water Electrolysis

The process uses current to drive the conversion; in this case, two platinum electrodes are placed in water and a potential difference is applied. Hydrogen appears in the cathode and oxygen in the anode.

$$\text{Cathode}: \ 2H^+ + 2e \rightarrow H_2 \qquad (6.6)$$

$$\text{Anode}: \ 2OH^- \rightarrow \frac{1}{2}O_2 + 2e \qquad (6.7)$$

The global reaction is:

$$H_2O \rightarrow H_2 + \frac{1}{2}O_2 \qquad (6.8)$$

Electrolysis is used to produce hydrogen anywhere in the world. To put it into operation, the only requirements are electricity and water. The attributes of water electrolysis represent a method capable of generating hydrogen in a wholly sustainable and environmentally friendly manner. However, achieving this objective is contingent upon obtaining the necessary electricity exclusively through emission-free methods, such as solar, wind, geothermal systems, ocean waves, or other renewable and environmentally conscious sources (Mazloomi and Gomes, 2012). This method does not produce a significant proportion of hydrogen. Additionally, high production costs are prevalent due to the method's low conversion efficiency and substantial electricity expenses. These factors may constitute the primary drawbacks of utilizing this approach (Mazloomi and Gomes, 2012).

6.2.4 Hydrogen Purification Processes

As mentioned above, different methods for the production of hydrogen based on steam reforming and gasification lead to a final phase of purification of H_2 in which CO_2, CO, and other impurities must be removed. The main objective of the hydrogen separation/purification is due to the importance for the implementation of the hydrogen economy because of different processes which require a very high-purity hydrogen (molar fraction >99.97% and CO <0.2 ppm) (Bernardo et al., 2020).

6.2.4.1 Pressure Oscillation Adsorption (PSA)

The PSA (Pressure Swing Adsorption) method is the most prevalent industrial process for hydrogen separation and purification, operating on the selective adsorption of impurities from a gas stream. PSA finds particular application in the chemical and petrochemical sectors. Its primary advantages include its capability to significantly reduce undesirable impurities to low levels, yielding high-purity hydrogen (up to 99.99%). However, the main drawback lies in its substantial energy demand, coupled with a considerable loss of hydrogen (around 20%) due to pressure release during desorption, leading to an approximate 80% recovery of hydrogen (Bernardo et al., 2020).

6.2.4.2 Cryogenic Distillation

Cryogenic distillation involves partially condensing gas mixtures at low temperatures and high pressures to separate them through distillation. Hydrogen's relatively high volatility compared to hydrocarbons enables its separation using this method. However, it demands a significant energy input and proves to be costly due to the need for numerous pieces of equipment. In cases where CO and CO_2 are present, the inclusion of a CH_4 washing column becomes necessary to lower their levels. The resulting hydrogen purity is moderate, typically reaching around 95% or less. Both PSA and cryogenic distillation methods result in elevated equipment expenses and substantial energy consumption (Bernardo et al., 2020).

6.2.4.3 Membrane Separation

A membrane is a barrier that, according to its physical nature, allows one of the components to selectively permeate only with a certain permselectivity through them. The driving force that compels various substances to pass through the membrane is termed as the "permeation driving force," often depicted by the variance in partial pressure of a specific compound between the membrane's two sides (Mulder., 1997)

6.3 MEMBRANE TECHNOLOGY

In gas separation procedures, employing membranes holds significant promise as it involves the interaction between the gas and the utilized membrane, contingent upon the permeability characteristics and the selectivity of a particular gas mixture species (Ismail et al., 2015). Membranes are materials that ensure we have greater performance in the gas treatment industry. Currently, gas separation processes have become one of the main methods to be applied in the industrial sector based on membrane technology; this field remains dynamic and swiftly expanding, proving its technical and economic superiority over previously utilized technologies. Membrane technology guarantees superior performance within the gas treatment industry. Presently, gas separation is one of the most critical industrial applications of membrane technology, continuing to evolve rapidly and demonstrating its technical and economic superiority over other emerging technologies (Ismail et al., 2015).

Membrane applications have gained a very important place in chemical industry and technology. They find extensive application across various laboratory and industrial scales, including but not limited to the chemical and petrochemical industry, water desalination, agro-food industrial refineries, and pre/post-treatment systems (Iulianelli et al., 2019). The primary characteristic of these membranes lies in their capacity to regulate the rate of permeation of a chemical species through their structure.

6.3.1 Gas Separation

Within membrane gas separation, the primary factors of concern encompass membrane permeability and selectivity. Permeability refers to the membrane's capability to allow substances to pass through, whereas selectivity denotes the membrane's ability to enable the passage of very few species (ideally just one). The balance between these two attributes epitomizes the recognized trade-off in membranes: higher permeability generally accompanies lower selectivity, and vice versa (Ismail et al., 2015).

6.3.1.1 Membrane-Based Gas Separation

As previously mentioned, membrane-based separation processes occur due to variations in the rates at which different chemical species move across the membrane interface. The rate of movement is dictated by the influencing factor or factors that affect the individual components, including their mobility and concentration within the interface (Basile et al., 2011) (Figure 6.2).

6.3.1.2 Membrane Classification

Membranes can be categorized into two groups: porous and non-porous (Ismail et al., 2015). Membranes may exhibit homogeneity in structure, but can also possess symmetric or asymmetric characteristics. They might exist as solid or liquid membranes, bearing positive or negative charges, or even be bipolar. Thickness varies greatly, ranging from less than 100 nm to over a centimeter, contingent upon the specific application. Membranes can be presented as flat sheets or in the form of hollow fibers. Typically, hollow fibers are favored due to their capacity to achieve a larger effective membrane area within a given module volume (Ismail et al., 2015).

6.3.1.2.1 Porous Membranes

Porous membranes are rigid structures, with pores interconnected and randomly distributed. The separation of mixtures using these types of membranes primarily relies

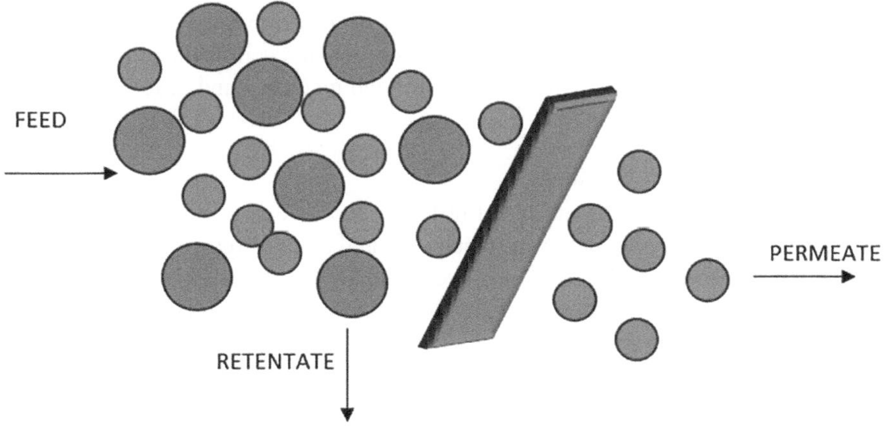

FIGURE 6.2 Schematic of membrane separation.

on the permeant's attributes and the membrane's properties, such as the molecular size of the membrane polymer and the dimensions and arrangement of its pores. Porous membranes employed for gas separation exhibit notably high levels of flow but relatively lower selectivity values (Ismail et al., 2015).

We can classify three types of porous membranes:

a. Relatively large porous membranes with pore sizes ranging from 0.1 to 10 μm: The permeation process occurs by convective mechanism, where no separation process occurs.
b. Porous membranes with pores measuring 0.1 μm: It has the diffusive mechanism of the Knudsen type due to the presence of pores smaller than the average free path of gas molecules.
c. Porous membranes with extremely small pores: 5–20 Å: The separation process takes place by molecular sieve; the transport mechanism in these membranes is very complex to describe, however if they highlight two mechanisms.
d. The first involves gaseous phase diffusion, while the latter pertains to the diffusion of adsorbed species along the pore surface (Porter, 1990).

6.3.1.2.2   Non-Porous Membranes

Non-porous membranes, also known as dense membranes, lack intentionally engineered microscopic pores in their structure. Instead, they possess openings formed by gaps between the molecular chains, often termed as "free volume," typically in the range of 6–9 Å. These membranes offer considerable selectivity in separating gases from a mixed stream, but they typically exhibit a lower gas transport rate. These materials possess an important property: permeants having similar dimensions can be separated if their solubility in the membrane differs significantly. The preparation methods are different, but there are two more relevant methods to mention: casting extrusion and solution casting method (Pandey and Chauhan, 2001).

6.3.1.3   Gas Separation Membrane Components

Selecting membrane materials is a crucial aspect in membrane separation technology. The efficacy of separating gas mixture components is influenced by the chemical interactions occurring between the membrane materials and gaseous penetrants (Ismail et al., 2015) (Table 6.2).

6.3.1.3.1   Polymer Membranes

Polymer membranes can be produced utilizing cost-effective materials and exhibit impressive performance capabilities when operated at lower temperatures. The operating temperature generally does not overcome 100°C. Typical polymer materials adopted for membrane preparation are cellulose acetate, polydimethylsiloxane, polycarbonate, polyvinyl difluoride, polyimide, etc. (Bernardo et al., 2020).

6.3.1.3.2   Inorganic Membranes

Inorganic membranes are distinguished into metal, zeolite, carbonic, ceramic, etc. Most of the metal membranes are palladium based. Being the predominant metallic

TABLE 6.2

Comparison between Membranes Made of Polymers and Those Made of Inorganic Materials (Ismail et al., 2015)

Characteristics	Inorganic Membrane	Polymeric Membrane
Material	Silica, glass, palladium, etc.	Glassy or rubbery (polyimide, cellulose, etc.)
Benefic	• High resistance to chemical and thermal deterioration • Mechanical resistance	• Soft and flexible for rubbery state • Firm and inflexible when in a glass-like state
Advantages	• Able to withstand harsh environmental conditions, high pressure • Long lifetime	• Low cost
Disadvantage	• Large capital cost • Low membrane surface per module volume • Fragile	• Short lifetime • Limited resistance to severe chemical and thermal breakdown

material compared to others, it exhibits superior hydrogen permselectivity when contrasted with organic materials. They operate between 300°C and 500°C, sometimes up to 600°C, even though they are sensitive to the presence of impurities (Voitic et al., 2018). While these metal membranes demonstrate outstanding performance and maintain reasonable chemical stability, they are typically fabricated from metals or oxides, including zeolites, palladium alloys, amorphous silica, and similar materials (Bernardo et al., 2020).

6.3.1.3.3 Hybrid Membranes

Heterogeneous membranes, also called MMMs, are composed of a separation layer that contains a continuous phase, typically a polymer, combined with a second dispersed phase containing inorganic fillers with different chemical properties. These membranes offer a promising option for enhancing the gas separation capabilities of polymer membranes. The incorporation of these two materials allows for the integration of the high permselectivity of the fillers, such as molecular carbon sieves, zeolites, and inorganic particles. The concept of hybrid matrix combines the advantages of each phase: high selectivity of dispersed fillers and desirable mechanical properties and economic advantages of polymers (Hashemifard, et al., 2010; Ismail et al., 2015).

6.4 HYBRID MEMBRANE

6.4.1 Summary

As stated previously, the advancement of polymer membranes in gas separation is among the rapidly progressing segments within membrane technology. Consequently, in recent years, there has been significant focus among scientists to maximize the distinct advantages offered by both polymer and inorganic membranes, leading to the creation of hybrid membranes that combine the properties of these two membranes in a single matrix (Cheng et al., 2017). This is due to the fact that these materials have advantages and disadvantages when being used alone. For example, polymers have advantages: ease of processing, low cost, good mechanical stability, and adjustable carrying property, but the main disadvantages include their poor resistance to the presence of impurities, low thermal and chemical stability and, mainly, low compromise between permeability and selectivity. This implies that enhancing permeability comes with a trade-off in selectivity, and vice versa. On the contrary, inorganic-based membranes show higher performance than polymeric in terms of thermal and chemical stability, as well as mechanical robustness, but they also possess some limitations. Typically, they are thick and exhibit limited film-forming characteristics (resulting in low permeation), high brittleness, and often come with a higher cost, especially in the case of metallic membranes (Cheng et al., 2017). Consequently, while polymer membranes remain appealing, alternative strategies are required to significantly enhance their gas separation properties beyond the Robeson upper bound (Cheng et al., 2017).

The combination of the advantages of these two materials leads to inherit attributes that are higher than both the polymer and inorganic membranes. Therefore, the idea of hybrid membranes holds promise in enhancing the selectivity, permeability, or both compared to current polymer membranes by incorporating inorganic particles. Blending these particles with a polymer matrix presents a promising strategy to surpass the limitations set by Robeson while preserving the polymer matrix's flexible mechanical properties (Gonzo et al., 2006).

Hybrid membranes exhibit the distinct 4M features: multiple interactions, multiscale structures, multiphase composition, and multiple functionalities. These membrane structures encompass a spectrum from nanoscale to macroscale. They encompass both polymer and inorganic phases, along with the polymer–inorganic interface phase. Diverse combinations of polymers and fillers provide hybrid membranes with a variety of functions (Cheng et al., 2017).

6.4.2 Theoretical Background

The concept of hybrid membranes combines the advantages of each phase present in the matrix, such as high selectivity of dispersed fillers and desirable mechanical properties and economic advantages of polymers (Figure 6.3).

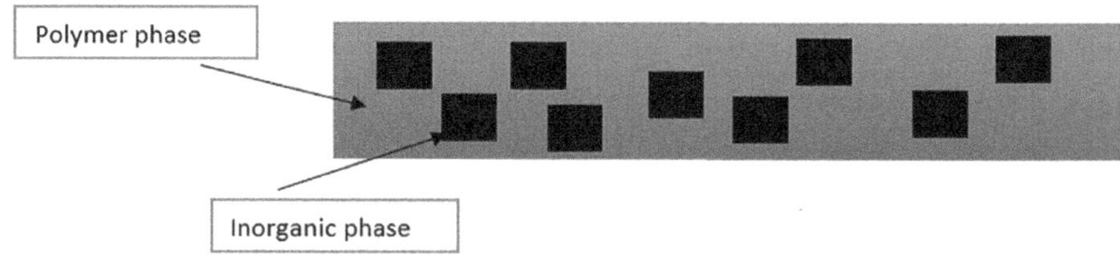

FIGURE 6.3 Schematic of hybrid membrane.

Such membranes find application in various gas separation processes, including but not limited to: segregating oxygen from nitrogen mixtures, purifying natural gas through carbon dioxide extraction, and separating hydrogen.

Figure 6.3 depicts the mass phase (phase A) as a polymer material, while the dispersed phase (phase B) showcases inorganic particles such as zeolite, carbon molecular sieves, or metallic nanoparticles.

Inorganic particles dispersed on the polymer can have three possible effects on gas permeability:

- Particles possess the capability to serve as molecular sieves, modifying permeability based on molecular size.
- Particles have the potential to disrupt the polymer matrix, resulting in an ensuing rise in micro-cavities and subsequently increasing permeability.
- Particles can function as a hindrance to gas transport, thereby reducing permeability (Ismail et al., 2015).

There are several inorganic fillers used in hybrid membranes; in general, they are porous materials of the molecular sieve type, among which zeolites and carbon molecular sieves, carbon silica and nanotubes can be mentioned. Incorporating molecular sieve-type (CMS) materials into the polymer matrix results in enhanced permeability, selectivity, or both when contrasted with the utilization of a mere polymer membrane. This is due to the molecular sieve-type materials' ability to distinguish between various molecules within the mixture, typically based on their size and configuration. Meanwhile, CMSs operate on the varying adsorption kinetics of distinct molecular species; for instance, oxygen molecules exhibit faster adsorption on CMS compared to nitrogen molecules, thus permeating more rapidly through the membrane (Hashemifard et al., 2010).

6.4.3 Transport Mechanism

6.4.3.1 Model of Maxwell

Current models explaining the permeation across hybrid membranes rely on thermal or electrical conductivity models. This is due to the close resemblance between thermal or electrical conduction and the permeation of substances through these compound materials. Hence, conductivity models serve as a tool to elucidate the penetration of substances into hybrid membranes (Aroon et al., 2010). Typically, the permeability of a gaseous penetrant in these membranes is anticipated to rely on the permeability of both the continuous phase (polymer) and the dispersed phase (inorganic filling particles), in addition to the volume fraction of the dispersed phase. Maxwell's model is the prominent equation utilized to forecast the permeability of the MMM (Aroon et al., 2010).

Maxwell's transport model, initially designed for the electrical conductivity of particle-based composites, can be adjusted and applied to aspects related to permeability:

$$P = Pc \left[\frac{Pd + 2Pc - 2\varnothing(Pc - Pd)}{Pd + 2Pc + \varnothing(Pc - Pd)} \right] \tag{6.9}$$

where

- P: Effective permeability of a gaseous penetrant
- $\varnothing$: Volume fraction of the dispersed phase
- Pd: Dispersed phase permeability
- Pc: Continuous phase permeability

The Maxwell equation is suitable for a sparse suspension and specifically for lower concentrations, typically around 20% fill particles. It assumes that the flow patterns around the dispersed phase remain unaffected by neighboring particles (Aroon et al., 2010). However, the model isn't capable of accurately foreseeing the permeability in hybrid membranes at higher concentrations of fill particles. Moreover, it overlooks considerations such as the distribution of particle size, shape, and aggregation (Aroon et al., 2010).

6.4.3.2 Model of Bruggeman

This model is employed for estimating the permeability of hybrid membranes characterized by a substantial volume fraction of fill material. Bruggeman's model, initially formulated to evaluate the dielectric constant of particulate composites, accounts for the impact of introducing extra particles into a sparse suspension, assuming random dispersion of spherical particles (Pandey and Chauhan, 2001). This model can be suitable for predicting the permeability according to the following equations:

$$\left[\frac{\left(\dfrac{P}{Pc}\right)-\left(\dfrac{Pd}{Pc}\right)}{1-\left(\dfrac{Pd}{Pc}\right)}\right]\left(\frac{P}{Pc}\right)^{-1/3}=1-\varnothing \tag{6.10}$$

This equation is an enhancement of the Maxwell model, specifically addressing higher concentrations. However, it does not accurately predict the permeability of hybrid membranes at the maximum fill volume fraction of filler particles, neglecting the consideration of particle size, shape, and distribution. To assess the actual permeability with this model, a trial-and-error method becomes necessary. Consequently, the Bruggeman model is an implicit correlation requiring numerical resolution to determine permeability (Pal, 2008).

6.4.3.3 Model of Lewis-Nielsen

This model is used for the elastic modulus of particulate composites (Baker, 2012). The equation, suitable for determining permeability, is defined as follows:

$$P = Pc\left[\frac{1+2\left(\left(\dfrac{Pd}{Pc}\right)-1\right)/\left(\left(\dfrac{Pd}{Pc}\right)+2\right)\varnothing}{1-\left(\left(\dfrac{Pd}{Pc}\right)-1\right)/\left(\left(\dfrac{Pd}{Pc}\right)+2\right)\varnothing\psi}\right] \tag{6.11}$$

where

$$\psi = 1 + \left(\left(\frac{1 + \varnothing_m}{\varnothing_m^2} \right) \varnothing \right) \tag{6.12}$$

- with $\varnothing_m$ being the maximum fraction of the packing volume of the filled particles. Generally, it has a value of 0.64 (random close-up packing of uniform spheres) (Aroon et al., 2010).

The Lewis–Nielsen model gives the correct behavior for maximum values of $\varnothing$ and the effects of morphology on permeability, since $\varnothing_m$ is a function of particle distribution, particle size and shape, and particle aggregation (Aroon et al., 2010).

When $\varnothing_m \to 1$, the Lewis–Nielsen model becomes Maxwell's model.

6.4.3.4 Model of Pal

This model was developed for the thermal conductivity of particulate composites, drawing from the concept of the differential effective medium approach; considering the challenge in compacting the filler particles (Hashemifard et al., 2010), the Pal equation serves to compute the effective permeability of hybrid membranes, even at the highest fill volume fraction of particles. Furthermore, it accounts for the influence of morphology on permeability by incorporating a specific parameter $\varnothing_m$ (Aroon et al., 2010).

$$\left(\frac{P}{Pc} \right)^{1/3} \left[\frac{\left(\frac{Pd}{Pc} \right) - 1}{\left(\frac{Pd}{Pc} \right) - \left(\frac{P}{Pc} \right)} \right] = \left(1 - \frac{\varnothing}{\varnothing_m} \right)^{-\varnothing_m} \tag{6.13}$$

The Pal model, like the Bruggeman model, is an implicit relationship that must be resolved numerically for P (Permeability).

When $\varnothing_m \to 1$, the Pal model becomes the Bruggeman model.

The Lewis–Nielsen model is a very attractive equation because:

- It's an explicit relationship that's easy to solve.
- It gives the correct behavior to $\varnothing \to \varnothing_m$.
- It affects morphology $\varnothing_m$ (Pal, 2008).

6.4.4 CLASS OF HYBRID MEMBRANES

The classification of these membranes into two major categories is facilitated by the nature of the bonds formed between the inorganic and organic phases:

A. Class I: The integration of various organic and inorganic compounds occurs through a relatively weak bond, which might involve Van der Waals, hydrogen, or ionic bonding.
B. Class II: These membranes exhibit a connection between the two phases established by robust chemical bonds, such as covalent or ion-covalent bonds (Judeinstein and Sanchez, 1996).

6.4.5 COMMON METHODS OF MANUFACTURING HYBRID MEMBRANES

The processes for the preparation of hybrid membranes can be categorized into three types: physical mixing, sol–gel, and in situ polymerization. The first is a kind of physical method, while the other two are chemical methods. The solution mixing and the sol–gel are the most used methods for the manufacture of this type of membrane (Cheng et al., 2017). The structure of the membrane can be categorized into two distinct groups: the first having a concrete-like structure and the second having a sandwich-like structure (Figure 6.4).

In the initial structure, the fillers are uniformly dispersed within the polymer's continuous phase, whereas in the second structure, multiple layers are formed by integrating inorganic and polymer layers through multiple iterations (Cheng et al., 2017). The distinct physicochemical properties between the polymer and inorganic filler phases, along with the pronounced inclination toward filler aggregation, contribute to the absence of non-ideal effects in hybrid membranes. For instance, interfacial voids and pore blockages are minimized (Cheng et al., 2017).

6.4.5.1 Mixture of the Solution

In the initial structure, the fillers are uniformly dispersed within the polymer's continuous phase, whereas in the second structure, multiple layers are formed by integrating inorganic and polymer layers through multiple iterations (Cheng et al., 2017). The distinct physicochemical properties between the polymer and inorganic filler phases, along with the pronounced inclination toward filler aggregation, contribute to the absence of non-ideal effects in hybrid membranes. For instance, interfacial voids and pore blockages are minimized (Cheng et al., 2017).

6.4.5.2 Mixture of the Solution

The polymer is dissolved in a solvent to form a solution into which fillers are introduced. The fillers are introduced in a dispersed manner inside the polymer matrix, which is mixed until a completely uniform mixture is obtained. This mixture is then poured onto a porous support layer, and the solvent is evaporated using conventional methods to form the hybrid membrane. The solution mixing technique is user-friendly

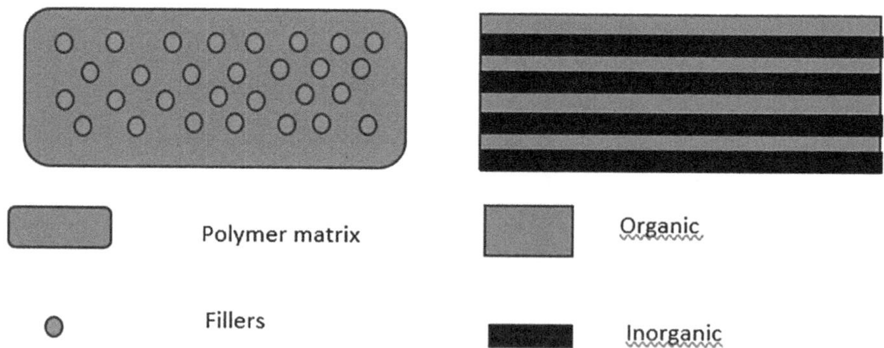

FIGURE 6.4 The two types of hybrid membrane structures.

and adaptable for various types of inorganic materials. It allows for easy control of the concentration of both polymeric and inorganic components. However, one drawback is the tendency of the inorganic constituents to aggregate within the membranes (Cong et al., 2007; Ahmad and Hägg, 2013; Cheng et al., 2017).

The solution mixing technique offers simplicity and versatility, allowing application across various inorganic materials, alongside convenient control over polymer and dispersed phase concentrations. Nonetheless, a notable challenge of this method is the tendency of inorganic fillers to aggregate within the resulting membranes (Cong et al., 2007). These fillers encompass a spectrum of types, extending from zero-dimensional materials to three-dimensional counterparts (Cong et al., 2007):

This mixture is then poured onto a porous support layer, and the solvent is evaporated using conventional methods to form the hybrid membrane. The solution mixing technique is user-friendly and adaptable for various types of inorganic materials. It allows for easy control of the concentration of both polymeric and inorganic components. However, one drawback is the tendency of the inorganic constituents to aggregate within the membranes (Cong et al., 2007; Ahmad and Hägg, 2013; Cheng et al., 2017).

The solution mixing technique offers simplicity and versatility, allowing for application across various inorganic materials, alongside convenient control over polymer and dispersed phase concentrations. Nonetheless, a notable challenge of this method is the tendency of inorganic fillers to aggregate within the resulting membranes (Cong et al., 2007). These fillers encompass a spectrum of types, extending from zero-dimensional materials to three-dimensional counterparts (Cong et al., 2007):

- Silica nanoparticles (SiO_2)
- Carbon nanotubes (CNTs)
- Nanoparticles of metal oxide
- Organic metal structures (MOF)
- Graphene, graphene oxides (GO)
- The zeolites

6.4.5.3 The Sol–Gel Method

It is an attractive and environmentally friendly method because it works at low temperature. In the sol–gel method, inorganic materials are blended into the polymeric solution at ambient temperature. When coupled with the solidification of the polymer chains, these processes culminate in the creation of hybrid membranes. Some advantages of this technique are (Cheng et al., 2017):

- The process takes place under mild conditions, typically at room temperature and normal pressure.
- Managing the concentration of organic and inorganic elements within the solution is straightforward.
- Both organic and inorganic substances are uniformly dispersed within the membranes at the molecular or nanometer scale, resulting in homogeneous membrane products (Cong et al., 2007; Ahmad and Hägg, 2013; Cheng et al., 2017).

This process involves the dispersion of inorganic materials in solvents through the growth of the metal–bone polymer. The first step of this method is the hydroxylation of metal alkoxide.

$$M - OR + H_2O \rightarrow M - OH + ROH \qquad (6.14)$$

where

- $M-(OR)z$: Metal alkoxides
- M: Sn, Zr, Si, Al, Mo, Ti, V, etc.
- OR: Alcox-group OC_nH_{2n+1}

First, hydroxy groups are formed, followed by two successive steps of polycondensation reactions.

$$\text{Oxolation } M - OH + M - OX \rightarrow M - O - M + XOH \qquad (6.15)$$

$$\text{Olation } M - OH + HO - M \rightarrow M(OH)_2 M \ X = H \text{ or alkyl group} \qquad (6.16)$$

These different reactions result in the formation of a structure where oligomers and polymers based on metal–bone are covered by residual hydroxy and alcohol groups. Once this structure becomes macroscopic, a gel can be formed in which the solvent and the polymer are trapped (Judeinstein and Sanchez, 1996).

6.4.5.4 In Situ Polymerization

In this technique, the process involves thoroughly blending inorganic nanoparticles with organic monomers in a solution, followed by polymerization of the monomers. The surface of inorganic particles contains various functional groups like hydroxyl and carboxyl, which create initiating radicals, cations, or anions. These entities are formed due to high-energy radiation, plasma, or other conditions, prompting the initiation of monomer polymerization on the particle surface. Inorganic nanoparticles possessing functional groups can establish links with polymer chains through covalent bonds. However, preventing the aggregation of inorganic nanoparticles in the end product remains a challenge (Cong et al., 2007; Ahmad and Hägg, 2013; Cheng et al., 2017).

6.4.6 Hybrid Membranes for Hydrogen Separation

6.4.6.1 Membranes Based on Zeolite

Zeolites, classified among inorganic materials, represent microporous and crystalline aluminosilicates connected through AlO_4 and tetrahedral SiO_4 primary units. Their exceptional physical–chemical stability and substantial specific area have made them a focus for gas separation and storage research. These zeolites can be synthesized in various shapes, pore diameters, and particle sizes, tailored for specific applications. In the realm of hydrogen separation, hybrid membranes exhibit H_2-selective or reverse-selective behaviors contingent on the choice of polymer and zeolite. For instance, Hu et al. investigated a MMM comprising Pebax, an ionic liquid (IL), and surface-modified SAPO-34. The SAPO-34 surface was enriched with $-NH_2$ groups to enhance its compatibility with the polymer matrix. The introduction of IL increased CO_2 affinity, resulting in a notable CO_2/H_2 selectivity of 22.1 (Hu et al., 2017; Chuah et al., 2021).

Zeolite membranes are generally used as selective H_2 membranes due to the molecular sieving effect. Table 6.3 shows some experimental result about the zeolite Matrimid5218/DDR, in which the pore size of zeolite DDR is 3.6–4.4 Å. This hybrid membrane has a high selectivity of about 130 for the gas mixture H_2/CH_4 (Peydayesh et al.); as a further example, 4A zeolites, with a pore diameter of 3.8 Å, exhibit better the H_2/N_2 and H_2/CH_4 selectivity than previously mentioned (Ahmad and Hägg, 2013).

Recent studies have indicated that the use of spherical particles of ordered mesoporous silica has been effective in mitigating zeolite agglomeration within

TABLE 6.3

Summary of the Notable Achievements Observed in Zeolite-Based Membranes for Hydrogen-Based Separation (Chuah et al., 2021)

MMM							
					Selectivity		
Filler	**Polymer**	**Test Condition**	**PH_2 (GPU)**	**H_2/CO_2**	**H_2/CH_4**	**H_2/N_2**	**References**
IL/SAPO-34	Pebax MH1657/ PEGDME with ceramic	1 bar, 20°C	4.9	0.11	1.2	–	Hu et al. (2017)
SAPO-34	Pebax MH1657/ PEGDME with ceramic	1 bar, 20°C	1.5	0.06	1.4	–	Hu et al. (2017)
IL/SAPO-34-NH_2	Pebax MH1657/ PEGDME with ceramic	1 bar, 20°C	2.2	0.05	1.9	–	Hu et al. (2017)
DD3R	Matrimid 5218	1 bar, 25°C	34.9	–	375	–	Peydayesh et al. (2017)
Zeolite A4 (25 wt%)	PVAc	–	3.8	1.6	–	156	Esmaeili et al. (2019)
Hydroxyl sodalite	PSF	–	21.8	1.1	–	1.1	Eden and Daramola (2021)
Modified Zeolite A4 (15 wt%)	PVAc	0.75 bar 30°C	5.6	6.1	–	143	Esmaeili et al. (2019)
Silica sodalite	PSF	—	22.8	1.1	–	1.0	Eden and Daramola (2021)
HZS	6FDA-DAM	2 bar 35°C	541	0.77	25	21	Zornoza et al. (2015)

Note: Permeability reported in the units of GPU.

hybrid membranes. This efficacy stems from the spherical shape and the surface area-to-volume ratio, which help reduce filler agglomeration (Chuah et al., 2021). The resulting hydrogen permeability is measured at 15, 38, and 541 Barrer, along with H_2/CH_4 selectivity values of 80, 180, and 25, respectively. These outcomes are attributed to the presence of hollow spheres, which facilitate low resistance to transport while allowing for a greater free volume by breaking the polymer chains (Chuah et al., 2021).

6.4.6.2 MOF-Based Membranes

MOFs, which are part of the organic–inorganic hybrid material family, are porous structures comprising metal ions/clusters and rigid organic ligands. Unlike inorganic zeolites, MOFs possess a remarkably high specific area, microporous structures, and customizable surface chemical properties. These characteristics render MOF membranes highly proficient in gas adsorption and molecular sieving, making them immensely appealing for gas separation applications (Seoane et al., 2015; Denny et al., 2016). An additional advantage lies in the excellent interfacial compatibility of MOF nanoparticles with the polymer matrix, which is a significant factor in enhancing gas separation efficiency. Functional ligands such as -OH, $-NH_2$, and $-NO_3$ can be introduced into MOF structures to improve compatibility with the utilized organic polymers. For example, Cao et al. employed $CAU-1-NH_2$ functionalized with NH_2 as a filler for poly(methyl methacrylate) (PMMA) membranes -OH, $-NH_2$, and $-NO_3$, thereby improving their compatibility with organic polymers (Cao et al., 2013). Hydrogen bond formation between $CAU-1-NH_2$ and PMMA achieved a fully compact and robust hybrid membrane with high hydrogen permeability ($>1 \times 10^4$ Barrer) and good selectivity (>10) for H_2/CO_2 separation (Cao et al., 2013). Table 6.4 summarizes various membranes developed in this classification (Chuah et al., 2021).

6.4.6.3 COF-Based Membranes

They fall under the category of microporous organic polymers, encompassing various porous aromatic frameworks such as porous aromatic frameworks (PAF), conjugated microporous polymers (CMP), and hyper-crosslinked polymers (HCP) (Chuah et al., 2021), formed by covalent bonds. Unlike the extensively studied zeolite and MOF structures, these COF materials consist solely of organic compounds and possess the advantage of exhibiting excellent interfacial compatibility with organic polymers (Zou and Zhu, 2018). Moreover, the covalent linkage in COFs confers high chemical stability compared to MOF materials, making them highly effective filling materials for gas separation applications, particularly under challenging conditions (Chuah et al., 2021). Most COFs tend to have pore sizes that are relatively large, similar to those found in zeolitic materials, making it difficult to differentiate between hydrogen molecules and other gases during gas separation processes. To address this issue and achieve discriminatory sieving of H_2 from other gases, COF membranes are often combined with other materials like MOF and GO, creating double-layered/composite membranes (Chuah et al., 2021). Alternatively, another approach involves creating vertically aligned COF membranes, facilitating gas molecule separation through narrow nanochannels delineated by the spacing between COF nanosheets and layers. COF fillers are commonly employed to enhance the low selectivity and

TABLE 6.4

Summary of Exemplary Performances Demonstrated by MOF-Based Membranes in H_2-Based Separation (Chuah et al., 2021)

MMM		Performance					
					Selectivity		
Filler	Polymer	Test Condition	PH$_2$ (GPU)	H_2/CO_2	H_2/CH_4	H_2/N_2	References
CAU-1-NH$_2$	PMMA	3 bar, 25°C	11.000	13	–	–	Cao et al. (2013)
CBMN	6FDA-Durene-DABA	3 bar, 25°C	410	1.2	29	42	Bi et al. (2020)
MIL-53AL-NH$_2$	PI-1388 VTEC™	5 bar, 35°C	5.4	5.4	–	–	Perez et al. (2017)
PSM-nZIF-7	PEI	2 bar, 35°C	8.6	9.9	23	13	Al-Maythalony et al. (2017)
UiO-66-NH$_2$	6FDA-DAM	3 bar, 35°C	1810	0.7	10	10	Ma and Urban (2019)
UiO-66-Hf-OH$_2$	PBI	2 bar, 35°C	8.2	12	–	–	Hu et al. (2016)
ZIF-7-I	6FDA-DAM with α-alumina	–	921	2	67	36	Park, Cho and Jeong (2020)
ZIF-8-DA	Matrimid 5218	3.5 bar, 35°C	400	4	52	44	Jiang et al. (2021)
ZIF-93	PBI	3 bar 180°C	128	5	–	–	Sánchez-Laínez et al. (2018)

Note: Permeability reported in the units of GPU.

physical aging resistance of glass polymers (Chuah et al., 2021), such as PI (polyimide), PIM-1 (polymer of intrinsic microporosity), PSF (polysulfone), PVAm (polyvinylamine), and TPIM (phenazine-containing triptycene ladder polymers) (Chuah et al., 2021) (Table 6.5).

6.4.6.4 Graphene-Based Membranes

These materials contain a layer of carbon atoms hybridized sp^2 in their structure and are placed in a hexagonal lattice that has the shape of a honeycomb. Graphene structures such as graphene oxide (GO) are used as membranes for the separation of gases due to their 2D morphology and monoatomic thickness (Chuah et al., 2021). Furthermore, these properties offer a low resistance to the gas transport. As shown in Table 6.6, Castarlenas et al. demonstrated that GO materials in polymeric matrices of type PSF and Matrimid 5218 may be adopted for separation and purification of H_2 (Castarlenas et al., 2017). Sometimes, the non-porous nature of the GO may be responsible for the decrease in both permeability and H_2/CH_4 selectivity. Indeed, the diffusion mechanism causes the gas to travel a tortuous path through the membrane. Recognizing that CH_4 exhibits higher polarizability compared to H_2, the reduction in H_2/CH_4 selectivity is

TABLE 6.5

Summary of Representative Performances for COF-Based Membranes in H_2-Based Separation (Chuah et al., 2021)

Filler	Polymer	Test Condition	PH_2 (GPU)	H_2/CO_2	H_2/CH_4	H_2/N_2	References
COF_L	PVAm/Mpsf	5 bar	90	0.1	–	–	Cao et al. (2020)
COF_M	PVAm/Mpsf	5 bar	85	0.085	–	–	Cao et al. (2020)
COF_P	PVAm/Mpsf	5 bar	58	0.052	–	–	Cao et al. (2020)
PAOPIM-1	PI-COOH	3 bar	1279	1	16.7	23.5	Huang et al. (2021)
NUS-2	Ultem	2 bar, 35°C	22.7	4.6	103	–	Kang et al. (2016)
NUS-2	PBI	2 bar, 35°C	4.1	31.4	–	–	Kang et al. (2016)
NUS-3	Ultem	2 bar, 35°C	33.4	2.2	63	–	Kang et al. (2016)
PAF-1	TPIM-2	–	4886	1	18.8	23.4	Hou et al. (2020)
PAF-1	PIM-1	–	5500	–	–	4.5	Lau et al. (2015)

TABLE 6.6

Overview of Exemplary Performances Demonstrated by Membranes Based on Graphene in H_2 Separation Applications (Chuah et al., 2021)

Filler	Polymer	Test Condition	PH_2 (GPU)	H_2/CO_2	H_2/CH_4	H_2/N_2	References
Cgo-76 C=cysteamine	Anodized AL_2O_3	1.5 bar 25°C	52	21	–	–	Cheng et al. (2020)
GO	PSF	35°C	4.7	–	29	–	Castarlenas et al. (2017)
GO	PI	35°C	14	–	82	–	Castarlenas et al. (2017)
GO/PEI	Porous Al_2O_3	25°C	1000	29	–	–	Shen et al. (2016)
GO/EDA-2	Porous Al_2O_3	25°C	73	23	–	–	Lin et al. (2018)

presumed to arise from the favorable interaction between CH_4 and the functional groups present in GO. Addressing this concern, various investigations have highlighted the potential of manufacturing GO materials in the form of ultra-thin layers on substrates of polymer membranes. This strategic application aims to prevent excessive compromise of membrane gas permeability. The configuration of these layered membranes enables molecular sieving by leveraging the interlayer spacing between the GO nanosheets, serving as nanochannels that facilitate molecular transport (Chuah et al., 2021).

6.4.6.5 Binary Fillers

As mentioned above, the use of GO as a filler for hybrid membranes results in a lower effectiveness in their use due to a significant and unwanted decrease in H_2 permeability. Therefore, by incorporating binary fillers, it becomes possible to introduce two distinct fillers into the polymer matrix or combine two 3D composite fillers (Chuah et al., 2021). These modifications are presumed to enhance the hydrogen separation performance. Castarlenas et al. (2017) conducted a study involving two different filler types: a composite UiO-66/GO and another comprised of individual fillers UiO-66 and GO physically mixed in the Matrimid 5218 polymer matrix. The resultant hybrid membrane from these fillers showed an improvement in both hydrogen permeability and selectivity, owing to the enhanced interfacial adhesion between the polymer and the filler (Chuah et al., 2021). Conversely, employing UiO-66 and GO separately in the polymer matrix did not yield similar enhancements as the individual fillers failed to establish effective interfacial adhesion (Chuah et al., 2021). Table 6.7 summarizes the diverse membranes created using binary fillers.

TABLE 6.7

Summary of Typical Performances Observed in Mixed Matrix Membranes (MMMs) Incorporating Binary Fillers for H_2-Based Separation (Chuah et al., 2021)

MMM		Performance					
					Selectivity		
Filler	Polymer	Test Condition	PH$_2$ (GPU)	H_2/CO_2	H_2/CH_4	H_2/N_2	References
UiO-66+GO	PSF	35°C	12	–	60	–	Castarlenas et al. (2017)
UiO-66+Go	PI	35°C	41	–	136	–	Castarlenas et al. (2017)
MCM-41+MIL-53-AL-NH2	PSF	35°C	20	–	67	–	Valero et al. (2014)
MCM-41+MIL-53-AL-NH2	PI	35°C	16	–	132	–	Valero et al. (2014)
MCM-41+JDF-L1	6FDA-based copolyimide	35°C	16	–	83	–	Galve et al. (2013)
ZIF8+Silicatile-1	PSF	35°C	17	–	74	–	Zornoza et al. (2011)

6.4.7 SUMMARY OF DIFFERENT HYBRID MEMBRANES DEVELOPED FOR HYDROGEN PURIFICATION

6.4.7.1 Graphene Oxide–Molybdenum Disulfide Hybrid Membranes

Hybrid membranes of GO and molybdenum disulfide (MoS_2) are prepared by vacuum filtration method, which are easily manufactured as thin films with high chemical strength. GO has high gas selectivity but its permeability is low, while molybdenum disulfide has high permeability but has low selectivity (Ostwal et al., 2018).

Li et al. (2013) have manufactured membranes GO with a thickness of 1.8–18 nm, achieving a H_2/CO_2 selectivity of 3,400. Chi et al. (2016) realized GO membranes with a thickness of 20 nm with H_2/CO_2 selectivity of 250 and H_2 permeability of 340×10^{-9} mol m^{-2}s^{-1} Pa^{-1}. Achari et al. (2016) obtained membranes of MoS_2 (500 nm), showing H_2/CO_2 selectivity of 8.2 and H_2 permeability of 780×10^{-9} mol m^{-2}s^{-1} Pa^{-1} (Achari et al., 2016), whereas Wang et al. (2015) demonstrated that membranes of MoS_2 (17–60 nm) may reach H_2/CO_2 selectivity of 4.4 and H_2 permeability of $9,186 \times 10^{-9}$ mol m^{-2}s^{-1} Pa^{-1}.

Mayur et al. used the GO as a packing agent to increase the performance of MoS_2-based membranes (Ostwal et al., 2018). In Table 6.8, hybrids of GM membranes are reported with their respective permeability and selectivity values.

6.4.7.2 TNT@CNT Membranes

Titanium dioxide (TNT) nanotubes and filled CNTs embedded within cellulose triacetate (CTA) hybrid membranes are commonly employed for natural gas purification. The combination of CNT@TNT hybrid fillers with the CTA polymer is facilitated through a process known as easy casting. Compared to the membrane with a single filler (CNT/TNT), this hybrid membrane exhibits a stronger inclination toward CO_2, thanks to the hybrid nature of the charges (CNT@TNT), which renders it notably selective for CO_2 over CH_4, as well as for H_2 over CH_4 (Regmi et al., 2021).

Porous fillers can separate the gaseous mixture through the physique-absorption or dimensional exclusion, allowing to obtain both high permeability and selectivity. These materials can be micro or nano size. Among the common materials used for this purpose, zeolite, CNT, fullerenes, silica, and MOF organic metal structures are dispersed on the polymer matrix to form hybrid membranes (Regmi et al., 2021).

TABLE 6.8

Hydrogen Permeance and H_2/CO_2 Selectivity Data for Hybrid Membranes (as well as Pure GO and MoS_2 Membranes) (Ostwal et al., 2018)

MMM	GO (%wt)	MoS_2 (% wt)	Permeance $H_2 \times 10^{-9}$ (mol m^{-2}s^{-1} Pa^{-1})	H_2 Permeability (Barrer)	Selectivity H_2/CO_2
GO	100	–	188	280	69.9
MoS_2	–	100	105	312.7	6.7
GM-60-1	72.46	27.53	686	120.1	37.0
GM-100-1	70.30	29.70	399	127.0	43.5
GM-150-1	71.15	28.84	287	126.5	442

TABLE 6.9
Permeability and Selectivity of the Membranes Produced through Synthesis (Regmi et al., 2021)

MMM	Permeability (Barrer)			Selectivity
	H_2	CH_4	CO_2	H_2/CH_4
CTA	4.39	0.12	3.01	36.58
CTA-TNT@CNT	22.28	0.46	19.77	48.43
CTA-CNT	15.83	0.41	14.03	38.61
CTA-TNT	12.83	0.30	8.24	42.77

CNT shows excellent selectivity for CO_2 as a condensing penetrant, and it is the most studied materials in gas separation. Nanometer dimensions provide a high surface area, high surface energy, and a frictionless surface, which allow interaction with most of the gases (Zeng et al., 2010). The transport of gases within these materials is extremely fast, generally superior to porous materials such as zeolite. Functionalizing the conjugated sp2 carbon framework with groups like carboxyl, hydroxyl, or amines serves to deter the agglomeration of CNTs within the polymer matrix. CNT was used as a filler in different polymer materials, which improved the gas separation performance. This can be attributed to good dispersion in the polymer matrix, higher mechanical strength, and pore size control (Ismail et al., 2008). Nevertheless, high energy consumption, post-purification stages, and costly manufacturing technique make these materials expensive.

The particles of TiO_2 are very interesting materials if used as fillers for the development of hybrid membranes because they offer favorable physical property, stability, ease of synthesis, and low cost, but above all, they show quite high selectivity absorption capacity, for example at CO_2 (Regmi et al., 2021). As reported in Table 6.9, CTA is derived from cellulose acetate wherein the hydroxyl groups of cellulose are replaced by acetyl groups. This particular compound finds application in gas separation processes, specifically in separating CO_2 from CH_4. Moreover, CTA exhibits a relatively high stability in water at pH levels between 3 and 7, ensuring increased permeability without compromising selectivity. Additionally, it maintains stability even in the presence of acidic gases like SO_2 (Regmi et al., 2021).

6.4.7.3 Zeolite/Carbon Hybrid Membranes

Hybrid membranes of zeolite/carbon present high gas permeability, and are generally prepared by adding zeolite Y and β to the precursor of polyamic acid and high temperature carbonization (Table 6.10). Incorporating zeolite into the carbon membrane results in enhanced gas permeability while maintaining the selectivity of the original polymeric membrane. This improvement is attributed to the internal pore structure within the incorporated zeolite, which offers rapid and continuous diffusion paths

TABLE 6.10

The Influence of Particle Concentration on the Gas Separation Capabilities of the Zeolite/Carbon Hybrid Membranes (Li et al., 2015)

MMM	Content (wt%)	Permeability (Barrer)			Selectivity	
		H_2	CO_2	N_2	H_2/N_2	CO_2/N_2
Pure carbon membrane		84.4	52.7	0.27	312.6	195.1
Y	5	561	250	5.6	100.2	44.6
	15	2280	1022	32.1	71	31.8
	25	4094	1783	67.1	61	26.6
B	5	543	255	5.6	97	45.5
	15	2108	1129	27.9	75.6	40.5
	25	3996	1644	59.4	67.3	27.7

TABLE 6.11

The Impact of Varying Carbonization Temperatures on the Gas Separation Efficiency of Both the Zeolite/Carbon Hybrid Membranes and the Pure Carbon Membrane (Li et al., 2015)

MMM	Carbonization T (K)	Permeability (Barrer)			Selectivity	
		H_2	N_2	CO_2	CO_2/N_2	H_2/N_2
Pure carbon	873	16.5	0.05	6.57	123	330
	973	84.4	0.27	52.7	195	312.6
	1073	36	0.11	25	234	327.3
Zeolite-Y/C	873	2,304	50.8	1,148	22.6	45.4
	973	1,717	20.7	761	36.7	82.9
	1073	102	0.7	42.6	59.3	145.7
Zeolite-β/C	873	2,234	51.9	1,303	25.1	43
	973	1,721	21.3	810	80.8	38
	1073	253	0.81	158	312.3	195

for gas molecules (Li et al., 2015). Carbon membranes are produced by subjecting a polymer membrane to pyrolysis, creating a disordered porous structure characterized by ultramicropores (>0.7 nm) and supermicropores (0.7–2 nm). This structure facilitates a molecular sieving mechanism, which significantly impedes the diffusion of gas molecules through the membrane. Introducing inorganic nanoparticles, such as zeolites like Y, β, and SAPO-34 (as listed in Table 6.11), into the carbon matrix is a strategy employed to augment gas permeability and/or achieve heightened selectivity due to the ordered pore structure they provide (Li et al., 2015).

Compared to pure carbon membranes, the developed zeolite/C membranes show superior gas permeability, implying the molecular diffusion effect of the gas, due to the hybridization of zeolite particles. With the increase in weight of zeolite, gas permeability may be improved due to more interfaced pores formed by more zeolites. However, the addition of more zeolites in the carbon-matrix is advantageous for the permeation by the molecule of larger gases, but this could be responsible for a reduction in selectivity (Li et al., 2015)

Pure carbon membranes show maximum gas permeability at 973 K. While the pure carbon membranes showcase a decrease in gas permeability with higher pyrolytic temperatures, the hybrid zeolite/carbon membranes exhibit a contrasting trend. In the latter, an escalation in temperature leads to heightened gas selectivity. This effect occurs because, at elevated pyrolytic temperatures, the rate of gas permeation diminishes more rapidly for larger molecules such as N_2 compared to smaller gas molecules like H_2, O_2, and CO_2 (Li et al., 2015).

6.5 CONCLUSION

This chapter has provided an overview of the utilization of hydrogen as a renewable and environmentally friendly energy source, as well as the methods of separation/purification adopting hybrid membranes. With regard to the existing technologies for the production and purification of hydrogen, membrane technology is a valid candidate for contributing to the future of the hydrogen economy, allowing the intensification of the conventional processes such as PSA and cryogenic distillation. In the current market, there are several membranes (organic and inorganic) used for hydrogen separation and purification, presenting several advantages but also exhibiting certain disadvantages. The main aim of the researchers all over the world is to develop solutions that can bring in a single structure with the different characteristics of both polymeric (ease of preparation, low cost, and wide materials availability) and inorganic membranes (large hydrogen separation capacity, high mechanical and thermal resistance), which has led to the development of the so-called "hybrid membranes."

In this chapter, we have described in detail different kinds of hybrid membranes used in the hydrogen separation/purification. Among different inorganic fillers that can be used for the preparation of hybrid membranes, zeolites, MOF, COF, and graphene-based materials can change the role of the pristine polymer membranes used as a substrate. Multiple discussions are presented regarding the application of these fillers, exploring their advantageous physical and chemical properties such as morphology, porosity, and chemistry. There is a particular focus on the interfacial interaction between the polymer and filler in the context of polymer–filler hybrids. Furthermore, the resulting performance of these hybrid membranes for hydrogen separation and purification is extensively deliberated upon. Furthermore, a critical analysis of the hydrogen separation performance of hybrid membranes present in the specialized literature is provided, assessing in particular the permeability and selectivity properties of various materials.

On the other hand, apart from different benefits that the hybrid membranes offer, the current difficulties are due to the preparation process, which is related to the aggregation of fillers in the polymer matrix during the sizing of the hierarchical structure, and the non-ideal interfacial effects have also been discussed, describing different methods that today exist for the preparation of hybrid membranes (solution blending, in situ polymerization, and sol–gel). Last but not least, different mechanisms that can explain the process of hydrogen permeation/separation by using the hybrid membranes have also been taken into account, describing in detail the Bruggeman, Maxwell's, and Lewis–Nielsen models.

In summary, the incorporation of inorganic fillers into the polymer matrix has emerged as an expanding area of research, but further efforts are still needed to develop hybrid membranes as a well-established technology to be scaled up at industrial level.

ACKNOWLEDGMENT

The project has been prepared in the frameworks of the project "Hydrogen demo Valley: Multifunctional infrastructures for the experimentation and demonstration of hydrogen technologies" Mission Innovation – Scientific Agreement (21A03302) (GU Serie Generale n.133 del 05-06-2021) between MISE and ENEA.

NOMENCLATURE

j : Permeating flow
C_i : Concentration of the component in the medium
k' : Coefficient that reflects the naturalness of the medium
dp : Gradient of pressure
Di : Diffusion coefficient
l : Membrane thickness
ki : Gas sorption coefficient
$\alpha_{i/j}$: Selectivity of the membrane
Pc : Continuous phase permeability
Pd : Dispersed phase permeability
$\varnothing$: Volume fraction on the dispersed phase
$\varnothing_m$: Maximum fraction of the packing volume of the filled particles

ACRONYMS

IGCC	Combined cycle of integrated gasification
PSA	Pressure swing adsorption
MMM	Mixed matrix membrane
SMR	Steam methane reforming
IL	Ionic liquid
DAM	2,4,6-Triphenyl-m-phenylenediamide
PSF	Polysulfone
PVA	Polyvinyl acetate
PMMA	Polymethyl methacrylate
PBI	Polybenzimidazole

PI	Polyimide
PAF	Aromatic frameworks
CMP	Microporous polymers
GO	Graphene oxide
CNT	Carbon nanotube

REFERENCES

Achari, A., Sahana, S. and Eswaramoorthy, M. (2016) 'High performance MoS2 membranes: Effects of thermally driven phase transition on CO2 separation efficiency', *Energy and Environmental Science*, 9(4), pp. 1224–1228. Available at: https://doi.org/10.1039/c5ee03856a.

Adhikari, S. and Fernando, S. (2006) 'Hydrogen membrane separation techniques', *Industrial and Engineering Chemistry Research*, pp. 875–881. Available at: https://doi.org/10.1021/ie050644l.

Ahmad, J. and Hägg, M.B. (2013) 'Preparation and characterization of polyvinyl acetate/zeolite 4A mixed matrix membrane for gas separation', *Journal of Membrane Science*, 427, pp. 73–84. Available at: https://doi.org/10.1016/j.memsci.2012.09.036.

Al-Maythalony, B.A. et al. (2017) 'Tuning the interplay between selectivity and permeability of ZIF-7 mixed matrix membranes', *ACS Applied Materials and Interfaces*, 9(39), pp. 33401–33407. Available at: https://doi.org/10.1021/acsami.6b15803.

Armaroli, N. and Balzani, V. (2011) 'The hydrogen issue', *ChemSusChem*, 4(1), pp. 21–36. Available at: https://doi.org/10.1002/cssc.201000182.

Aroon, M.A. et al. (2010) 'Performance studies of mixed matrix membranes for gas separation: A review', *Separation and Purification Technology*, 4(1), pp. 229–242. Available at: https://doi.org/10.1016/j.seppur.2010.08.023.

Baker, R.W. (2012) *Membrane Technology and Applications*. John Wiley & Sons.

Bernardo, G. et al. (2020) 'Recent advances in membrane technologies for hydrogen purification', *International Journal of Hydrogen Energy*, 45(12), pp. 7313–7338. Available at: https://doi.org/10.1016/j.ijhydene.2019.06.162.

Basile A., Gugliuzza A., Iulianelli A., Morrone P. (2011) 'Membrane technology for carbon dioxide (CO_2) capture in power plants', In: *Advanced Membrane Science and Technology for Sustainable Energy and Environmental Applications*. Elsevier (UK), pp. 113–159. Available at: https://doi.org/10.1533/9780857093790.2.113

Bi, X. et al. (2020) 'MOF nanosheet-based mixed matrix membranes with metal-organic coordination interfacial interaction for gas separation', *ACS Applied Materials and Interfaces*, 12(43), pp. 49101–49110. Available at: https://doi.org/10.1021/acsami.0c14639.

Cao, L. et al. (2013) 'A highly permeable mixed matrix membrane containing CAU-1-NH_2 for H_2 and CO_2 separation', *Chemical Communications*, 49(76), pp. 8513–8515. Available at: https://doi.org/10.1039/c3cc44530e.

Cao, X. et al. (2020) 'Preparation of high-performance and pressure-resistant mixed matrix membranes for CO_2/H_2 separation by modifying COF surfaces with the groups or segments of the polymer matrix', *Journal of Membrane Science*, 601. Available at: https://doi.org/10.1016/j.memsci.2020.117882.

Castarlenas, S., Téllez, C. and Coronas, J. (2017) 'Gas separation with mixed matrix membranes obtained from MOF UiO-66-graphite oxide hybrids', *Journal of Membrane Science*, 526, pp. 205–211. Available at: https://doi.org/10.1016/j.memsci.2016.12.041.

Cheng, L. et al. (2020) 'Cysteamine-crosslinked graphene oxide membrane with enhanced hydrogen separation property', *Journal of Membrane Science*, 595. Available at: https://doi.org/10.1016/j.memsci.2019.117568.

Cheng, X. et al. (2017) 'Hybrid membranes for pervaporation separations', *Journal of Membrane Science*, 541, pp. 329–346. Available at: https://doi.org/10.1016/j.memsci.2017.07.009.

Chi, C. et al. (2016) 'Facile preparation of graphene oxide membranes for gas separation', *Chemistry of Materials*, 28(9), pp. 2921–2927. Available at: https://doi.org/10.1021/acs.chemmater.5b04475.

Chuah, C.Y. et al. (2021) 'Recent progress in mixed-matrix membranes for hydrogen separation', *Membranes*. 11(9), p. 666. Available at: https://doi.org/10.3390/membranes11090666.

Cong, H. et al. (2007) 'Polymer-inorganic nanocomposite membranes for gas separation', *Separation and Purification Technology*, 55(3), pp. 281–291. Available at: https://doi.org/10.1016/j.seppur.2006.12.017.

Dunbar, Z.W. (2015) 'Hydrogen purification of synthetic water gas shift gases using microstructured palladium membranes', *Journal of Power Sources*, 297, pp. 525–533. Available at: https://doi.org/10.1016/j.jpowsour.2015.08.015.

Denny, M.S. et al. (2016) 'Metal-organic frameworks for membrane-based separations', *Nature Reviews Materials*, 1(12), pp. 1–17. Article ID: 16078.

Eden, C.L. and Daramola, M.O. (2021) 'Evaluation of silica sodalite infused polysulfone mixed matrix membranes during H_2/CO_2 separation', In: *Materials Today: Proceedings*. Elsevier Ltd, pp. 522–527. Available at: https://doi.org/10.1016/j.matpr.2020.02.393.

Esmaeili, N. et al. (2019) 'Improving the gas-separation properties of PVAc-zeolite 4A mixed-matrix membranes through nano-sizing and silanation of the zeolite', *ChemPhysChem*, 20(12), pp. 1590–1606. Available at: https://doi.org/10.1002/cphc.201900423.

Galve, A. et al. (2013) 'Combination of ordered mesoporous silica MCM-41 and layered titanosilicate JDF-L1 fillers for 6FDA-based copolyimide mixed matrix membranes', *Journal of Membrane Science*, 431, pp. 163–170. Available at: https://doi.org/10.1016/j.memsci.2012.12.046.

Gernot Voitic, Birgit Pichler, Angelo Basile, Adolfo Iulianelli, Karin Malli, Sebastian Bock, Viktor Hacker, Chapter 10 - Hydrogen Production, Editor(s): Viktor Hacker, Shigenori Mitsushima, *Fuel Cells and Hydrogen*, Elsevier, 2018, Pages 215–241, ISBN 9780128114599, https://doi.org/10.1016/B978-0-12-811459-9.00010-4.

Ghasemzadeh, K., Morrone, P., Iulianelli, A., Liguori, S., Babaluo, A.A., Basile, A., H2 production in silica membrane reactor via methanol steam reforming: Modeling and HAZOP analysis, *International Journal of Hydrogen Energy*, 38 (2013) 10315–10326.

Gonzo, E.E., Parentis, M.L. and Gottifredi, J.C. (2006) 'Estimating models for predicting effective permeability of mixed matrix membranes', *Journal of Membrane Science*, 277(1–2), pp. 46–54. Available at: https://doi.org/10.1016/j.memsci.2005.10.007.

Hashemifard, S.A., Ismail, A.F. and Matsuura, T. (2010) 'A new theoretical gas permeability model using resistance modeling for mixed matrix membrane systems', *Journal of Membrane Science*, 350(1–2), pp. 259–268. Available at: https://doi.org/10.1016/j.memsci.2009.12.036.

Hou, R. et al. (2020) 'Highly permeable and selective mixed-matrix membranes for hydrogen separation containing PAF-1', *Journal of Materials Chemistry A*, 8(29), pp. 14713–14720. Available at: https://doi.org/10.1039/d0ta05071g.

Hu, L. et al. (2017) 'Composites of ionic liquid and amine-modified SAPO 34 improve CO_2 separation of CO_2-selective polymer membranes', *Applied Surface Science*, 410, pp. 249–258. Available at: https://doi.org/10.1016/j.apsusc.2017.03.045.

Hu, Z. et al. (2016) 'Mixed matrix membranes containing UiO-66(Hf)-$(OH)_2$ metal-organic framework nanoparticles for efficient H_2/CO_2 separation', *Industrial and Engineering Chemistry Research*, 55(29), pp. 7933–7940. Available at: https://doi.org/10.1021/acs.iecr.5b04568.

Huang, M. et al. (2021) 'In-situ generation of polymer molecular sieves in polymer membranes for highly selective gas separation', *Journal of Membrane Science*, 630, p. 119302. Available at: https://doi.org/10.1016/j.memsci.2021.119302.

Ismail, A. F. et al. (2008) 'A review of purification techniques for carbon nanotubes.', *Nano*, 3, pp. 127–143.

Ismail, A.F. et al. (2015) *Gas Separation Membranes Polymeric and Inorganic*. New York, Dordrecht, London: Springer Cham Heidelberg.

Iulianelli, A. et al. (2019) 'A supported Pd-Cu/Al$_2$O$_3$ membrane from solvated metal atoms for hydrogen separation/purification', *Fuel Processing Technology*, 195, p. 106141. Available at: https://doi.org/10.1016/j.fuproc.2019.106141.

Jiang, X. et al. (2021) 'Aqueous one-step modulation for synthesizing monodispersed ZIF-8 nanocrystals for mixed-matrix membrane', *ACS Applied Materials and Interfaces*, 13(9), pp. 11296–11305. Available at: https://doi.org/10.1021/acsami.0c22910.

Judeinstein, P. and Sanchez, C. (1996) 'Hybrid organic-inorganic materials: A land of multidisciplinarity', *Journal of Materials Chemistry*, 6(4), pp. 511–525.

Kang, Z. et al. (2016) 'Mixed matrix membranes (MMMs) comprising exfoliated 2D covalent organic frameworks (COFs) for efficient CO$_2$ separation', *Chemistry of Materials*, 28(5), pp. 1277–1285. Available at: https://doi.org/10.1021/acs.chemmater.5b02902.

Lau, C.H. et al. (2015) 'Gas-separation membranes loaded with porous aromatic frameworks that improve with age', *Angewandte Chemie*, 127(9), pp. 2707–2711. Available at: https://doi.org/10.1002/ange.201410684.

Li, H. et al. (2013) 'Ultrathin, molecular-sieving graphene oxide membranes for selective hydrogen separation', *Science*, 342(6154), pp. 95–98.

Li, L. et al. (2015) 'The preparation and gas separation properties of zeolite/carbon hybrid membranes', *Journal of Materials Science*, 50(6), pp. 2561–2570. Available at: https://doi.org/10.1007/s10853-015-8819-1.

Lin, H. et al. (2018) 'Permselective H$_2$/CO$_2$ separation and desalination of hybrid GO/rGO membranes with controlled pre-cross-linking', *ACS Applied Materials and Interfaces*, 10(33), pp. 28166–28175. Available at: https://doi.org/10.1021/acsami.8b05296.

Ma, C. and Urban, J.J. (2019) 'Hydrogen-bonded polyimide/metal-organic framework hybrid membranes for ultrafast separations of multiple gas pairs', *Advanced Functional Materials*, 29(32). Available at: https://doi.org/10.1002/adfm.201903243.

Mazloomi, K. and Gomes, C. (2012) 'Hydrogen as an energy carrier: Prospects and challenges', *Renewable and Sustainable Energy Reviews*, 16(5), pp. 3024–3033. Available at: https://doi.org/10.1016/j.rser.2012.02.028.

Mulder, M. (1997) *Basic Principles Membrane Technology*. Second Edition. Amsterdam: Kluwer Academic Publishers.

Ostwal, M. et al. (2018) 'Graphene oxide - molybdenum disulfide hybrid membranes for hydrogen separation', *Journal of Membrane Science*, 550, pp. 145–154. Available at: https://doi.org/10.1016/j.memsci.2017.12.063.

Pal, R. (2008) 'Permeation models for mixed matrix membranes', *Journal of Colloid and Interface Science*, 317(1), pp. 191–198. Available at: https://doi.org/10.1016/j.jcis.2007.09.032.

Pandey, P. and Chauhan, R.S. (2001) 'Membranes for gas separation', *Progress in Polymer Science*, 26(6), pp. 853–893. Available at: www.elsevier.com/locate/ppolysci.

Park, S., Cho, K.Y. and Jeong, H.K. (2020) 'Polyimide/ZIF-7 mixed-matrix membranes: Understanding thein situ confined formation of the ZIF-7 phases inside a polymer and their effects on gas separations', *Journal of Materials Chemistry A*, 8(22), pp. 11210–11217. Available at: https://doi.org/10.1039/d0ta02761h.

Perez, E.V. et al. (2017) 'Amine-functionalized (Al) MIL-53/VTECTM mixed-matrix membranes for H2/CO2 mixture separations at high pressure and high temperature', *Journal of Membrane Science*, 530, pp. 201–212. Available at: https://doi.org/10.1016/j.memsci.2017.02.003.

Peydayesh, M., Mohammadi, T. and Bakhtiari, O. (2017) 'Effective hydrogen purification from methane via polyimide Matrimid(r) 5218-deca-dodecasil 3R type zeolite mixed matrix membrane', *Energy*, 141, pp. 2100–2107. Available at: https://doi.org/10.1016/j.energy.2017.11.101.

Regmi, C. et al. (2021) 'CO_2 /CH_4 and H_2 /CH_4 gas separation performance of CTA-TNT@ CNT hybrid mixed matrix membranes', *Membranes*, 11(11), p. 862. Available at: https://doi.org/10.3390/membranes11110862.

Sánchez-Laínez, J. et al. (2018) 'Synthesis of ZIF-93/11 hybrid nanoparticles via post-synthetic modification of ZIF-93 and their use for H_2/CO_2 separation', *Chemistry - A European Journal*, 24(43), pp. 11211–11219. Available at: https://doi.org/10.1002/chem.201802124.

Seoane, B. et al. (2015) 'Metal-organic framework based mixed matrix membranes: A solution for highly efficient CO_2 capture?', *Chemical Society Reviews. Royal Society of Chemistry*, 44(8), pp. 2421–2454. Available at: https://doi.org/10.1039/c4cs00437j.

Shen, J. et al. (2016) 'Subnanometer two-dimensional graphene oxide channels for ultrafast gas sieving', *ACS Nano*, 10(3), pp. 3398–3409. Available at: https://doi.org/10.1021/acsnano.5b07304.

Valero, M. et al. (2014) 'Mixed matrix membranes for gas separation by combination of silica MCM-41 and MOF NH2-MIL-53(Al) in glassy polymers', *Microporous and Mesoporous Materials*, 192, pp. 23–28. Available at: https://doi.org/10.1016/j.micromeso.2013.09.018.

Voitic G., Pichler B., Basile A., Iulianelli A., Malli K., Bock S., Hacker V. (2018) 'Hydrogen production', in *Fuel Cells and Hydrogen: From Fundamentals to Applied Research*, pp. 215–241, UK: Elsevier. https://doi.org/10.1016/B978-0-12-811459-9.00010-4

Wang, D. et al. (2015) 'Ultrathin membranes of single-layered MoS2 nanosheets for high-permeance hydrogen separation', *Nanoscale*, 7(42), pp. 17649–17652. Available at: https://doi.org/10.1039/c5nr06321c.

Yin, H. and Yip, A.C.K. (2017) 'A review on the production and purification of biomass-derived hydrogen using emerging membrane technologies', *Catalysts*, 7(10), p. 297. Available at: https://doi.org/10.3390/catal7100297.

Zeng, Y. et al. (2010) 'Increasing the electrical conductivity of carbon nanotube/polymer composites by using weak nanotube-polymer interactions', *Carbon*, 48(12), pp. 3551–3558. Available at: https://doi.org/10.1016/j.carbon.2010.05.053.

Zornoza, B. et al. (2011) 'Combination of MOFs and zeolites for mixed-matrix membranes', *ChemPhysChem*, 12(15), pp. 2781–2785. Available at: https://doi.org/10.1002/cphc.201100583.

Zornoza, B. et al. (2015) 'Mixed matrix membranes based on 6FDA polyimide with silica and zeolite microsphere dispersed phases', *AIChE Journal*, 61(12), pp. 4481–4490. Available at: https://doi.org/10.1002/aic.15011.

Zou, X. and Zhu, G. (2018) 'Microporous organic materials for membrane-based gas separation', *Advanced Materials*, 30(3). Available at: https://doi.org/10.1002/adma.201700750.

7 Silica Membrane for Hydrogen Separation

*Sina Mosallanezhad and
Mohammad Reza Rahimpour*

7.1 INTRODUCTION

Hydrogen is crucial in manufacturing fuels, various chemicals (Daneshmand-Jahromi et al., 2017), and fuel cell applications (Baptista et al., 2010). Consequently, the purification of hydrogen is a significant process in industrial operations. Hydrogen separation via the use of membranes has become an important technology. While polymeric membranes have been used to some extent, they cannot allow substances to pass through and selectively separate hydrogen. As a result, there has been a focus on inorganic materials. Palladium is the most well-recognized material, although it is expensive and prone to metallic failure due to hardening, as well as being vulnerable to poisoning by sulfur and other foreign elements. This chapter explicitly examines silica-based membranes fabricated via chemical vapor deposition (CVD). These membranes have the potential to be cost-effective, possess high thermal stability, and are resistant to toxins.

The practical utility of membranes hinges on achieving high permeation rates and superior selectivities, prerequisites realized through membranes characterized by low thickness, devoid of cracks and pinholes. Furthermore, these membranes must exhibit mechanical robustness, extended lifespan, and resilience against poisons for effective implementation in practical equipment. Addressing the dual requirements of mechanical strength and high permeability involves the production of ceramic membranes, typically achieved by applying a thin film of the selective material onto a thick porous support, often in tubular form. The field of membrane science leverages film deposition technology for this purpose. Inorganic silica membranes offer distinct advantages in industrial applications, enabling simultaneous reaction and separation and high-temperature hydrogen separation. These advantages encompass notable selectivity and stability, even under elevated temperatures and in chemically challenging environments, rendering them a focal point of substantial research attention (Sea et al., 1997; Gavalas et al., 1989; Okubo and Inoue, 1989a). Furthermore, inorganic membranes in membrane reactors, whether catalytically active or passive, have shown promise. This is because they have achieved yields higher than the equilibrium by continuously separating the hydrogen product from the reaction system (Akamatsu et al., 2008; Prabhu and Oyama, 2000). Figure 7.1 depicts a typical membrane reactor used for the steam-reforming process of ethanol.

DOI: 10.1201/9781003382522-9

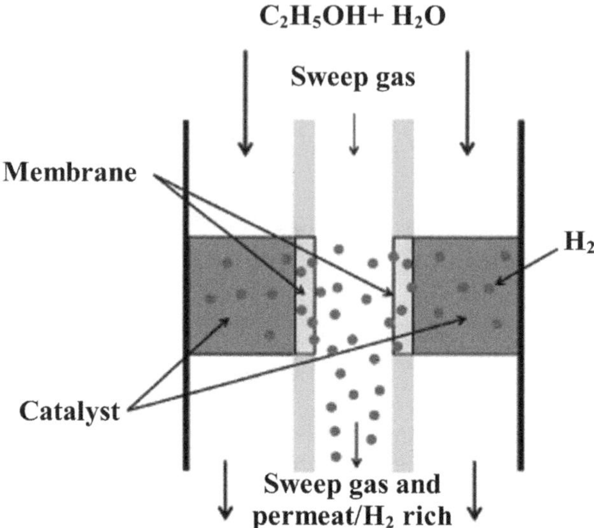

FIGURE 7.1　Diagram illustrating the use of a membrane reactor in steam reforming ethanol. (Adapted from Khatib et al., 2011.)

The supported silica membranes are created by applying the separation layer using appropriate precursors suspended in a liquid or gaseous medium. Deposition from the liquid phase may be accomplished using dip coating in polymeric or particle suspensions. On the other hand, deposition from the gas phase is often performed using CVD because of its flexibility. Therefore, silica membranes typically consist of silica layers applied onto ceramic supports like porous Vycor glass (Okubo and Inoue, 1989a) and alumina (Nair et al., 1997). These layers are commonly deposited using sol–gel techniques (Nair et al., 1997) or CVD (Okubo and Inoue, 1989a) with silica precursors. The sol–gel modification technique offers excellent gas penetration rates, mainly due to fragile top layers, typically measuring 50–100 nm. Additionally, this procedure ensures strong selectivity, unlike CVD techniques, which result in reduced permeability but improved selectivity. The sol–gel approach, however, needs to be more repeatable.

7.2　SILICA MEMBRANE LAYER SYNTHESIS

As discussed in Section 7.1, silica membranes are primarily synthesized using two distinct methods: sol–gel modification and CVD (Nomura et al., 2005). The sol–gel modification gives high selectivity and permeability, as opposed to CVD procedures, where there is an associated loss of permeability, albeit the selectivity is boosted. The sol–gel process, however, needs more repeatability. CVD procedures typically need sizeable financial investment and regulated conditions of deposition. More extensive explanations of each approach are provided below.

7.2.1 SOL–GEL PROCESSING

There are three distinct synthetic ways of processing sol–gel: particulate-sol, silica polymers, and template approaches. The silica polymer approach involves the controlled hydrolysis and condensation of alkoxysilane precursors such as tetraethyloxysilane (TEOS). In addition, de Lange demonstrated the existence of ultrathin microporous membranes of 60 nm in thickness, with pores ranging from 0.5 to 0.7 nm in size (Nair et al., 2000). The gas transportation for H_2 was initiated with an activation energy (E_{act}) of 21.7 kJ mol^{-1}. Additionally, there were molecular sieve-like separation factors 200 for combinations of H_2/C_3H_6 at a temperature of 260°C (Asaeda and Yamasaki, 2001). The particulate sol approach relies on the arrangement of nanoparticles to form a structure with a high degree of porosity. Various silica particles are densely packed into the support substrate to produce membranes with varying pore diameters. Incorporating a binder substance or utilizing hierarchical size packing may enhance the packing of particles, hence preventing any potential flaws. The template approach employs organic molecules as templates inside the sol matrix, which are then eliminated by calcination. The size and form of the organic molecule may be imprinted in the sol to achieve a specific porosity. Surfactants, chemical ligands, and polymers have been documented as templates. Silica membranes are produced by the sol–gel deposition method, which involves applying an aqueous silica polymer sol onto a mesoporous support surface and then drying and calcining it at temperatures ranging from 400°C to 800°C (Brinker et al., 1993). The silica polymers are created by acid-catalyzed hydrolysis and polymerization of TEOS and MTES at a pH lower than 7. Strict control of reaction parameters, including time, temperature, pH, and mixing, is necessary during all process phases. Nevertheless, the impact of these characteristics on the ultimate microporous configuration is restricted. The micropores in sol–gel silica are believed to be created near the first solvent molecules, such as H_2O and C_2H_5OH. Instances of preexisting templates have been documented with various alcohols, namely HTEAB and methacryl oxypropyl trimethoxy silane (Lai et al., 2003). Amorphous silica is composed of many Si-O bonds that are not connected to other atoms, and these bonds encircle the micropores. Typically, they are ended with protons to create Si-OH; however, when MTES is added before hydrolysis, Si-CH$_3$- terminated groups are formed. The latter enhances the stability and openness of the structure. Upon heating below 600°C, the terminal groups undergo condensation and carbonization, resulting in their progressive disappearance. At 800°C, a silica structure with high density is produced. In this process, polysilazane was applied by spin coating, then cross-linked in a nitrogen environment at a temperature of 270°C, and finally subjected to pyrolysis in an air environment at 600°C (Ockwig and Nenoff, 2007).

7.2.2 CVD OF A MEMBRANE LAYER

Thin films are applied onto a substrate using the CVD process, which alters the porosity of support membranes by reacting gas phase precursors near, inside, or around the mesoporous pores of the substrate. Although these films have lower gas permeances than sol–gel alternatives, they are more stable, last longer, and show

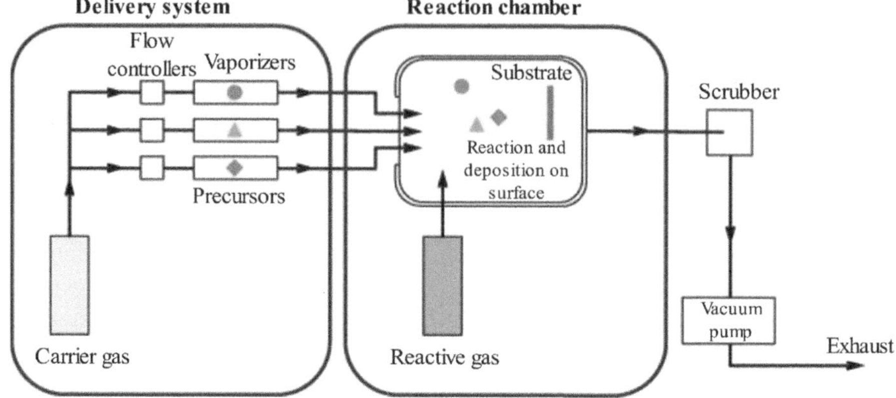

FIGURE 7.2 Constituents of a typical CVD system. (Adapted from Khatib and Oyama, 2013.)

better selectivities. Furthermore, because the CVD process eliminates the need for a repeated coating step, it is more convenient than the sol–gel approach. The gas delivery system of a CVD system usually combines reactive and carrier gases (e.g., H_2, N_2, or Ar) with volatile, reactive chemicals (e.g., metal halides, carbonyls, or alkoxides). Deposition of the target film occurs when these gases move over a substrate in a reaction chamber at a controlled temperature. Figure 7.2 shows the parts of a standard CVD system (Khatib and Oyama, 2013).

Films may be deposited using CVD via heat breakdown, oxidation, and hydrolysis processes. The reactions between different components may start in the gaseous phase above the heated substrate, followed by the formation of a thin layer on the surface. Alternatively, a surface layer may be formed via a chemical interaction between the substrate surface and one or more vapor phase components. Hence, the processes of CVD include both reactions occurring in the gaseous phase and reactions taking place on the surface. The techniques for membrane production using CVD have yet to be extensively researched. However, some empirical criteria for producing high-quality membranes have been established (Pattanaik and Sarin, 2001).

Qualitative information has also been proposed to understand the CVD process for film creation better (Tsapatsis et al., 2000). The size and shape of the film-making molecules, as well as the modifications introduced by subsequent drying and heat treatments, are the two main determinants of the membrane pore structure. Either the unconverted feed or species formed by the feed molecules dissolving, such as small clusters or oligomers, could make up the growth species. These creatures may go through further steps once connected to the surface.

A key component in dictating the size and shape of developing species is the distribution of pore sizes inside the porous support. Depositions on the surface and inside the pores of the support material are the outcomes of postponing or avoiding particle growth. When the growth species responsible for deposition cannot squeeze into the pores any longer, deposition stops; the limiting pore size is about equal to the kinetic diameter of the species. However, this pore size may be changed by reactions and diffusion during a future heat treatment. After the pores are shut to prevent species formation, the support outside becomes a deposition site. The size and location of the support pores determine how long it takes for the pores to close. Based on the correlation between the two variables, it seems that the deposit penetration into the pores is approximately equal to the square root of the ratio of the response rate constant to the pore diffusion coefficient. A more consistent density of the inner deposit layer must be attained; this layer must have its maximum density close to the pore entrance and its lowest density further from the surface (Tsapatsis and Gavalas, 1992; Lin and Burggraaf, 1992).

7.3 SILICA MEMBRANE SYNTHESIS VIA CVD

Extensive research has been conducted on silica membranes produced through CVD for their potential use in selective hydrogen separations (Lee and Oyama, 2002). However, as the achieved selectivities are relatively low, these membranes are less commonly studied for other separations, such as nitrogen/oxygen or air/ hydrocarbon separations. This chapter will focus only on the process of hydrogen separation. The primary goal of most of the SiO_2-CVD methods documented is to effectively seal the pores of the support material within a limited range, hence facilitating the selective separation of hydrogen. In recent decades, several practical demonstrations of this concept have been documented. However, there is a need for improvements in flux and stability, mainly when dealing with water vapor and high temperatures, to make it suitable for commercial applications. Under diverse circumstances, various research groups have used distinct iterations of the CVD technique to fabricate silica membranes on a porous substrate that exhibits high efficiency and stability.

7.3.1 PRECURSORS

The gaseous silica precursor undergoes a chemical reaction with the reactive agent (such as oxygen, air, water, and ozone, depending on the specific response used for chemical CVD), forming the deposited silica layer. This chapter presents the results of many experiments conducted by multiple research groups, including $SiCl_4$ hydrolysis, SiH_4 oxidation, TEOS thermocracking, and oxidation. Various settings and geometries have been studied for different precursors to optimize the stability and selectivity of the membranes produced. The deposition of the silica film occurs when the gaseous precursors react.

7.3.2 SUPPORT

Two kinds of supports have been employed in the literature on CVD-made silica membranes: alumina and Vycor glass. Since porous Vycor glass is known to have a high selectivity for hydrogen penetration due to its narrow mesopore size distribution, this material was initially the focus of the investigation. Additionally, Vycor glass is thermally compatible with SiO_2, which means it won't shatter as easily when subjected to thermal cycling (Tsapatsis et al., 2000). Other materials, such as α-alumina and γ-alumina-coated α-alumina tubes, were used as supports since Vycor glass has a low natural permeability and does not mix well with silica.

7.3.2.1 Vycor Glass Support

A different kind of membrane, composed of silica and supported by Vycor glass, separates hydrogen. Vycor glass contains high-purity fused silica and has a porous structure and remarkable heat stability. The selective permeability of hydrogen via silica membranes based on Vycor glass has been extensively studied. The production of silica membranes based on Vycor glass involves many distinct processes. The Vycor glass undergoes an initial cutting process and subsequent cleaning to remove any impurities. Next, the sol–gel method is used, including the hydrolysis of a silica precursor in a solution to generate a gel. This gel then creates a thin silica coating on the Vycor glass substrate. Afterward, the gel is desiccated and subjected to heat to eliminate any organic matter and increase the density of the silica layer (Okubo and Inoue, 1989a).

The result is the creation of a silica membrane with many tiny pores, allowing for the specific separation of hydrogen gas. The pore size may be controlled by adjusting the sol–gel synthesis parameters, including pH, temperature, and precursor concentration. Vycor glass-based silica membranes provide several advantages for hydrogen separation, such as excellent selectivity, high permeance, and remarkable thermal and chemical resistance. The membranes often exhibit susceptibility to cracking or delamination during use, while the production process may provide challenges and consume significant time. Scientists are now enhancing the mechanical properties of these membranes to enhance their reliability and durability for use in industrial settings. Several organizations have tried modifying the absorbent composition of Vycor glass by using the CVD of silicon precursors. Nevertheless, only a limited number of materials have effectively attained significant selectivity and permeability when exposed to temperatures beyond 573 K. Okubo and Inoue (1989a) were the first to attempt the fabrication of silica membranes using CVD. The scientists injected TEOS into the holes of tubular porous glass support with a pore size of 2 nm at a temperature of 473 K and a pressure of 1 atm. It was found that TEOS decomposed in the holes at the surface of the porous glass tube, leading to an irregular membrane structure and improved selectivity for gas separation. After a thorough investigation of the membranes (Okubo and Inoue, 1989b), it was found that they exhibited permeance to H_2 of $5.0 \times 10^{-9}\,\mathrm{m^{-2}\,s^{-1}\,Pa^{-1}}$ at a temperature of 473 K, with an H_2/N_2 ratio of 11. The original glass substrate has a permeance that is tenfold lower than this.

Furthermore, it was found that the permeability rose with temperature instead of decreasing as $T^{-1/2}$, which goes against the prediction of Knudsen diffusion. The temperature relationship had an Arrhenius signature, suggesting that permeation was an active process, even if the activation energy was lower than that of quartz or vitreous glass. Currently, gas and surface diffusion work together to regulate the permeation, which is in a state of transition.

Following Okubo and Inoue's publication, Gavalas and his team conducted a series of comprehensive investigations on the fabrication of silica membranes using CVD (Gavalas et al., 1989). The researchers used porous Vycor glass tubes obtained from Corning Glass as a support material. These tubes had a pore size of 4 nm.

The tubes were sintered at both ends to create impermeable sections that may be connected to the gas lines using fittings. Their first publication (Gavalas et al., 1989) documented the formation of SiO_2 coatings with H_2-permselective properties inside the porous support tubes by oxidizing SiH_4. This process, referred to as the counter diffusion geometry or opposing reactants geometry, included the deposition of SiO_2 within the tube walls. SiH_4 was introduced into the tube, while O_2 was introduced outside. The two reactants exhibited diffusion in opposing directions at a total pressure of 1 atm. They underwent a chemical reaction inside a confined region along the inner surface of the tube, resulting in the formation of a thin SiO_2 layer.

Upon the occlusion of the pores, the reaction ceased as the reactants could not establish contact with one another. The reaction reached completion within 15 minutes at a temperature of 723 K and with SiH_4 and O_2 pressures of 0.1 and 0.33 atm, respectively. Conversely, the reaction exhibited sluggishness at lower temperatures, while at higher temperatures, the silane broke into silicon, forming an impenetrable layer. The membranes fabricated at a temperature of 723 K showed a remarkable capacity to allow the passage of H_2 gas selectively. The permeability ratio of H_2 to N_2 reached an impressive value of 3,000, while the H_2 permeance was measured at 1.4×10^{-8} mol m^{-2}s^{-1} Pa^{-1}. The diffusion process was discovered to occur in the transitional area between Knudsen and microporous materials. They remained stable after further subjecting the membranes to heat treatment at 723 K. However, when exposed to a higher temperature of 873 K, the membranes experienced densification, significantly decreasing the film's permeability and selectivity. The penetration of H_2 was thought to take place via an activated diffusion process known as the solution–diffusion mechanism. The activation energy was determined to be 35 kJ mol^{-1}, indicating the presence of a molecular sieving action that separated the gases. In this study, the researchers also performed deposition using the "one-sided geometry", but they discontinued it due to their rapid success utilizing the counter-diffusion geometry (Khatib and Oyama, 2013).

This group performed experiments in which they applied Al_2O_3, TiO_2, SiO_2, and B_2O_3 films into the pores of Vycor tubes using the respective chlorides as precursors. This was done to address the issue of densification that occurs with silica films at high temperatures, which leads to a decrease in permeability and selectivity. In the case of silica films, they conducted experiments using a different deposition process: the hydrolysis of $SiCl_4$ to deposit SiO_2 (Tsapatsis et al., 1991). This reaction may be performed at temperatures up to 1,073 K, forming denser sheets that exhibit enhanced thermal stability. While one-sided deposition resulted in membranes with thinner deposited layers, higher H_2 penetration coefficients, and a quicker deposition rate than counter-diffusion deposition, both approaches yielded SiO_2 membranes that selectively allow the passage of hydrogen.

7.3.2.2 Alumina Support

The permeance of the membranes previously mentioned, which was around 10^{-8} mol m^{-2}s^{-1} Pa^{-1}, was limited by the small pore size (2–4 nm) of the Vycor glass support. Making membranes out of alumina with larger pores increases permeability and decreases inherent resistance. The porous supports, which have pore sizes varying from 110 to 180 nm, are made by compressing α-alumina particles. On top of being more robust and more resistant to high pressure, these supports are more cost-effective than Vycor glass tubes. Given the large pore sizes of alumina supports, it was proposed that by adding a layer between the active layer and the support tube, selectivity may be enhanced by preventing pinhole defects from forming on the active layer. As described in this section, various research groups integrated an γ-alumina intermediate layer into their membranes. Results from other research confirmed that the Knudsen diffusion mechanism controlled the permselectivity of these sol–gel modified supports with ultrafine γ-alumina particles (Okubo et al., 1991).

Megiris and Glezer pioneered altering alumina tubes, both α- and γ-alumina, using low-pressure CVD by oxidizing triisopropylsilane (TPS) (Megiris and Glezer, 1992). While specific information on permeance measurements was not provided, the authors said their membranes exhibited hydrogen permeance of 4×10^{-7} mol $m^{-2}s^{-1}$ Pa^{-1} and selectivities above 40. Although the specific gas against which the selectivity was assessed is not specified, it was likely studied against N_2.

Kin and Sea (2001) conducted further permeance studies of H_2, CH_4, N_2, CO_2 i-C_4H_{10}, and C_3H_8 using these membranes, and their findings further confirmed that the underlying support influences the effectiveness of the upper selective layer. When applied over an γ-alumina layer, the silica membranes exhibited a molecular sieving effect that separated H_2, N_2, and CO_2. On the other hand, the permeation of CH_4, i-C_4H_{10}, and C_3H_8 occurred by a Knudsen diffusion process. Nevertheless, in the case of an α-alumina support, no gases bigger than CH_4 were seen, suggesting the lack of mesopores in these membranes. Consequently, all gases were segregated based on their molecular size, showing the prevailing molecular sieve mechanism. The poor selectivity may be attributed to bigger micropores than membranes with an intermediate layer. However, the decreased H_2 permeance can be attributed to fewer micropores. These findings align with the anticipated outcomes when using a middle layer. Sea and colleagues conducted a more comprehensive study and confirmed that improved membranes were achieved by evacuation and including an intermediate γ-alumina layer (Seo and Lee, 2001).

Research conducted by Hwang et al. (Hwang and Choi, 2012; Hwang et al., 2000; Hwang et al., 1999) evaluated the capabilities of silica membranes produced by CVD for hydrogen separation. For the thermochemical iodine-sulfur (IS) process, these membranes were designed to be used in gaseous mixtures of H_2-H_2O-HI. It was tested in several research. The top layers of the support tubes supplied by Noritake, namely the γ-alumina or α-alumina layers, were replaced using the TEOS decomposition with the evacuation process. Hydrogen permeance of 6×10^{-9} mol $m^{-2}s^{-1}$ Pa^{-1} was attained at 873 K in their first investigation (Hwang et al., 1999). When both membranes were made on supports with α-alumina top layers and γ-alumina top layers were used at the same temperature, the selectivity of H_2/N_2 was 5.2 and 160, respectively. The use of an intermediate layer to improve permselectivity was further shown in this case. The top-tier membrane showed the highest separation factor between hydrogen and hydrogen iodide (650 at 723 K) with a hydrogen permeance of about 10^{-7} mol $m^{-2}s^{-1}$ Pa^{-1}.

On the other hand, it displayed Knudsen diffusion and had the lowest selectivity for hydrogen relative to nitrogen. The researchers also tested how long silica membranes made by CVD on porous α- and γ-alumina support tubes might last. The support tubes had 100 and 10 nm pore sizes, respectively. The membranes were heated to 723 K in a gaseous mixture of HI and H_2O for a long time (Hwang et al., 2003). One way the CVD treatment was carried out was by leaving it running continuously until the membrane showed a certain amount of, He/N_2 selectivity; another way was to run it for a certain amount of time, then let the membrane cool to room temperature and repeat the process until it reached the same level of selectivity. The results of the stability tests performed in a gaseous combination of $HI-H_2O$ showed that the first method, which did not include interruptions, produced better membranes. Moreover, after the stability test, the membrane—supported by α-alumina—showed a remarkable H_2/HI selectivity (ranging from 240 to 2,600 at temperatures between 573 and 873 K) and showed extraordinary stability in the $HI-H_2O$ gaseous combination.

An amorphous silica membrane was created by Nagano et al. (2008) using the counter-diffusion CVD process at 873 K. The α-alumina substrate was coated with γ-alumina before the membrane was made. The precursors for the CVD procedure were O_2 and tetramethyl orthosilicate. Improved permselectivity was seen in membranes with thicker γ-alumina layers and were calcined at higher temperatures (1,073 K). The significance of the middle layer was highlighted by the fact that optimizing it was critical in lowering the permeability of gases with large kinetic diameters. At temperatures 798 K and above, the best membrane showed an H_2/N_2 ratio above 10,000 and a much greater H_2 permeance at $3.0 \times 10^{-7}\,mol\ m^{-2}s^{-1}\ Pa^{-1}$. In agreement with the Oyama solubility site mechanism, activated diffusion was shown to be the principal means of He and H_2 permeation within the temperature range of 373–873 K. Knudsen diffusion was noted below 673 K for CO_2, Ar, and N_2. In contrast, activated permeance only happened beyond this temperature.

7.4 THEORETICAL FOUNDATION OF HYDROGEN-SELECTIVE SUPPORTED SILICA MEMBRANE GAS PERMEATION

A silica membrane called Nanosil was found by Lee and Oyama (2002) that is very permeable to hydrogen. A thin layer of SiO_2 was deposited over a porous Vycor glass substrate using CVD to create a membrane. For small gas molecules like He, Ne, and H_2 at 873 K, the composite membrane showed an outstanding permeability of about $10^{-8}\,mol\ m^{-2}s^{-1}\ Pa^{-1}$. The selectivity toward other giant gas molecules, such as CO_2, CO, and CH_4, coming in at about 104, was also very impressive. As shown below, the gas diffusion model was used to investigate gas transport characteristics on the Nanosil and Vycor membranes.

Fick's first law pertains to the macroscopic depiction of transport phenomena:

$$J = -D(c)\nabla c \tag{7.1}$$

7.4.1 SORPTION

Gas transmission through dense or impermeable materials, such as solid oxides or zeolites, requires molecular adsorption and subsequent diffusion. Several sorption models have been published, and they all depend on different assumptions about the adsorbed gas's state (Ma, 1996). The Langmuir adsorption model (Langmuir, 1915) is the most applicable to gas separation membrane applications since adsorption is often not multilayer and is often far below a monolayer.

$$\theta = qq_s = bP1 + bP \tag{7.2}$$

The symbol θ represents the proportion of occupied adsorption sites, while q represents the number of gas molecules adsorbed per adsorbent unit (measured in mol g^{-1}). q_s represents the maximum number of adsorbed molecules that may reach saturation. P is the pressure in pascals (Pa), and b is the equilibrium adsorption constant measured in Pa^{-1}. The relationship between the equilibrium adsorption constant and temperature is expressed by

$$b = b_0 \exp \Delta H_a RT \tag{7.3}$$

ΔH_a represents the heat of adsorption. Under conditions of low pressure and high temperature (with small values of b), the Langmuir equation may be reduced to:

$$q = q_s bP = KP = K_0 \exp \Delta H_a RTP \tag{7.4}$$

K represents the Henry's constant, measured in units of mol g^{-1} Pa^{-1}. Under Henry's law, the quantity of adsorbed molecules exhibits a linear relationship with the applied pressure.

7.4.2 DIFFUSION

The process of molecules spreading across a membrane may occur in several ways, depending on the contact between the gas molecules and the membrane. Burggraaf (1999) examined several gas diffusion mechanisms about the energy potential wells formed inside the membrane's pores. The separation between the pore walls and the diffusing gas molecules mainly determines the morphology of these energy potential wells. Hence, the proportion between the molecular size of the diffusing gas and the pore width significantly influences the selection of the diffusion mechanism. We will examine four distinct gas diffusion processes (Knudsen, gas translational (GT), solid state, and surface diffusion) that often play a role in the transportation of gases across membranes.

Knudsen diffusion arises when the average distance traveled by the gas molecules during their motion is much greater than the size of the pores (Knudsen, 1909). During this regime, the gas molecules move through the pores and collide randomly with the walls of the pores. The gas's kinetic velocity and the membrane's geometric properties determine the Knudsen diffusivity:

$$D = \varepsilon d_p 3\tau 8RT\pi M^{0.5} \tag{7.5}$$

The variables in the equation are as follows:

- ε represents the porosity of the membrane
- d_p represents the pore diameter
- τ represents the tortuosity

- R represents the gas constant
- M represents the molecular weight of the diffusing gas

GT diffusion process has been used in microporous materials, namely zeolites (Xiao and Wei, 1992). The resulting expression resembles the Knudsen equation. However, in this model, the gas molecules are assumed to undergo translational motion between sorption sites (cages) by surmounting the obstacles created by the narrow channels connecting neighboring sorption sites. The following equation determines the GT diffusivity:

$$D = g_d d_p 8RT\pi M^{0.5} \exp(-\Delta ERT) \tag{7.6}$$

The geometric factor, g_d, represents the relationship between d_p, which is the material's pore size, and ΔE, which is the activation energy of diffusion. The activation energy is defined as the disparity in potential energy between the sorption sites and the channels of the zeolites.

The gas molecules will change their motion from translational to vibrational when the size of the holes in the membrane to the diffusing gas molecule decreases, leading to an energy potential within the steeply parabolic pores. Vibrational frequency, v_e, and hop length, λ, determine the movement of a chemical bond from one sorption site to another. Gas diffusion in this regime is quite similar to solid-state diffusion. The following equation may be used to find the diffusivity (Burggraaf, 1999):

$$D = g_d \lambda^2 v_e \exp(-\Delta ERT) \tag{7.7}$$

where g_d represents a constant in geometry, and ΔE represents the activation energy required for diffusion. A statistical technique may be used to derive a solid-state diffusion model. The diffusivity of monatomic gas in fused silica glass may be described by a statistical model, as shown in the following equation (Masaryk and Fulrath, 1973):

$$D = 16kThd^2 \left(e^{\frac{hv}{2kT}} - e^{-\frac{hv}{2kT}} \right)^3 \left(e^{\frac{hv^*}{2kT}} - e^{\frac{hv^*}{2kT}} \right)^2 e^{-\frac{\Delta E}{RT}} \tag{7.8}$$

where k represents Boltzmann's constant, h represents Planck's constant, d represents the distance between sorption sites in the structure, T represents the absolute temperature, v represents the vibrational frequency of gas molecules on the sorption sites, v^* represents the vibrational frequency on the doorway sites, R represents the gas constant, and ΔE represents the activation energy of diffusion.

The physical meaning of the obtained model equation for surface diffusion is in two dimensions; however, it may be defined similarly to solid-state diffusion. According to the theory of surface diffusion, gas molecules are thought to be adsorbed onto surfaces and then move along those surfaces by hopping between the surface's lowest points of potential energy (Burggraaf, 1999). However, if the energy of motion is greater than that of attachment, the gas molecules dispersed over the surface may change from their attached condition to the gaseous state.

Therefore, the sorption energy of gas molecules determines the temperature range in which surface diffusion is proper. In addition, surface diffusion requires less activation energy than sorption.

7.5 CONCLUSION

Since its first discovery in the late 1980s, silica hydrogen separation membranes have been the subject of much investigation. Their performance has been significantly improved. The initial membranes, supported by Vycor glass, had a selectivity for H_2/N_2 of around 10 and a permeance ranging from 10^{-9} mol m^{-2}s^{-1} Pa^{-1}. Porous alumina, a more realistic material, was found in the 1990s. This substance greatly enhanced the permeance, rising from 10 to about 10^{-8} mol m^{-2}s^{-1} Pa^{-1}.

The selectivity for separating H_2 from N_2 also reached values in the hundreds. Advancements in membrane technology throughout the new century included incorporating graded intermediate layers and using inert gas CVD. These advances resulted in a tenfold increase in permeance, reaching a 10^{-7} mol m^{-2}s^{-1} Pa^{-1} level. Additionally, the selectivity for H_2 over other gases reached values in the thousands. This has now entered the realm of commercial significance. The problems include stability, affordability, and the capacity to manufacture modules with a high surface area-to-volume ratio (Khatib et al., 2011).

ACRONYMS

CVD   Chemical vapor deposition
TEOS Tetraethyloxysilane
TPS   Triisopropylsilane
IS   Iodine-sulfur
GT   Gas translational

REFERENCES

Akamatsu, K., Nakane, M., Sugawara, T., Hattori, T. & Nakao, S.-I. 2008. Development of a membrane reactor for decomposing hydrogen sulfide into hydrogen using a high-performance amorphous silica membrane. *Journal of Membrane Science*, 325, 16–19.

Asaeda, M. & Yamasaki, S. 2001. Separation of inorganic/organic gas mixtures by porous silica membranes. *Separation and Purification Technology*, 25, 151–159.

Baptista, P., Tomas, M. & Silva, C. 2010. Plug-in hybrid fuel cell vehicles market penetration scenarios. *International Journal of Hydrogen Energy*, 35, 10024–10030.

Brinker, C. J., Ward, T. L., Sehgal, R., Raman, N. K., Hietala, S. L., Smith, D. M., HUA, D. W. & Headley, T. J. 1993. "Ultramicroporous" silica-based supported inorganic membranes. *Journal of Membrane Science*, 77, 165–179.

Burggraaf, A. J. 1999. Single gas permeation of thin zeolite (MFI) membranes: theory and analysis of experimental observations. *Journal of Membrane Science*, 155, 45–65.

Daneshmand- Jahromi, S., Rahimpour, M. R., Meshksar, M. & Hafizi, A. 2017. Hydrogen production from cyclic chemical looping steam methane reforming over yttrium promoted Ni/SBA-16 oxygen carrier. *Catalysts*, 7, 286.

Gavalas, G. R., Megiris, C. E. & NAM, S. W. 1989. Deposition of H_2-permselective SiO_2 films. *Chemical Engineering Science*, 44, 1829–1835.

Hwang, G.-J. & Choi, H.-S. 2012. Multiple tube preparation characteristics of silica hydrogen permselective membrane. *Korean Journal of Chemical Engineering*, 29, 1796–1801.

Hwang, G.-J., Kim, J.-W., Choi, H.-S. & Onuki, K. 2003. Stability of a silica membrane prepared by CVD using γ-and α-alumina tube as the support tube in the HI-H$_2$O gaseous mixture. *Journal of Membrane Science*, 215, 293–302.

Hwang, G.-J., Onuki, K., Shimizu, S. & Ohya, H. 1999. Hydrogen separation in H$_2$-H$_2$O-HI gaseous mixture using the silica membrane prepared by chemical vapor deposition. *Journal of Membrane Science*, 162, 83–90.

Hwang, G. J., Onuki, K. & Shimizu, S. 2000. Separation of hydrogen from a H$_2$–H$_2$O–HI gaseous mixture using a silica membrane. *AIChE Journal*, 46, 92–98.

Khatib, S. J. & Oyama, S. T. 2013. Silica membranes for hydrogen separation prepared by chemical vapor deposition (CVD). *Separation and Purification Technology*, 111, 20–42.

Khatib, S. J., Oyama, S. T., De Souza, K. R. & Noronha, F. B. 2011. Chapter 2- Review of silica membranes for hydrogen separation prepared by chemical vapor deposition. In: Oyama, S. T. & Stagg- Williams, S. M. (eds.) *Membrane Science and Technology*. Netherlands: Elsevier.

Kin, S.-S. & Sea, B.-K. 2001. Gas permeation characteristics of silica/alumina composite membrane prepared by chemical vapor deposition. *Korean Journal of Chemical Engineering*, 18, 322–329.

Knudsen, M. 1909. The law of molecular flow and viscosity of gases moving through tubes. *Annals of Physics*, 28, 75.

Lai, Z., Bonilla, G., Diaz, I., Nery, J. G., Sujaoti, K., Amat, M. A., Kokkoli, E., Terasaki, O., Thompson, R. W. & Tsapatsis, M. 2003. Microstructural optimization of a zeolite membrane for organic vapor separation. *Science*, 300, 456–460.

Langmuir, I. 1915. Chemical reactions at low pressures. *Journal of the American Chemical Society*, 37, 1139–1167.

Lee, D. & Oyama, S. T. 2002. Gas permeation characteristics of a hydrogen selective supported silica membrane. *Journal of Membrane Science*, 210, 291–306.

Lin, Y. S. & Burggraaf, A. J. 1992. CVD of solid oxides in porous substrates for ceramic membrane modification. *AIChE Journal*, 38, 445–454.

Ma, Y. H. 1996. Adsorption phenomena in membrane systems. In: Burggraaf, A. J. & Cot, L. (eds.) *Membrane Science and Technology*. Netherlands: Elsevier.

Masaryk, J. S. & Fulrath, R. M. 1973. Diffusivity of helium in fused silica. *The Journal of Chemical Physics*, 59, 1198–1202.

Megiris, C. E. & Glezer, J. H. E. 1992. Preparation of silicon dioxide films by low-pressure chemical vapor deposition on dense and porous alumina substrates. *Chemical Engineering Science*, 47, 3925–3934.

Nagano, T., Fujisaki, S., Sato, K., Hataya, K., Iwamoto, Y., Nomura, M. & Nakao, S. I. 2008. Relationship between the mesoporous intermediate layer structure and the gas permeation property of an amorphous silica membrane synthesized by counter diffusion chemical vapor deposition. *Journal of the American Ceramic Society*, 91, 71–76.

Nair, B. N., Okubo, T. & Nakao, S. I. 2000. Structure and separation properties of silica membranes. *Membrane*, 25, 73–85.

Nair, B. N., Yamaguchi, T., Okubo, T., Suematsu, H., Keizer, K. & Nakao, S.-I. 1997. Sol-gel synthesis of molecular sieving silica membranes. *Journal of Membrane Science*, 135, 237–243.

Nomura, M., Ono, K., Gopalakrishnan, S., Sugawara, T. & Nakao, S.-I. 2005. Preparation of a stable silica membrane by a counter diffusion chemical vapor deposition method. *Journal of Membrane Science*, 251, 151–158.

Ockwig, N. W. & Nenoff, T. M. 2007. Membranes for hydrogen separation. *Chemical Reviews*, 107, 4078–4110.

Okubo, T., Haruta, K., Kusakabe, K., Morooka, S., Anzai, H. & Akiyama, S. 1991. Preparation of a sol-gel derived thin membrane on a porous ceramic hollow fiber by the filtration technique. *Journal of Membrane Science*, 59, 73–80.

Okubo, T. & Inoue, H. 1989a. Introduction of specific gas selectivity to porous glass membranes by treatment with tetraethoxysilane. *Journal of Membrane Science*, 42, 109–117.

Okubo, T. & Inoue, H. 1989b. Single gas permeation through porous glass modified with tetraethoxysilane. *AIChE Journal*, 35, 845–848.

Pattanaik, A. K. & Sarin, V. K. 2001. Basic principles of CVD thermodynamics and kinetics. In: Jong-Hee Park, T. S. Sudarshan (eds.) *Chemical Vapor Deposition*. Geauga County, OH: ASM International Materials Park.

Prabhu, A. K. & Oyama, S. T. 2000. Highly hydrogen selective ceramic membranes: application to the transformation of greenhouse gases. *Journal of Membrane Science*, 176, 233–248.

Sea, B. K., Kusakabe, K. & Morooka, S. 1997. Pore size control and gas permeation kinetics of silica membranes by pyrolysis of phenyl-substituted ethoxysilanes with cross-flow through a porous support wall. *Journal of Membrane Science*, 130, 41–52.

Seo, B. G. & Lee, G. H. 2001. Molecular sieve silica membrane synthesized in mesoporous γ-alumina layer. *Bulletin of the Korean Chemical Society*, 22, 1400–1402.

Tsapatsis, M. & Gavalas, G. R. 1992. A kinetic model of membrane formation by CVD of SiO_2 and Al_2O_3. *AIChE Journal*, 38, 847–856.

Tsapatsis, M., Gavalas, G. R. & Xomeritakis, G. 2000. Chemical vapor deposition membranes. *Membrane Science and Technology*, 6, 397–416.

Tsapatsis, M., Kim, S., Nam, S. W. & Gavalas, G. 1991. Synthesis of hydrogen permselective SiO_2, TiO_2, Al_2O_3, and B_2O_3 membranes from the chloride precursors. *Industrial and Engineering Chemistry Research*, 30, 2152–2159.

Xiao, J. & Wei, J. 1992. Diffusion mechanism of hydrocarbons in zeolites – I. Theory. *Chemical Engineering Science*, 47, 1123–1141.

8 Mixed Matrix Membranes for Hydrogen Separation

Henry Bryan Trujillo Ruales, Alberto Figoli, and Adolfo Iulianelli

8.1 INTRODUCTION

Currently, global energy production does not come from sustainable sources; this is due to the fact that the resources used come from fossil hydrocarbons; of course it is a great resource, but unfortunately it is not renewable; this has led to a profound search to find new and more efficient forms of energy, which leads us to develop sustainable energy processes.

The processes to have sustainable energy are research processes developed in recent years, where the main factor is to obtain a cleaner and alternative form of energy; this means that they are friendly with the environment. The important example is hydrogen, which is the resource used in these processes and is generally derived from the production and purification of syngas. This element significantly contributes to the prolonged transition, thereby preventing the reliance on hydrocarbon sources that contribute to environmental pollution.

Membrane technology is the current process that allows us to have an economical and efficient solution for the separation/purification of the hydrogen current coming from a gaseous mixture; other advantages during their use are low energy consumption, environmentally friendly, and easy installation and operation.

Mixed matrix membranes (MMMs) are heterogeneous materials, which have the presence of inorganic particles and organic polymers in their structure; this type of membrane has been developed as an alternative to intensify the separation performance of gases that offer the organic and polymer membranes to be connected (Kamble et al., 2021).

In general, organic and polymer membranes have advantages and disadvantages when used for separation in gaseous separation processes; through the development of the structure of the MMMs, we can say that the combination of these two materials allows to have a greater permeability and selectivity, due to the dispersed inorganic fillers, which facilitates production, desirable properties (mechanical, chemical, and thermal) and economic advantages that the polymer membrane offers.

DOI: 10.1201/9781003382522-10

In conclusion, we can say that the combination of inorganic fillers in a polymer matrix becomes an excellent proposal to take advantage of these materials used while also exceeding the limit established by Robeson, maintaining the flexible mechanical characteristics of polymer membranes.

8.2 HYDROGEN

The element hydrogen (H_2) is the most abundant element on the planet, accounting for approximately 15% in moles (Dunn, 2002). It can be found in fossil fuels, water, biomass, etc. This element is generally not found in its natural gaseous form on Earth; it is always bound to other components (Momirlan and Veziroglu, 2005). One of the main advantages of this element is that it is non-toxic, colorless, tasteless, and odorless. At ambient conditions, molecular hydrogen becomes one of the lightest substances on Earth. However, a disadvantage is its difficulty condensing it in liquid form because of its very low critical point Tc = –240°C and Pc = 13 bar (Møller et al., 2017). Hydrogen has a high energy content (143 MJ kg^{-1}), and when it burns, it produces no pollutants (Momirlan and Veziroglu, 2005).

In the coming years, hydrogen will become one of the most important energy carriers. For this reason, several researchers indicate that hydrogen will establish a large-scale energy infrastructure that will supply consumers worldwide (Momirlan and Veziroglu, 2005). In a way to replace the current petroleum, natural gas, and coal-based infrastructures, this futuristic vision is called the "hydrogen economy." Purified hydrogen is applied in various industrial fields, including metallurgy, chemistry, refinery engineering, and environmental engineering.

There are various mechanisms for producing hydrogen; among them, the most important are the conversion of biomass and fossil fuels, resulting in a hydrogen-rich gaseous mixture. Therefore, hydrogen requires another phase of separation and purification (Conde et al., 2017). Currently, these processes (separation and purification) have a high cost, attributed to the energy consumption and investments involved in the processes (heat exchange, compression, pressure swing adsorption, and cryogenic distillation). However, the use of membranes as an alternative has become an ideal candidate to replace the aforementioned processes, allowing for cost reduction in the separation process (Aberg, 2006; Adams and Chen, 2011; Conde et al., 2017).

In addition to the mentioned processes, membrane gas separation (MS) has become an interesting alternative due to the membrane's ability to separate and purify hydrogen. MS offers several advantages, including lower energy consumption, continuous operation, lower operating costs, smaller plant sizes, operational versatility, and environmental friendliness (Iulianelli and Drioli, 2020).

The U.S. Department of Energy has established five objectives that materials must meet to deliver excellent hydrogen separation performance and advance the hydrogen economy (U.S. Department of Energy, 2005):

I. Higher hydrogen flux rates
II. Low material costs
III. Increased durability
IV. Lower parasitic power requirements
V. Low membrane production/fabrication cost

In recent times, membrane separation has received more attention because of its inherent advantages over traditional separation methods (Chiappetta et. al, 2006). Since its first application in 1980, this technology has experienced significant growth, with a market worth $150 million annually in 2020 and a projected annual growth rate of 10% for the next 20 years (Baker, 2002).

Depending on the material used, there are several selective membranes for hydrogen separation. These include polymeric membranes, inorganic membranes (zeolite, carbon, and silica), and metallic membranes (Pt, Pd, etc.) (Shiraz and Shiraz, 2017; Weber et al., 2020; Pal and Agarwal, 2021). Polymeric membranes offer excellent mechanical strength, ease of molding, and cost-effectiveness (Goh et al., 2011; Pal and Agarwal, 2021). However, they need improvements in permeability and selectivity to be applied in industrial settings.

Inorganic membranes show excellent performance in hydrogen separation and can be divided into two groups: non-porous (metallic) and porous (zeolite, carbon, and silica) membranes (Cardoso et al., 2018; Pal and Agarwal, 2021). They have advantages such as high chemical, mechanical, and thermal stability, reduced plasticization, and pore uniformity (Cardoso et al., 2018). They are favorable for high-temperature and high-pressure applications (Caro et al., 2000; Balachandran et al., 2006).

While metal inorganic membranes are mainly composed of Pd and its alloys (Weber et al., 2020), platinum (Kajiwara et al., 2000), and other metals belonging to Group 10 of the periodic table, there is also the presence of other metals from Groups 3 and 5, which exhibit the ability to dissociate and dissolve hydrogen (Yun et al., 2011). Group 5 metals show high hydrogen permeability, but their excessive solubility makes the membrane susceptible to hydrogen work hardening, leading to the formation of holes and cracks (Pal et al., 2020).

Another technology is MMMs have been suggested as a solution for separation and purification of H_2. These membranes consist of inorganic particles dispersed within continuous polymer matrices. Achieving optimal particle size and dispersion is crucial for the fabrication of MMMs, particularly when considering industrial membrane configurations like hollow fibers and spiral-wound membranes, which feature thin selective layers (Chung et al., 2007; Choi et al., 2008; Yang et al., 2011; Yang and Chung, 2013).

8.3 MEMBRANE ENGINEERING IN GAS SEPARATION

8.3.1 POLYMERIC MEMBRANES

Polymer membranes are materials that have particular importance in the fields of gaseous separation; these membranes are generally characterized to possess excellent permeability and selectivity to be used as non-porous membranes; the gas is separated due to the effect of solubility–diffusivity through the polymer membrane (Kamble et al., 2021).

In essence, all available polymers can be utilized as materials for barriers or separation in membrane processes. However, their diverse chemical and physical properties vary significantly, leading to the utilization of only a limited selection of membranes. Polymer membranes can be categorized into porous and non-porous types. Porous membranes find application in microfiltration and ultrafiltration

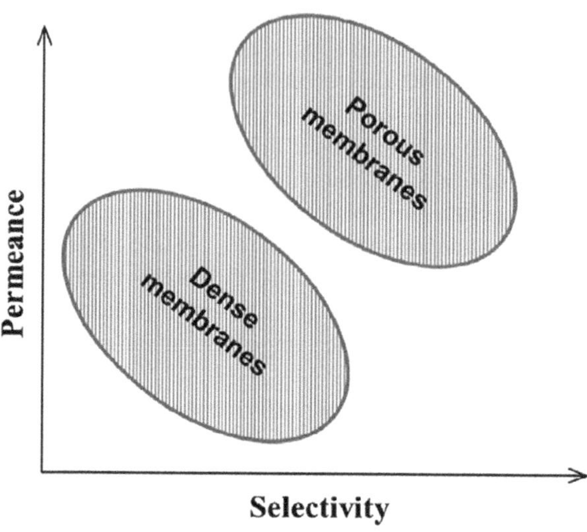

FIGURE 8.1 The trade-off for permeance and selectivity of polymeric membranes (Ghasemzadeh et al., 2020).

processes, while dense (non-porous) membranes are employed in gas separation processes and pervaporation (Mulder, 1997). The determination or choice of polymer material is based on their structure with the following criteria: with regard to porous membranes, the materials are chosen mainly by the requirements of processing (manufactured membranes), chemical and thermal stability, and tendency to soiling, whereas for the porous membrane, the material used is directly determined by the performance of the membrane (Mulder, 1997).

As we can highlight in Figure 8.1, the performance of the membrane can be determined by parameters such as selectivity and permeance. In general, permeance measures the extent to which the gas passes through the film with a known area and thickness per unit of time and pressure gradient, whereas selectivity is determined by the relationship that exists between the permeance of the gas under study in contrast to the other gas of interest (Ghasemzadeh et al., 2020).

Currently, there are several polymeric membranes developed; among the most known, we can mention: Polyimide, Polysulfone, Polystyrene, Cellulose Acetate, and Polyetherimide; these are membranes that have presented a better performance in hydrogen separation (Ghasemzadeh et al., 2020). In gas separation processes, polymer membranes are generally used in industrial processes due to their characteristics (Ghasemzadeh et al., 2020).

The permeation process in dense polymer membranes takes place through the mechanism of solution–diffusion, while the porous membranes may present the following mechanisms: According to diffusion of Knudsen, the flow within the membrane can be either convective or follow a molecular sieving mechanism, depending on the pore diameter of the membrane (Amin et al., 2023).

Polymeric membranes pose some issues, with plasticization being the main concern. To address this problem, highly crystalline polymeric membranes have been developed. These membranes exhibit a high molecular weight and long polymer chains (Amin et al., 2023).

Among the main advantages of using these membranes are their low cost, ease of processing, high film processing capacity, as well as their high permeability with moderate selectivity and vice versa (Amin et al., 2023). Hard and rigid glass is soft and flexible in the rubbery state. Among the disadvantages, we have low thermal stability that generally operates at temperatures $\leq 100°C$, and they are very sensitive to the presence of impurities (HCl, SOx, CO$_2$), thus reducing the life of the membrane (Amin et al., 2023).

8.3.1.1 Classification

In practice, only a limited group of polymers are used for this purpose (Kamble et al., 2021). Polymeric membranes are classified into two groups: rubbery and glassy. Some examples of rubbery polymers include poly(dimethylsiloxane), while glassy polymers include cellulose acetate, polysulfone, polyimides, and poly(phenylene oxide).

In general, we can say that the glassy polymer has higher selectivity than the rubbery polymers; due to that, the permeation mechanism that exists in glassy polymers is controlled by the diffusivity of the penetrating species, but they have low permeability. In gummy polymers, the permeation mechanism depends on the solubility of the penetrants and the solubility difference, which is an important factor in determining polymer selectivity (Amin et al., 2023). In Table 8.1, we can highlight a summary of the classification of polymer membranes.

Glassy polymers find common usage in gas separation due to their capacity to permit the passage of the smallest gaseous component in a mixture, thereby enabling high selectivity (Kamble et al., 2021).

New glass polymer materials developed in recent years, such as thermally rearranged (TR) polybenzoxazole and intrinsically microporous polymers (PIM), are able to overcome Robeson's upper limit compromise limitations (Budd and McKeown, 2010; Kamble et al., 2021). These substances demonstrated remarkably high rates of gas permeation while preserving acceptable selectivity values, as well as excellent chemical and thermal stability (Swaidan et al., 2014; Ricci and De Angelis, 2019).

TABLE 8.1

Classification of Polymer Membranes (Kamble et al., 2021)

Polymer Membranes	
Glassy Polymers	**Rubbery Polymers**
Polycarbonates (PC)	Poly(dimethylsiloxane) (PDMS)
Cellulose acetate (CA)	Ethylene oxide
Polyimides (PI)	Propylene oxide amide
Poly(phenylene oxide) (PPO)	Copolymers
Polysulfone (PS)	
Polyetherimide (PEI)	

TABLE 8.2

Important Polymer Membranes for Gas Separation (Sidhikku et al., 2021)

Membrane	Material	Gas Separation
Prism	Polysulfone	H_2/CO, H_2/Ar, H_2/N_2
IMS	Polyimide	H_2O/Air, O_2/N_2
Medal	Polyimide/polyaramide	CO_2/CH_4
Separex	Cellulose acetate	H_2O/CH_4, CO_2/CH_4
Cynara	Cellulose acetate	CO_2/CH_4

Table 8.2 summarizes different polymer membranes most used in the fields of gaseous separation.

8.3.1.2 Gas Transport Model

In polymer membranes, the driving force responsible for gas separation is the partial pressure gradient, which is the result of the molar fraction multiplied by the total pressure. However, the effectiveness of gas separation through the membrane primarily relies on its morphological properties. Various transport mechanisms using porous polymer membranes include Poiseuille flow, Knudsen diffusion, molecular sieving, capillary condensation, and the solution–diffusion mechanism (Sidhikku et al., 2021).

8.3.1.2.1 Hagen–Poiseuille Flow

This mechanism is observed in membranes with larger pores, where the pore diameter exceeds the average free path of molecules. In such cases, the predominant transport of molecules occurs through the flow of bulk liquid across the membrane's large pores (Sidhikku et al., 2021). The gas permeability according to this mechanism is as follows:

$$P = \frac{\varepsilon \eta r^2}{8 \mu RT} P_{avg} \left(mol\ m^{-1}s^{-1}Pa^{-1} \right) \tag{8.1}$$

where

- ε: Porosity
- μ: Viscosity (Pa s)
- η: Shape factor
- P_{avg}: Mean pressure (Pa)
- r: Pore radius (m)

8.3.1.2.2 Knudsen Diffusion

This mechanism is observed in porous membranes where the pore diameter is larger than the size of the molecule but smaller than the average free path of the molecule. In this process, each molecule moves independently of others, interacting with the walls of the pores (Sidhikku et al., 2021) It often happens that there are collisions between molecules, but these collisions do not have any participation in interacting with the surface because they are elastic in nature. Generally, the mechanism takes place with pore diameter ranging between 50 and 100 Å (Javaid, 2005; Sidhikku et al., 2021). Permeability is given as follows:

$$P = \frac{2\varepsilon\eta rv}{3RT} \left(\text{mol m}^{-1}\text{s}^{-1}\text{Pa}^{-1} \right) \tag{8.2}$$

where v is the molecular velocity (m/s):

$$v = \sqrt{\frac{8RT}{M\pi}} \tag{8.3}$$

8.3.1.2.3 Molecular Sieving

As the name suggests, this method involves gas separation through molecular sieving, permitting the passage of species smaller than the feed gas through the pores while impeding the passage of larger species. The pore diameter utilized in this process is a crucial determinant in the membrane, which must be between the diameter of the gas molecule that we want to separate, which is generally <2 nm (Sidhikku et al., 2021).

8.3.1.2.4 Capillary Condensation

This process is accomplished through the partial condensation of one of the gaseous species in the feed mixture. The membrane's pores become filled with the condensed gas, effectively excluding the other gas species. This occurs at a specific kinetic pressure, typically with mesoporous pores having a diameter greater than 3 nm, resulting in exceptionally high selectivity (Pandey and Chauhan, 2001; Sidhikku et al., 2021).

8.3.1.2.5 Solution–Diffusion Mechanism

The mechanism that exists for the movement of gases through the polymer membrane grid is called the solution–diffusion mechanism. The mechanism describes the separation that results from the variations between the diffusion rate of gas molecules in the permeating feed and the amount of penetrating gases that dissolve in the membrane (Kamble et al., 2021; Sidhikku et al., 2021). The process is described in three steps:

a. The adsorption of gas molecules onto the membrane surface, typically exposed to elevated pressure.
b. The active diffusive mechanism occurring through the membrane.
c. The desorption of gas molecules on the downstream side of the membrane at lower pressure (Kamble et al., 2021; Sidhikku et al., 2021).

To evaluate membrane performance, it is essential to calculate the permeability and selectivity of gas passing through the membrane. Permeability represents the ability the membrane possesses to allow gas molecules to pass through their grid:

$$P = K_i D_i \quad (\text{Barrer}) \tag{8.4}$$

where

- K_i: Sorption coefficient
- D_i: Diffusion coefficient

Selectivity is the membrane's capability to differentiate between two gas molecules (i and j), expressed as the ratio of the permeabilities of the gases being analyzed:

$$\alpha_{i.j} = {P_i}/{P_j} \tag{8.5}$$

Permeant flow can be described through Fick's diffusion law:

$$J = -DS\,\frac{P_1 - P_2}{l} \tag{8.6}$$

where D is the coefficient of diffusion, S is the coefficient of solubility, $P_1 - P_2$ is the difference of pressure across the membrane, and l is the thickness of membrane (Kamble et al., 2021).

8.3.1.3 Polymeric Membrane-Based for Gas Separation

Currently, polymer membranes find application in the separation processes of virtually all conceivable gas mixtures. At an industrial scale, membranes have been designed for use in processes such as oxygen and nitrogen enrichment, as well as the removal of acid gases, the recovery of ammonia purge gas, and refinery gas purification (Yampolskii, 2012; Sidhikku et al., 2021)

These processes are examined below:

- Air separation (nitrogen or oxygen enrichment)
- Hydrogen separation (H_2/N_2, H_2/CH_4 and H_2/CO)
- Natural gas purification (Separation CO_2/CH_4)
- Treatment of flue gas (CO_2/N_2) (Yampolskii, 2012).

a. Hydrogen separation

There are various polymers available in the market that are employed for the separation and purification of hydrogen streams. Polymer membranes exhibit elevated selectivity and permeability for hydrogen in gas mixtures when compared to other gases such as N_2, CH_4, CO_2, and CO. Hydrogen is recognized as one of the gases with high permeability in polymers, often surpassing the permeability of other gases. Membranes find applications in various gas separation processes, including but not limited to, ammonia purge gas recovery, refinery gas, and oxo-chemical synthesis (Sanders et al., 2013).

Among different membranes used for the hydrogen separation of the gaseous mixture, we have polysulfone, polyimide, cellulose acetate, poly(methyl methacrylate) (PMMs), poly(1-trimethylsilyl-1-propyne) (PTMSP), and polydimethylsiloxane (PDMs) (Ockwig and Nenoff, 2007).

The polymeric membranes for H_2/CO_2 are generally manufactured using glassy polymers, because they have the characteristic of being able to discriminate the gas molecule according to the variation of their size; this difference is fundamental to separate the molecules, for example, the difference in the kinetic diameter between H_2 and CO_2 is 0.41 Å. While if we want to work with membranes selective for the CO_2, we must use rubbery polymers membranes; in these membranes, the solubility of the CO_2 is greater than H_2; this is due to the critical temperature which is very high than that of the CO_2 (Shao et al., 2009).

b. Natural gas purification

The application of membrane separation in the purification processes of natural gas is a substantial area within industrial practices. Typically, natural gas comprises methane, carbon dioxide, hydrogen sulfide, inert gases, and minimal traces of additional components such as BTEX aromatics (benzene, toluene, xylene, and ethylbenzene) (Baker and Lokhandwala, 2008; Sanders et al., 2013; Sidhikku et al., 2021).

Gas treatment is very important to prevent corrosion of the pipes; to adjust the dew point of the fuels and the desired heating value, treatment is carried out using polymer membranes for the removal of acidic gases (H_2S, CO_2) present in natural gas. Among different membranes used, we have hollow fiber membranes of cellulose acetate, which is the most used membrane and the membrane based on polyimides (Sanders et al., 2013).

8.3.1.3.1 *Polymer Materials Used for Gas Separation*

Developed polymer membranes are widely used in the fields of gas separation processes, due to different characteristics they present such as being economical, easy to work, and environmentally friendly. Polymeric materials generally used for gas separation include polycarbonates, polysulfones, aramids, cellulose acetate, polyphenyl oxide, and polyamides (Baker and Lokhandwala, 2008; Sanders et al., 2013; Sidhikku et al., 2021).

Table 8.3 shows different polymers that are most relevant for the separation of gas currents. Values are reported as a function of common selectivity of gases in commercial polymer membranes.

8.3.2 INORGANIC MEMBRANES

8.3.2.1 Overview

The first inorganic membranes were prepared in 1940, which was a homogeneous porous membrane of glass called Vycor with a diameter of 20–40 Å (Pandey and Chauhan, 2001). Inorganic membranes are classified according to their structure (Figure 8.2): membranes of dense structures and membranes of porous structures, as shown in the following graphic (Ismail and David, 2001).

Porous membranes typically have pores with a diameter greater than 0.3 nm, and they usually function as sieves for larger molecules and particles (Ismail and David, 2001). Among different porous membranes used, we have metal, glass, alumina,

TABLE 8.3

Common Gas Selectivity for Commercially Significant Polymers (Sanders et al., 2013)

Polymer	H_2/CH_4	N_2/CH_4	H_2/N_2	CO_2/N_2	O_2/N_2	CO_2/CH_4	CO_2/H_2
CA-2.45	80	1	80	32	5.5	32	0.4
Matrimid®	64	1.1	56	31	6.6	36	0.56
PPO	14	0.95	15	15	4.1	14	1
PSF	56	1	56	22.4	5.6	22.4	0.4
TB-BisA-PC	–	1.4	–	23	7.8	32	–
Selected aramid	245	–	–	–	–	–	–

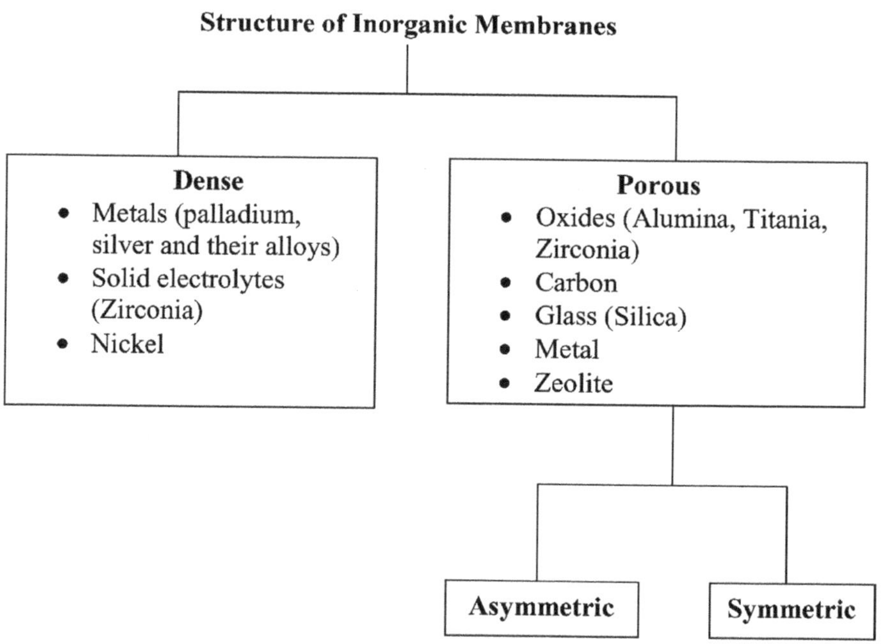

FIGURE 8.2   Structure of inorganic membranes (Ismail and David, 2001).

zeolite, zirconia, and carbon membranes (Ismail and David, 2001). While dense membranes that are generally Pd and its alloys (nickel, silver and zirconium) are structures used for the separation of gaseous components, they are highly selective for the separation of oxygen and hydrogen (Ismail and David, 2001). But unfortunately, this type of membrane has a very limited application at an industrial level due to its low permeability compared to porous (inorganic) membranes (Ismail and David, 2001). Pd membranes, with a thin film of the metal deposited on porous media, find applications in hydrogenation, dehydrogenation, and dehydrogenation–oxidation reactors (Ismail and David, 2001).

Inorganic membranes are more expensive materials with respect to polymer membranes, but they have some advantages: high temperature resistance, stable and well-defined pores, chemical and thermal stability, much higher gas flows, and greater selectivity (Pandey and Chauhan, 2001; Ismail and David, 2001).

Palladium-based materials and their alloys have been thoroughly investigated as prospective membrane materials for gas separation. Recent advancements in palladium membrane development involve the creation of composite membranes, where a dense metal layer is deposited as a thin film onto porous media, aiming to achieve optimal hydrogen permeability (Pandey and Chauhan, 2001).

Some important examples in the field of inorganic membranes application are (Keizer and Verweij, 1996):

- Separation of H_2 from coal-derived gases.
- Separation of H_2O from chemical reactions.
- Separation of CO_2 from gas natural and coal.
- Separation of O_2 from air using a petrochemical process.
- Removal contaminants from water.

8.3.2.2 Structure of Inorganic Membranes

Inorganic membranes are classified according to their structure: porous and non-porous inorganic membranes.

With regard to porous structures, these can be ceramic membranes, zeolite, and carbon, while dense membranes are characteristic metal-based membranes (palladium and its alloys) (Ismail and David, 2001).

Dense membranes have no discrete pores or voids and are well defined (Hsieh, 1996). The properties of these membranes strongly depend on the chemical species to be separated, its material, and mainly their interactions with the membrane (Hsieh, 1996).

The structure of the porous membranes can be described according to the scheme in Figure 8.3, and the shape of the pores is determined through the preparation method. Membranes that have straight pores across the thickness of the membrane are called straight pore membranes. Generally, in porous membranes, the pores that are interconnected through tortuous pathways are called tortuous pore membranes (Hsieh, 1996). When both the separation layer and the bulk support are constructed with a focus on mechanical strength, they typically exhibit a uniform structure, and their composition is consistent throughout the membrane thickness. These membranes are referred to as symmetric membranes or isotropic membranes (Hsieh, 1996). The flow through the membrane is inversely proportional to the thickness of the membrane in use, which leads to work with the thinnest possible layer of membrane (Hsieh, 1996). Unfortunately, very thin membranes do not have a high mechanical strength, so that they can withstand during their use; a proposed solution to this problem is the use of symmetrical or composite membranes, where the thin layer of separation and their support mechanical structure (open cell) are completely different (Hsieh, 1996).

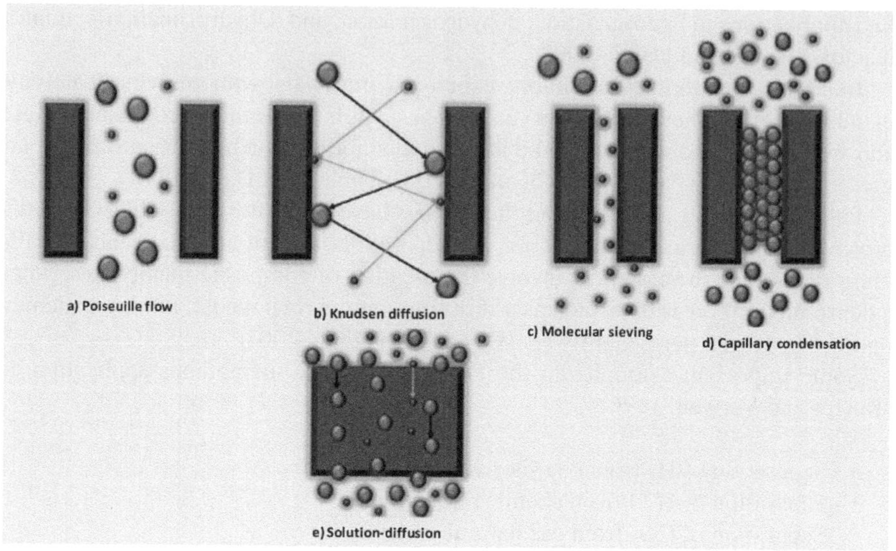

FIGURE 8.3   Structure of inorganic membranes (Hsieh, 1996).

8.3.2.2.1   Dense Inorganic Membranes

a. Metal membranes

They are represented by the palladium membrane and its alloys; the main feature of these structures is that they are only permeable to hydrogen. They are characterized by low permeability but high selectivity (De, 2017; Hsieh, 1996). The field of application of dense membranes is for the separation–purification of gaseous streams of hydrogen and oxygen, because their structure only allows the step of selected atoms or ions of gases (De, 2017). Unfortunately, these membranes have a limitation in the field of industrial scale application due to the fact that they have a low permeability and high cost (for example, Pd and its alloys) in comparison with porous inorganic membranes (Ismail and David, 2001).

8.3.2.2.2   Porous Inorganic Membranes

They are structures that have better selectivity, greater chemical and thermal stability with regard to polymer membranes (De, 2017; Ismail and David, 2001). Porous membranes are more attractive due to the fact that they have the main characteristic of molecular sieving, as do membranes such as zeolites, silica, and carbon (De, 2017), which are materials applied in the field of gaseous separation. Silica membranes basically work by selectively separating the hydrogen from other gases, but the problem of this membrane when working with gaseous molecules that have similar dimensions (oxygen and nitrogen) is that it is not able to identify and separate only one component (De, 2017; Ismail and David, 2001). Zeolite membranes have a high selectivity for separating isomers, but in this case, the problem is producing large membranes without the presence of defects; in the end, carbon membranes are easily obtained (De, 2017).

a. Silica membranes

Silica is a chemical compound that has a silicon oxide (SiO_2) in their structure; this compound can be found in nature as sand or quartz. They are obtained by the sol–gel or chemical vapor deposition (CVD) method, where the membrane manufactured possesses an amorphous film of silica that is homogeneous and free of defects (De, 2017; Ismail and David, 2001; Ahmadd Fauzii Ismaill, 2015). Permeation for different gaseous components (He, H_2, CO_2, N_2, and CH_4) is determined as a function of temperature and pressure (Ahmadd Fauzii Ismaill, 2015), while selectivity is given as a function of the difference in the kinetic diameter of the gases (Ahmadd Fauzii Ismaill, 2015).

There are several silica membranes; among them, we have PTES (phenyltriethoxysilane), TEOS (tetraethoxysilane), and DPDES (diphenyl-diethoxysilane). The TEOS membrane permeates He and H_2, while the membranes PTES and DPDES can permeate large molecules like CO_2 (Morooka and Kusakabe, 1999).

b. Zeolites membrane

They are crystalline aluminosilicate structures, which have a 3D configuration and pores of uniform molecular size 2–20 Å (Ahmadd Fauzii Ismaill, 2015). There are two mechanisms that can be obtained from zeolites: natural and synthetic. Synthetic materials are made from chemicals that consume energy, while natural materials are made from natural minerals (Ahmadd Fauzii Ismaill, 2015). The ratio of silica–alumina in synthetic zeolite membranes is 1:1, while in natural membranes, the ratio is 5:1 (Ahmadd Fauzii Ismaill, 2015). Natural zeolite membranes are more resistant to acids because their structure has more silica (Ahmadd Fauzii Ismaill, 2015).

Among the most used zeolite membranes, we have LTA (Linde-type A), which has a pore size of about 0.4 nm; it is ideal for separating small molecules such as H_2 (Ahmadd Fauzii Ismaill, 2015). And the membrane SAPO-34 is used to peer separate CO_2 from the following mixtures: CO_2/H_2 and CO_2/CH_4.

c. Carbon membrane

These membranes have higher selectivity values for mixtures: C_3H_6/C_3H_8 and C_2H_4/C_2H_6 with regard to polymer membranes. They have micropores capable of differentiating between alkane and alkene molecules according to dimensional difference; the pore structure is controlled through the method of CVD (Morooka and Kusakabe, 1999).

8.3.2.3 Transport Mechanism

Similar to polymer membranes, inorganic membranes can be categorized into porous and dense membranes, each with distinct separation mechanisms. There are four mechanisms commonly employed to elucidate the gas transport through porous membranes (Pandey and Chauhan, 2001):

- Knudsen diffusion
- Surface diffusion
- Capillary condensation
- Laminar flow
- Molecules sieving

Permeation of gases through porous solids is a complex phenomenon to describe the interaction between pore walls and gas molecules. In general, the mechanism describing the phenomenon depends on the properties of the membrane, the operating conditions (temperature and pressure), and the permeating gas (De, 2017; Pandey and Chauhan, 2001), but conditions mainly affect only the temperature and pressure, which determine the average free path of the molecule λ:

$$\lambda = \frac{3\eta(\pi RT)}{2P2M} \tag{8.7}$$

where

- η: Viscosity of gas
- T: Temperature
- R: Universal gas constant
- M: Molecular weight
- P: Pressure

The mechanism through porous inorganic membranes is summarized below:

a. Knudsen diffusion: It occurs in the Knudsen flow (Figure 9.10) when the r/λ ratio is much less than 1. This implies that there are fewer collisions with the walls of the pores compared to interactions with other gas molecules. During collisions, the gas is absorbed and then reflected, causing separation because different species of gas move at different speeds (De, 2017; Ismail and David, 2001). The prevailing flow mechanism is viscous flow (Ahmadd Fauzii Ismaill, 2015).

b. Surface diffusion: Gas molecules adsorbed on the pore wall move along the surface, leading to increased surface diffusion (see Figure 8.4). This enhances gas permeability while concurrently reducing the effective pore diameter, resulting in decreased non-absorbed gas and increased selectivity (Ahmadd Fauzii Ismaill, 2015).

c. Molecular sieving: The mechanism occurs when the pore diameters become sufficiently small (3–5 Å), which allows to separate the molecules that have different kinetic diameters, which means that only the smallest gases can cross the membrane (Ahmadd Fauzii Ismaill, 2015).

d. Capillary condensation: It happens when a phase is partially condensed and occupies the pores of the membrane. When the pores are completely filled with the condensed phase, more gases pass through the membrane.

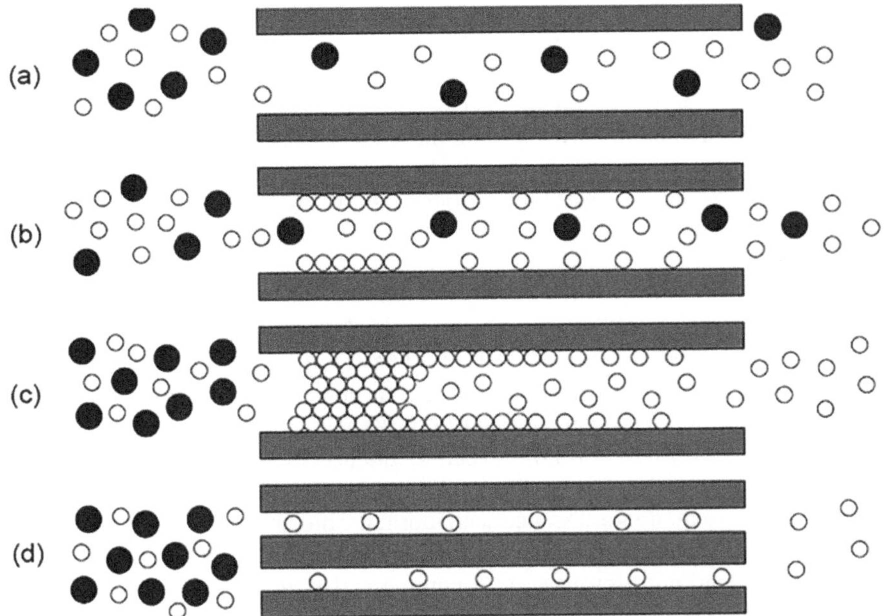

FIGURE 8.4 Four types of diffusion mechanism (Ahmadd Fauzii Ismaill, 2015).

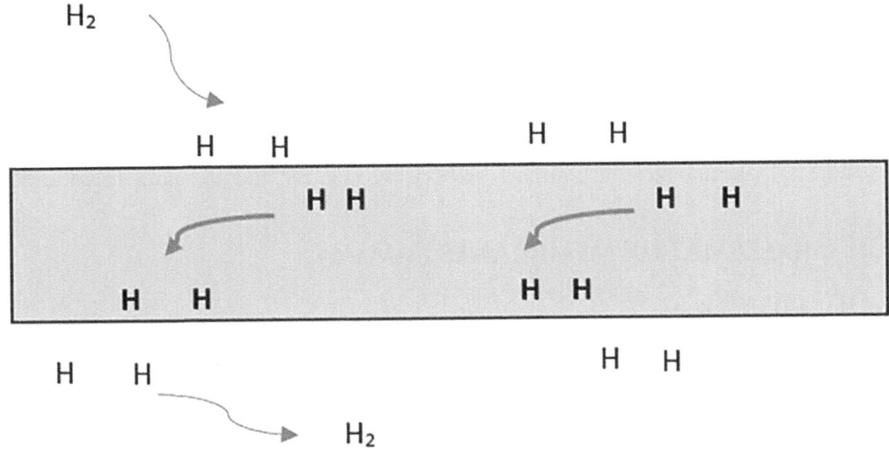

FIGURE 8.5 Adsorption of hydrogen in palladium (Brisotto et al., 2013).

The effectiveness of the process is highly dependent on the gas composition, pore size, and uniformity (Ahmadd Fauzii Ismaill, 2015).

The mechanism through dense membranes is generally described of hydrogen permeation in palladium membranes. The process casings several stages in series (Figure 8.5), ranging from the high-pressure zone to the base pressure zone, as described below:

a. Molecular transport of H_2 ions to the palladium surface.
b. Dissociative adsorption of H_2 molecules on the palladium surface.
c. Dissolution of hydrogen atoms (H) in bulk metal.
d. Diffusion of atomic hydrogen (H) through bulk metal.
e. Association of atomic hydrogen on the surface of palladium.
f. Molecular H_2 desorption from palladium surface.
g. Diffusion of molecular H_2 away from the palladium surface.

8.3.3 Comparisons of Organic and Inorganic Membranes

Inorganic membranes exhibit exceptional permeability and selectivity, demonstrating elevated chemical and thermal stability. They are resilient against microbiological degradation and can be effectively cleaned through repeated washing, surpassing the performance of polymer membranes (Sidhikku et al., 2021), but the manufacturing cost is high; they have a fragile structure and processing difficulties make these structures uncomfortable to manufacture it in large sizes (Sidhikku et al., 2021).

Typically, the expenses associated with inorganic membranes are 100–1,000 times greater than those of polymeric membranes, making them less viable for large-scale industrial applications. However, on a commercial scale, certain inorganic membranes, such as the palladium membrane and its alloys, find extensive use in the separation and purification of hydrogen (Sidhikku et al., 2021).

Inorganic membranes are more resistant to organic solvents, chlorine, and other chemical compounds, thus ensuring that it can work under chemical conditions and high temperatures; in general, inorganic membranes are applied at separation scale (Hsieh, 1996).

Polymer membranes have acceptable mechanical properties and are economical and easy to work. They are the most popular materials in gaseous separation processes (Sidhikku et al., 2021).

Table 8.4 summarizes the comparison between organic and inorganic membranes.

8.4 MIXED MATRIX MEMBRANES (MMMS)

8.4.1 Overview

They are new materials developed that have gained much interest in the field of research in order to overcome the performance compromise highlighted polymer membranes. They are membranes that combine the advantages that organic and polymer materials have at the same time show very high performance without having a significant increase in manufacturing cost (Carreon et al., 2017).

Polymer membranes separate gas mixtures based on the difference in permeation rates of each species of gas across the membrane; Robeson established a graph where he shows the limitations of polymer membranes called "Upper bound," which is supported by a theoretical framework provided by an empirical report (Carreon et al., 2017).

Inorganic membranes, such as zeolite and metal, with porous structures, have high performance in the field of gas separation, but also represent materials that

TABLE 8.4

Comparison between Organic and Inorganic Membranes (Sidhikku Kandath Valappil, Ghasem and Al-Marzouqi, 2021)

Properties	Organic	Inorganic
Material	Rubbery or glassy properties depend on the operational temperature (such as polysulfone, polyimide, and cellulose acetate)	Created using inorganic materials like glass, silica, palladium alloys, etc.
Characteristics	Hard and inflexible in the glassy state, soft and pliable in the rubbery state	High resistance to chemical and thermal deterioration Mechanical resistance
Advantages	• Economical production and purification processes • Lower expenses compared to inorganic materials membranes • Appreciable selectivity	• Tolerant to aggressive environmental conditions • Tolerant to high pressure • Well-defined and enduring pore arrangement • Extended lifespan
Disadvantages	• Short life time • Reduced resistance to severe chemical and thermal deterioration • Vulnerable to fouling	• Large capital cost • Fragile • Low membrane surface • Modeling the sealing of the membrane at high temperatures can be challenging

have high gas transport properties, excellent chemical and mechanical thermal properties, and are expensive and difficult to work with; so there are few commercial inorganic membranes (Carreon et al., 2017). This explains why membranes are more used in the industrial field but still need alternative materials to improve performance (Carreon et al., 2017).

Several researchers focused on the idea of being able to overcome the limitation of the "upper bound" established by Robeson for polymeric materials, but also the problems of processing and cost of inorganic materials; the reasons shifted led to the development of MMM (Carreon et al., 2017).

MMMs are innovative materials that combine different advantages of organic–inorganic materials (Carreon et al., 2017). In general, the structure of a MMM comprises inorganic fillers, often in the form of micro- or nanoparticles (discrete phase), dispersed within a polymer matrix (continuous phase), as illustrated in Figure 8.6 (Tanh et al., 2012; Carreon et al., 2017). The addition of these inorganic fillers allows the polymer membrane to inherit the superior separation performance that this structure offers, while the presence of the polymer matrix allows to solve the fragility of these fillers (Carreon et al., 2017).

The presence of inorganic fillers allows to create preferential permeation paths for certain gases, allowing a selective permeability by placing a barrier for unwanted gaseous compounds (Carreon et al., 2017).

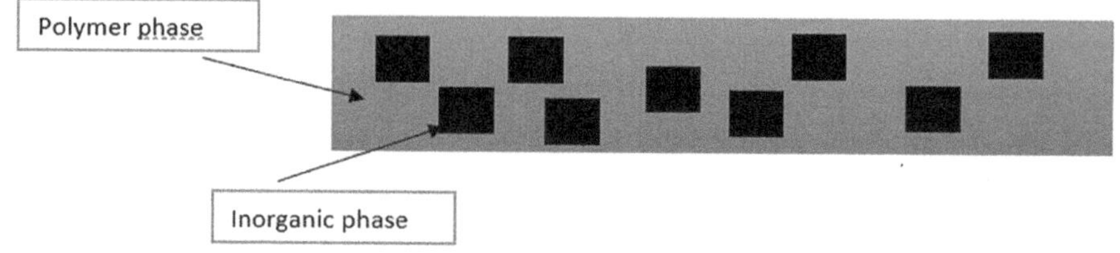

FIGURE 8.6   Schematic representation of a mixed matrix membrane.

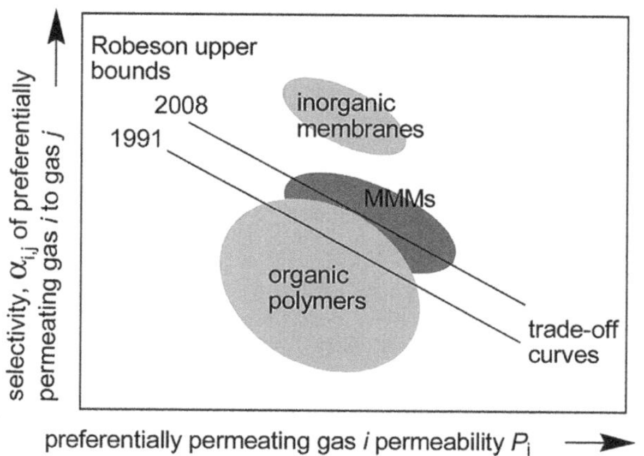

FIGURE 8.7   Schematic representation of 1991 and 2008 Robeson upper bound (Carreon et al., 2017).

The main concept of MMM was introduced to overcome the upper bound (Figure 8.7) of both polymeric and inorganic membranes (Hamid and Jeong, 2018).

With MMM, the separation properties (permeability–selectivity) can be well above the upper limit of Robeson, where the use of porous inorganic fillers helps to counteract the compromise between selectivity and permeability characteristic of pure polymers (Tanh Jeazet, Staudt and Janiak, 2012). Moreover, we can say that MMMs have properties that are physical and thermally stable, easier to produce, and cheaper than inorganic membranes (Tanh Jeazet, Staudt and Janiak, 2012).

8.4.2   MMMs for Gas Separation

MMMs use porous solids with higher permeation and selectivity as additives than polymers. There are several particles such as metal–organic framework (MOF), zeolite, carbon, porous organic polymer (POP), covalent organic structure (COF), and nanoparticles that are used in MMM to improve the performance of these structures (Carreon et al., 2017). The interaction between the polymer and the porous particle, the increase in free volume, the chain structure, and the presence of interfacial voids are responsible for the transport of gas in MMM (Carreon et al., 2017).

In broad terms, permeability is the result of multiplying diffusivity (D) by solubility (S). In MMMs, changes in permeability are elucidated by considering the diffusion and solubility coefficients. The impact of the polymer's free volume on the diffusion coefficient of penetrants is elucidated through the mechanical static model introduced by Cohen and Turnbull, which characterizes the diffusion of spheres in a liquid. Consequently, the penetrant diffusion coefficient (D) can be expressed as follows (Carreon et al., 2017):

$$D = A\, e^{\frac{-\gamma V^*}{Vf}} \tag{8.8}$$

where

- A: Pre-exponential factor (f(T))
- γ: Overlap factor
- Vf: Average free volume in the media accessible to penetrants transport
- V^*: Minimum free volume element size of penetrant molecule

The equation describes that an increase in free volume improves the diffusion of penetrants (Carreon et al., 2017).

The solubility coefficient of the penetrating species depends on the interaction between the inorganic fillers and the polymer. Solubility is dependent on temperature (T) and adsorption enthalpy (ΔH), which is described by Van't Hoff's law (Carreon et al., 2017):

$$S = So\, e^{\frac{-\Delta H}{RT}} \tag{8.9}$$

where

- So: A constant
- R: Ideal gas constant
- T: Absolute temperature

Enhanced interaction between the functional groups of the membrane and the polar penetrating species results in a decrease in ΔH and, consequently, an increase in the solubility coefficient of the gas (Carreon et al., 2017).

8.4.3 MORPHOLOGIES OF THE MMMS

The main objective in the manufacture of MMM is to improve and facilitate the transport properties of gas in inorganic particles (Dong, Li and Chen, 2013). This implies that an ideal MMM should possess a morphology that facilitates the preferential transport of species through the inorganic phase rather than the polymeric phase (Dong, Li and Chen, 2013).

8.4.3.1 MMMs Have Two Types of Morphology: Symmetrical and Asymmetric

Symmetrical (dense) membranes are the most studied due to simplicity in the synthesis process. The manufacture of this structure has the characteristic of having a good dispersion of inorganic fillers in the membrane structure (Dong, Li and Chen, 2013).

The problem is that the particle load must not exceed 50% by weight, because it leads to an agglomeration of particles; the difficulty is called percolation. The threshold and the transport of gas in this structure are very dominated by the polymer phase (Dong, Li and Chen, 2013), which presents another problem; the membrane thickness (>50 µm) is necessary to have good mechanical stability but reduces gas permeability (Dong, Li and Chen, 2013).

Asymmetric membranes have a thin layer (<1 µm) with a dense top layer and porous support layer (Dong, Li and Chen, 2013). These structures are generally used in the field of industrial scale gas separation (Dong, Li and Chen, 2013; Carreon et al., 2017). When this morphology is presented, the mechanical resistance is reduced compared to the symmetrical membranes (Dong, Li and Chen, 2013). To exploit the load of inorganic particles in the upper layer, the alternative that has been proposed is to prepare asymmetric composite membranes (Dong, Li and Chen, 2013). The main feature of composite membranes is that they consist of a thin polymer layer to which nanoparticles of similar size as the thickness of the upper layer are incorporated and a porous substrate to provide physical support. The developed structure allows to have a preferential flow path through the particles, improving the transport properties due to a higher particle load (Dong, Li and Chen, 2013).

MMMs hold promise as applicable materials in gas separation. However, addressing various challenges is essential during their design and fabrication (Carreon et al., 2017).

8.4.4 Membrane Materials

The selection of the material (polymer–inorganic particles) is fundamental for the manufacture of MMM, in order to ensure the desired structure, separation properties, and compatibility between the components used (Dong, Li and Chen, 2013).

The most used polymeric materials for MMM are:

- Polysulfones (PSF)
- Polyarylates
- Poly(arylketones)
- Polycarbonates
- Polyimides
- Poly(arylethers)

The inorganic materials:

- Zeolite
- Carbon nano-tubes
- Mesoporous materials
- Silica
- Graphite
- Carbon molecular sieves
- MOFs

8.4.4.1   Inorganic Fillers Materials-Based MMMs

Inorganic materials can be agglomerated in terms of their size such as:

- Macroporous: It has a pore size greater than 50 nm and is not used as particles in MMM.
- Mesoporous: The pore range between 2 and 250 nm is one of the materials used to manufacture MMM due to selective surface flow.
- Microporous: These particles have pore sizes smaller than 2 nm and are the most used materials in MMM (Dong, Li and Chen, 2013).

Non-porous inorganic materials can also be used as silica fumed. Table 8.5 shows the pore sizes of different particles used to make MMM (Dong, Li and Chen, 2013).

8.4.4.1.1   Metal Oxides

Inorganic metal oxide nanoparticles are materials considered for applications in the gas separation field (Dong, Li and Chen, 2013). But studies with these materials are still poor. There are several researchers who have developed different membranes for example: Hu et al. (1997), created fluorinated poly(amide-imide) that is mixed with particles TiO_2 to obtain nanocomposite membranes using the sol–gel technique (Hu et al., 1997); the results obtained from the membrane demonstrate that for the mixture H_2/CH_4 to obtain nanocomposite membranes using the sol–gel technique (Hu et al., 1997; Dong, Li and Chen, 2013), as well as for the mixture CO_2, H_2, and TiO_2. Hosseini et al. developed MgO nanocomposite membranes in a Matrimid matrix, thanks to the large MgO pores (30 Å) that had increased permeability (H_2/N_2, CO_2/CH_4) compared to the pure polymer membrane (Hosseini et al., 2007; Cheng et al., 2018).

TABLE 8.5
Pore Size of Inorganic Membranes

	Particles	Pore Size (Å)
Zeolites	Zeolite-A	3.2–4.3
	Zeolite-13x	73.
	Silicalite-1	5.2–5.8
	ZSM-5	5.1–5.6
Mesoporous	MCM-41,48	>25
	ZSM-5	27
	Activate carbon	20–30
	MgO	30
MOF	MIL-96	2.5–3.5
	MIL-100	5.5×8.6
	MOF-177	7.1–7.6
	ZIF-8	3.4

8.4.4.1.2   Metal–Organic Framework

They are materials of increasing interest as inorganic porous fillers to manufacture MMM (Tanh Jeazet, Staudt and Janiak, 2012). MOFs are composed of porous crystalline compounds that can form one-two-three-dimensional structures that contain metal ions or clusters to organic ligands (Carreon et al., 2017).

The main feature of MOFs that makes them more attractive for MMM application is due to its compatibility with polymers due to the presence of organic linker and their functionality (Carreon et al., 2017).

Another important advantage is that its porosity and the surface area are much higher than that of zeolite which makes it the most attractive adaptive for MMM (Carreon et al., 2017). Unlike pure zeolite, organic binders in MOF offer a high interaction with the polymer matrix, having a more intimate contact between the filler and the polymer (Hamid and Jeong, 2018).

One of the first studies of MMM using MOF was done by Yehia et al., using copper-based crystals (Cu-MOF) that are mixed with a gummy polymer called poly(3-acetoxyethylthiophene) to form MMM (Hamid and Jeong, 2018); the test is conducted at 308 K and 2 bars, which shows that an increase in CH_4 causes a decrease in O_2 permeation rate due to pore size and hydrophobic Cu-MOF crystals in the polymer matrix (Hamid and Jeong, 2018).

In Table 8.6, some MMMs are identified using MOF used for the purification and separation of the hydrogen current.

8.4.4.1.3   Carbon Molecular Sieve CMS

Carbon-based materials such as activated carbon, carbon nanotubes (CNTs), fullerenes, and graphene oxides are used as inorganic fillers for MMM (Carreon et al., 2017).

They have a high porosity and a fine distribution of the pores; the opening of the pores is of the same order as the size of the gas molecule (Carreon et al., 2017).

CMS is obtained by pyrolysis/carbonization of polymeric precursors under controlled conditions (mm7). Unlike zeolite and MOF, CMSs have an amorphous structure due to the distribution of ultramicropores and micropores (Carreon et al., 2017).

TABLE 8.6

MMM with MOFs in Hydrogen Separation (Hamid and Jeong, 2018)

Polymer	MOFs	Gas Pair	Permeability	Selectivity	Ref
Polybenzimidazole	ZIF-8	H_2/CO_2	41	4.7	Sánchez-Laínez et al. (2016)
6FDA-Durene	ZIF-8	H_2/N_2	284	141	Wijenayake et al. (2013)
Matrimid	ZIF-8	H_2/CH_4	24	50	Diestel et al. (2015)
Matrimid	ZIF-11	H_2/CO_2	95.9	4.4	Sánchez-Laínez et al. (2015)
6FDA-DAM	ZIF-11	H_2/CH_4	272	32.8	Safak Boroglu and Yumru (2017)
VTEC PI-1388	MIL-53 (Al)	H_2/CO_2	5.4	5.4	Perez et al. (2017)

TABLE 8.7
MMM with CMS in Hydrogen Separation (Chuah et al., 2021)

Membrane				Separation Performance			
					Selectivity		
Filler	Polymer/ Support	Test Condition	PH_2 (GPU)	H_2/CO_2	H_2/CH_4	H_2/N_2	Ref.
Cgo-76 C=cysteamine	Anodized AL_2O_3	1.5 bar 25°C	52	21	–	–	Cheng et al. (2020)
GO	PSF	35°C	4.7	–	29	–	Castarlenas, Téllez and Coronas, (2017)
GO	PI	35°C	14	–	82	–	Castarlenas, Téllez and Coronas (2017)
GO/PEI	Porous Al_2O_3	25°C	1000	29	–	–	Shen et al. (2016)
GO/EDA-2	Porous Al_2O_3	25°C	73	23	–	–	Lin et al. (2018)

Further, Table 8.7 includes some membranes that use CMS fillers for H_2 purification.

8.4.4.1.4　Zeolite

They are microporous crystalline solid structures that contain silicon, oxygen, and aluminum. They are widely used as additives in MMM and have a large diversity of zeolite-based membranes with desirable physical and chemical properties (Carreon et al., 2017). Zeolitic materials have a significantly high diffusivity and selectivity with regard to polymeric materials due to having the property of molecular sieve (Carreon et al., 2017).

The performance of zeolite-based MMs depends on the percentage weight load of particles, particle size, shape, and surface of zeolites, which are factors that affect the separation performance (Carreon et al., 2017). For example, an increase in zeolite load improves MMM permeability in some cases, while in other cases permeability is decreased (Carreon et al., 2017). The interaction between zeolite and polymer is determined by the glass transition temperature (Tg) of MMM. Table 8.8 shows some zeolite-based MMs for hydrogen purification.

8.4.4.1.5　Carbon Nanotubes (CNTs)

They are unique tubular structures of carbon nanotubes with a nanoscale diameter with a very large length/diameter ratio (Dong, Li and Chen, 2013). CNTs are synthesized through two models: single-wall nanotubes (SWCNT) and multiple-wall nanotubes (MWCNT).

TABLE 8.8

MMM with Zeolite-Based in Hydrogen Separation (Chuah et al., 2021)

Membrane				Separation Performance			
					Selectivity		
Filler	Polymer/Support	Test Condition	PH_2 (GPU)	H_2/CO_2	H_2/CH_4	H_2/N_2	Ref.
IL/ SAPO-34	Pebax MH1657/ PEGDME with ceramic	1 bar, 20°C	4.9	0.11	1.2	–	Hu et al. (2017)
IL/SAPO-34-NH$_2$	Pebax MH1657/ PEGDME with ceramic	1 bar, 20°C	2.2	0.05	1.9	–	Hu et al. (2017)
SAPO-34	Pebax MH1657/ PEGDME with ceramic	1 bar, 20°C	1.5	0.06	1.4	–	Hu et al. (2017)
DD3R	Matrimid 5218	1 bar, 25°C	34.9	–	375	–	Peydayesh et al. (2017)
Zeolite A4 (25 wt%)	PVAc	–	3.8	1.6	–	156	Ahmad and Hägg (2013)
Modified Zeolite A4 (15 wt%)	PVAc	0.75 bar 30°C	5.6	6.1	–	143	Ahmad and Hägg (2013)
Hydroxyl sodalite	PSF	–	21.8	1.1	–	1.1	Eden and Daramola (2021)
Silica sodalite	PSF	–	22.8	1.1	–	1.0	Eden and Daramola (2021)
HZS	6FDA-DAM	2 bar 35°C	541	0.77	25	21	Zornoza et al. (2015)

Problems in using these materials are given to particle dispersion and nanotube alignment (Dong, Li and Chen, 2013). Among the CNTs, there is the presence of Van der Waals forces that hinder the good despair of the particles inside the polymer matrix. One of the mechanisms applied to solve this challenge is the so-called covalently and not covalently mechanism (Dong, Li and Chen, 2013).

The covalent method introduces carboxyl and hydroxyl groups at the ends of the CNT; adding these groups leads to a better dispersion of the CNT in the solvent, increases the interaction with the polymer, and serves as anchor point for bonding other functional groups (Dong, Li and Chen, 2013). The non-covalent method uses low molecular weight surfactants and room-temperature ion liquids (RTILs).

TABLE 8.9

Effect of Multi-Walled Carbon Nanotube (MWCNT) Concentration on the Permeability of MWCNTs/PBNPI Nanocomposite Membranes (Weng et al., 2009)

Polymer	MWCNT (wt%)	$P(H_2)$ Barrer
PBNPI 15 wt%	0	4.71 ± 0.77
PBNPI 15 wt%	1	4.95 ± 0.31
PBNPI 15 wt%	2.5	6.49 ± 1.94
PBNPI 15 wt%	5	6.42 ± 1.62
PBNPI 15 wt%	10	12.06 ± 3.16
PBNPI 15 wt%	15	14.31 ± 3.07

The RTIL has shown high attention due to the efficient ability to disperse CNT particles in the polymer matrix (Dong, Li and Chen, 2013).

The problem of the alignment of a CNT inside the polymer matrix is solved by using a filling between the CNT spaces with polystyrene (PS) by a spin coating approach (mmm5 20), which gives a vertically aligned structure.

Table 8.9 shows several carbons' nanotube-based MMMs for H_2 current separation.

8.4.4.2 Polymers Materials-Based MMM

It uses two types of polymers for gaseous separation: glass polymers and gummy polymers. The rubbery materials have a flexible chain structure that allows the chain segments to rotate freely (Dong, Li and Chen, 2013). Glass polymers have a rigid chain that causes limited segmental movements (Dong, Li and Chen, 2013).

Polymer materials have disadvantages when using MMM because they must ensure excellent adhesion with inorganic particles (Dong, Li and Chen, 2013). The gummy polymers characterized by a significant degree of flexibility enable robust interactions with inorganic particles, resulting in a flawless structure. However, their high flexibility contributes to increased permeability of the membrane. Consequently, gas transport through the membrane is governed by the polymer matrix (Dong, Li and Chen, 2013). Glass polymers show superior separations due to the rigid chain structure, but this decreases the interaction between the polymer and inorganic particles (Dong, Li and Chen, 2013).

The "reverse-selective polymers" such as PMP, PTB and PTMSP are highlighted in Table 8.10, which present a unique transport for the most condensed molecules (CO_2) on the smaller ones (H_2) (Dong, Li and Chen, 2013). The permeability of CO_2 increases with the size of the lateral functional groups (increased free volume) (Dong, Li and Chen, 2013). This transport behavior has an opposite mechanism than what occurs in glass polymers with a low free volume, which is the reason why they are so dominated (Dong, Li and Chen, 2013).

TABLE 8.10

Permeability to Gas for Certain Widely Used Polymers with Reverse-Selective Characteristics (Dong, Li and Chen, 2013)

Polymer Types	PTBA (Tert-butylacetylene)	PMP (4-Methyl-2-pentyne)	PTMSP (Poly 1-trimethylsilyl-1-propyne)
Chemical structure			

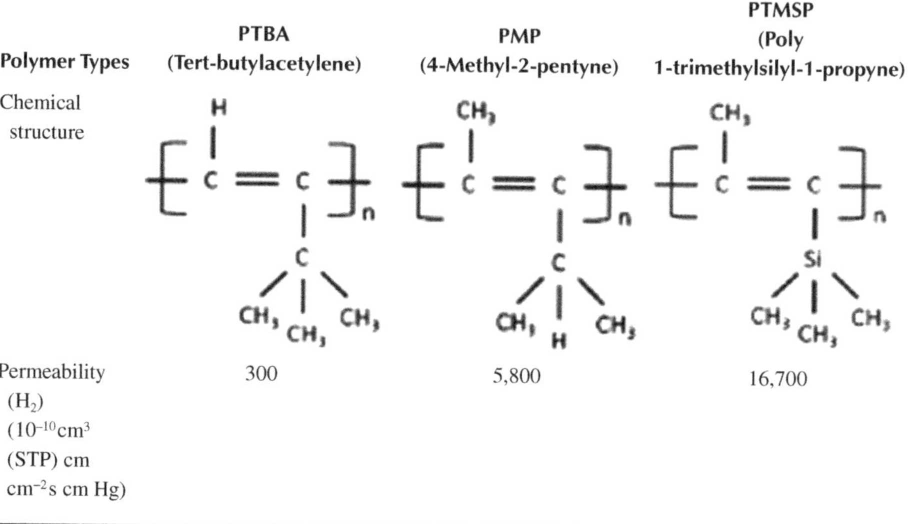

Permeability (H_2) $(10^{-10}cm^3$ (STP) cm $cm^{-2}s$ cm Hg)	300	5,800	16,700

8.4.5 FABRICATION AND CHARACTERIZATION

8.4.5.1 Flat Sheet-Hollow Fiber Configurations

MMMs are manufactured either as a flat sheet membrane or hollow fiber membrane (Carreon et al., 2017). The first MMMs were made of flat membranes, which were used in the gas separation industry. It has several disadvantages, including a dense selective layer and low permeant flow (Carreon et al., 2017). Dense flat symmetrical membranes are 50 μm thick, while asymmetric membranes are 5–25 μm thick. Most MMMs are dense flat-sheet membranes (Carreon et al., 2017). There is little study of flat asymmetric membranes, so Basu et al. prepared an asymmetric membrane Matrimid/(Cu$_3$ (BTC)$_2$) MMM using the phase inversion method to increase membrane performance in the field of gas separation (mm1). The obtained membrane has a greater permeability and selectivity with the increase of the filler load compared to the pure polymer membrane (Carreon et al., 2017).

Currently, there is a focus on the development of hollow fiber membranes due to their advantageous characteristics, including a high surface-to-volume ratio, robust mechanical strength, excellent selectivity, resistance to chemicals, thermal stability, absence of defects, and good compatibility (Carreon et al., 2017). But there are several challenges related to the manufacture of hollow MMM, for example: the thickness of the membrane (which must be thin), density, and ability to maintain the pressure during operation are important characteristics of asymmetric hollow fiber MMM (Carreon et al., 2017).

Hollow fiber MMMs are obtained by single-layer spinning, where the phase reversal method is used, while the inorganic filler is introduced into the spinning polymer (Husain and Koros, 2007; Carreon et al., 2017).

8.4.5.2 Polymer–Filler Compatibility

In an ideal world, MMM would show the best properties of gas transport, but in the real world, the presence of the dispersed phase affects the polymer matrix between the two phases (Carreon et al., 2017). Moore and Koros described several non-ideal cases that may occur during MMM production; the interfacial defects are classified into five cases (Moore and Koros, 2005):

i. Rigidified polymer layers lying around inorganic fillers.
ii. They have large voids (>~10–100 nm) around the particle–polymer interface.
iii. Presence of small gaps due to the rigidity of the polymer chains, as well as a less attractive interaction between the polymer and the particles.
iv. Fillers can be blocked by gas molecules due to the strong tightness that presents the inner surface of fillers.
v. The presence of a bottleneck on the filler side of the interfacial region of the structure (Moore and Koros, 2005; Carreon et al., 2017).

In order not to have the presence of these non-ideal behaviors, there must be a compatibility between the polymer and the filling both in terms of chemical and transport properties; it means that one must work with polymers and inorganic particles of the same polarity (Carreon et al., 2017)

In order to obtain a good interface, the interaction force between the polymer and the inorganic particles must be superior for both the filler–solvent interaction and the polymer–solvent interaction (Carreon et al., 2017). It is necessary that the polymer be flexible during the manufacturing; this is little difficult to achieve because the polymers at the upper limit are rigid, but it can have a solution by applying suitable plasticization (Carreon et al., 2017).

To prevent the occurrence of undesired voids at the interface, it is crucial to prepare the membrane in a manner that minimizes the stress between the polymer–particle interface (Carreon et al., 2017); other solutions applied are: membrane melting at higher temperatures (Tg), addition of surface coupling agents, and the preparation of membranes using melt processing (Jiang et al., 2005; Jiang, Chung and Kulprathipanja, 2006; Pechar et al., 2006; Ismail, Kusworo and Mustafa, 2008; Carreon et al., 2017).

The correct choice of both materials (organic–inorganic) greatly affects the morphology and performance of the membrane, requiring a clear compression of the transport mechanisms (Carreon et al., 2017). The characteristics of filling particles such as pore size and surface properties are necessary to understand to achieve high gas transport properties in MMM. The fillers used are generally microporous materials such as zeolite, carbon molecular sieves, and metal–organic structures (Carreon et al., 2017), but they can also include silica, activated carbon, graphene, covalent organic structures, and CNTs (Funk and Lloyd, 2008; Zhao et al., 2015; Carreon et al., 2017).

The use of microporous materials allows to have a very high selectivity, thanks to the ability of molecular sieving. The use of non-porous particles improves separation, increases the tortuous patter but decreases the spread of larger molecules (Carreon et al., 2017).

8.4.5.3 Filler Size And Morphology

8.4.5.3.1 *Dispersion of Particles*

In MMM, the thickness of the asymmetric membrane layer is <1 µm to be applied at industrial scale; we know that the layer largely depends on the particle size (Carreon et al., 2017). To have a very thin layer, the particles that are used must be much less than the layer thickness (Carreon et al., 2017).

Smaller inorganic particles have a higher surface-to-volume ratio, which leads to greater mass flow between phases (Carreon et al., 2017). The use of nanoscale inorganic materials causes an increase in the diffusion mechanism of the gas, which means an increase in permeability of penetrants due to the fact that these particles alter the packaging of the polymer chain, producing an increase in the free volume between the chains (Dong, Li and Chen, 2013; Carreon et al., 2017).

Another characteristic in the manufacture of MMMs is the particle load limit, which must not exceed 50%, to avoid agglomeration problems (Carreon et al., 2017). The amount of load is determined by the particle size, meaning that as the particle size decreases, the percentage of load being added decreases. Upon overcoming the established load, it results in a particle agglomeration, then reduction of separation properties; the problem is very common at high temperatures which causes large gaps between the polymer and the filler (Carreon et al., 2017); it can also be highlighted because of the surface tension and buoyancy of the processing fluid (Bastani et al., 2013; Carreon et al., 2017).

Particle distribution in the polymer matrix must be homogeneous, which can be achieved by dispersing particles in the solvent using the ultrasonication horn technique, before adding the polymer. Another technique is called "priming" which keeps particles dispersed in the polymer solution (Vu et al., 2003; Carreon et al., 2017). It has been observed that adding a small amount of polymer to the dispersed solution avoids agglomeration of particles (Carreon et al., 2017). Also, we can add the inorganic particles in the solvent (instead of the solution); the resulting diluted suspension has a low viscosity which allows to have a high cutting speed and vigorous stirring to avoid agglomeration (Dong, Li and Chen, 2013).

Another factor affecting agglomeration is sedimentation, but this effect can be counteracted by using ultrafine crystalline particles that reduce the rate of sedimentation and increase the Viscosity dope or by working with polymeric and inorganic materials of the same polarity (Merkel et al., 2002; Mahajan et al., 2002; Dong, Li and Chen, 2013).

8.4.5.3.2 *Interfacial Morphology*

The morphology of the polymer–particle interface is a crucial factor that influences the gas transport properties. A weak interaction between the polymer and particles can result in a reduction in the separation performance of the membrane. There are three factors that affect poor interface morphology (Dong, Li and Chen, 2013).

    i. Low polymer–particle adhesion
    ii. Partial pore block of particles from polymer chains
    iii. Polymer chain rigidification (Dong, Li and Chen, 2013).

A lack of adhesion between the materials used results in the formation of non-selective voids in the interface region (Dong, Li and Chen, 2013); this is attributed to de-wetting of the polymer chains on the outer surface of the inorganic filler (Dong, Li and Chen, 2013).

The introduction of coupling agents (silane) to modify the particles can cause the blocking of the pores by the addition of additional agents. To avoid the problem, it can use the precursor of polyimide 3-amino-propyltriethoxysilane (APTES), which has improved interfacial morphology and separation performance (Fryčová et al., 2012).

Furthermore, a heat treatment to eliminate the tensile strength and the addition of wet particles and liquids to polymers are proposed techniques to improve the adhesion between materials (Dong, Li and Chen, 2013). Another approach involves establishing a nanoscale separation between the particles and the polymer to prevent direct contact between them (Dong, Li and Chen, 2013).

The rigidity of the polymer chain at the interface can be improved by better lamination of the polymer matrix, thus increasing the chain mobility (Dong, Li and Chen, 2013).

8.4.6 MODELS FOR GAS TRANSPORT

8.4.6.1 Gas Transport in Ideal MMMs

Most models describing transport in mixed membranes are treated as membranes that do not have defects or distortions in their structure. The classic models they describe are Maxwell, Bruggeman, Lewis–Nielsen, and Cussler models (Dong, Li and Chen, 2013).

8.4.6.1.1 Low Particle Loading Maxwell Model

The equation was initially formulated to characterize the electrical conductivity of composite materials, but it has been adapted to describe gas permeability in MMMs, as the gas transport in MMMs is analogous to the dielectric properties of composite materials (Dong, Li and Chen, 2013).

The equation describing the mechanism is given as follows:

$$P_{\text{eff}} = Pc\left[\frac{nPd + (1-n)Pc + (1-n)(Pd - Pc) + \phi_d}{nPd + (1-n)Pc - n(Pd - Pc)\phi_d}\right] \tag{8.10}$$

where

- P_{eff}: Effective MMM permeability
- n: Shape factor

- Pd: Permeability of the dispersed inorganic particles
- ϕ_d: Volume fraction of the dispersed phase
- Pc: Permeability of the continuous polymer phase

The equation is utilized when there are no morphological defects in the individual phases. The form factor "n" ranges from 0 to 1, representing transport through parallel plates ($n=0$), cylinder fillers ($n=1/6$), and ideal spherical particles ($n=1/3$). The Maxwell equation is applied for transport through series laminates ($n=1$) (Dong, Li and Chen, 2013).

The model doesn't consider the interaction between the polymer chain and inorganic nanofillers. In many cases, particularly with nanocomposite membranes, the interaction is robust, markedly altering the diffusivity and solubility of penetrants.

8.4.6.1.2 Medium Particle Loading (Bruggeman Model)

The equation was initially formulated to describe the dielectric constant of composite materials and is applicable for elucidating gas transport in MMMs with moderate to high particle loads. The following equation describes permeation through a dispersion of spherical particles (Dong, Li and Chen, 2013):

$$\frac{P_{\text{eff}}}{Pc} = \frac{1}{\left(1-\phi_d\right)^3}\left(\frac{\dfrac{P_{\text{eff}}}{Pc}-\dfrac{Pd}{Pc}}{1-\dfrac{Pd}{Pc}}\right)^3 \tag{8.11}$$

The model explains situations where the local flow around a particle is influenced by the presence of other particles, a scenario commonly encountered in high particle load conditions (Dong, Li and Chen, 2013). The limitations of this model are:

i. It does not estimate transport behavior when the particle load is approaching the maximum level.
ii. It requires a numerical solution for permeability to being implicit in nature.
iii. It does not consider particle shape and particle size distribution.

8.4.6.1.3 Maximum Particle Loading (Lewis–Nielsen and Pal Models)

It was used to predict transport behavior at maximum particle load, and the model was developed to initially describe the elastic modulus of polymer compounds reinforced with particulates. The equation is (Dong, Li and Chen, 2013):

$$\frac{P_{\text{eff}}}{Pc} = \frac{1+2\phi_d\dfrac{\left(\dfrac{Pd}{Pc}\right)-1}{\left(Pd/Pc\right)+2}}{1-\phi_d\psi\dfrac{\left(\dfrac{Pd}{Pc}\right)-1}{\left(Pd/Pc\right)+2}} \tag{8.12}$$

where

$$\psi = 1 + \left(\frac{1 - \phi_m}{\phi_m^2} \right) \phi_d \tag{8.13}$$

- ϕ_m: Maximum achievable volume fraction of the inorganic particles

In general

- $\phi_m = 0.59$ for loose random packing of uniform spheres
- $\phi_m = 0.64$ for random close packing of uniform spheres
- $\phi_m = 0.86$ for binary packing with spherical particles
- $\phi_m \approx 1$ the equation if reduced to that of Maxwell

Similar to the Lewis–Nielsen model, Pal's model is used to describe MMM permeation with very high particle load (Dong, Li and Chen, 2013):

$$\frac{P_{\text{eff}}}{Pc} = \frac{1}{1 - \phi_d + \left(\dfrac{1}{\delta \phi_d + \dfrac{1 - \phi_d}{\alpha^2 \phi_d^{\,2}}} \right)} \tag{8.14}$$

where

- δ: Ratio of the diffusion coefficients between polymer and flake
- α: Aspect ratio of the flakes

8.4.6.2 Transport in Non-Ideal Mixed Matrix Membranes

In real-world scenarios, the absence of interaction between a polymer matrix and inorganic particles in MMMs implies that the models mentioned earlier do not provide accurate insights into the gas transport behavior across the membrane (Dong, Li and Chen, 2013). Defects at the interface between the polymer and inorganic particles include:

   i. Formation of voids at the interface.
  ii. Rigidification of polymer chains.
 iii. Particle pore blockage by the polymer chain.

Table 8.11 describes the models used for simulating gas transport in MMMs.

8.4.7 MMMs in Gas Separation

8.4.7.1 For Air Separation

The production of oxygen from the use of membranes is still not much explored, due to two main factors:

   i. Small difference in kinetic diameters between O_2 (3.46 Å) and N_2 (3.64 Å).

TABLE 8.11

Summary of Models for Simulating the Gas Permeation Behavior in Non-Ideal Mixed Matrix Membranes (Dong, Li and Chen, 2013)

Description	Derivation of the Model	Equation	Ref.
Model developed to consider the formation of voids at the interface	The mixed matrix membrane consists of a continuous polymer phase and a pseudo phase containing voids and particles. The Maxwell equation is applied twice to develop the simulation model	$\dfrac{P_{\text{eff}}}{P_v} = \dfrac{Pd + 2Pv - 2\phi_d\left(Pv - Pd\right)}{Pd + 2Pv + \phi_d\left(Pv - Pd\right)}$	Mahajan and Koros (2002)
Model developed to consider the rigidification of the polymer chains	Incorporating a "chain immobilization factor" to consider the stiffening of polymer chains	$Pv = {Pc}/{\beta}$	Felske (2004)
Model developed to consider the blockage of the particle pores	Introducing a "permeability reduction factor" to consider the obstruction of particle pores by polymer chains	$Pv = {Pc}/{\beta'}$	Felske (2004)

TABLE 8.12

Promising Mixed Matrix Membranes with Potential for Air Separation (Dong, Li and Chen, 2013)

Material	Permeability O_2 (Barrer)	Permeability N_2 (Barrer)	Ref.
Matrimid+MOF-5	4.12	0.52	Perez et al. (2009)
PAN+Silica	25.00	1.67	Iwata et al. (2003)
Pure PSf	1.30	0.22	Jeong et al. (2004)
PEBAX+silica	11.30	3.52	Kim and Lee (2001)
Pure PAN	4.00	1.14	Iwata et al. (2003)

ii. Because there are conventional air separation techniques that offer a purity of O_2 of 99.999% for the distillation process and 95% for the vacuum oscillation adsorption process (Dong, Li and Chen, 2013).

However, there are several MMMs used in the field of air separation (Table 8.12).

8.4.7.2 CO_2 Separation

The separation of CO_2 from natural gas is very important in order to meet the established characteristics of the pipes. It does the separation of $CO_2/$ CH_4, thanks to the

TABLE 8.13

Promising Mixed Matrix Membranes with Potential Natural Gas Purification (Dong, Li and Chen, 2013)

Material	Permeability CO_2 (Barrer)	Permeability CH_4 (Barrer)	References
Pure PSf	6.30	0.22	Jeong et al. (2004)
Pure ABS	2.87	0.12	Anson et al. (2004)
Pure Matrimid	7.29	0.21	Vu, Koros and Miller (2003)
ABS+AC-2	20.50	0.41	Anson et al. (2004)
PSf+AlPO flakes	51.00	1.30	Jeong et al. (2004)

fact that CO_2 has a high condensability and the moment of quadrupole allows to have different separation mechanisms. Table 8.13 shows several MMMs that are used to purify and separate CO_2.

8.4.7.3 H_2 Separation

It is the primary area of industrial-scale application for membrane gas separation. One of the key characteristics is that the majority of polymer membranes enable the passage of hydrogen while retaining larger molecules present in the feed stream. However, this necessitates rearranging the H_2. The issue described with traditional polymer membranes has led to the development of membranes known as reverse-selective membranes such as PMP (Poly(4-methyl-2-pentyne)) and PTMSP (Poly(1-trimethylsilyl-1-propyne)), which are structures that allow the most condensed gases (CO_2) to pass by keeping the smallest molecules (H_2), reducing pressure loss at the H_2 outlet (Dong, Li and Chen, 2013).

Table 8.14 shows several MMMs used for hydrogen separation.

TABLE 8.14

MMMs for Hydrogen Separation

MMMs	P (Bar)	T (°C)	$P(H_2)$ (Barrer)	Selectivity H_2/CO_2	H_2/CH_4	CO_2/H_2	H_2/N_2	References
VTEC(Polyimides)	5	35	5.0	5.0	–	–	–	Perez et al. (2017)
VTEC(Polyimides)	30	35	4.5	6.0	–	–	–	Perez et al. (2017)
VTEC(Polyimides)	5	300	86	4.0	–	–	–	Perez et al. (2017)
VTEC(Polyimides)	30	300	85	4.0	–	–	–	Perez et al. (2017)
NH$_2$-MIL-53/VTEC	5	35	5.4	5.4	–	–	–	Perez et al. (2017)

(Continued)

TABLE 8.14 (*Continued*)
MMMs for Hydrogen Separation

MMMs	P (Bar)	T (°C)	P(H$_2$) (Barrer)	H$_2$/CO$_2$	H$_2$/CH$_4$	CO$_2$/H$_2$	H$_2$/N$_2$	References
NH$_2$-MIL-53/VTEC	30	35	5.1	7.0	–	–	–	Perez et al. (2017)
NH$_2$-MIL-53/VTEC	5	300	157	5.7	–	–	–	Perez et al. (2017)
NH$_2$-MIL-53/VTEC	30	300	144	5.8	–	–	–	Perez et al. (2017)
Matrimid/Cu-BPY-HPFS	3.6	35	26.74	1.78	45.3	–	–	Zhang, Balkus, et al. (2008)
Matrimid/MOF-5	2	35	53.8	2.66	120	–	–	Perez et al. (2009)
Matrimid/ZIF-8	5	25	44.7	3.33	126.27	–	–	Kamble, Patel and Murthy (2021)
Matrimid/Pd@ZIF-8	5	25	68.85	5.05	201.1	–	–	Kamble, Patel and Murthy (2021)
6FDA-DAM-ZIF-11	–	–	106.70	0.97	30.48	–	–	Safak Boroglu and Yumru (2017)
6FDA-DAM	–	–	21.40	1.03	33.96	–	–	Safak Boroglu and Yumru (2017)
Polybenzimidazole/ZIF-8	-	–	64.5	12.3	–	–	–	Yang, Shi and Chung (2012)
Polyethersulfone/MWCNTs	-	–	70	4.4	–	–	–	Ghomshani et al. (2016)
Cu-BPY-HFS/Matrimid	–	35	17.50	2.40	83.32	–	–	Zhang, Musselman, et al. (2008)
PSf/Cu$_3$(BTC)$_2$ MOF	10	25	876	10.0	128	–	–	Hu et al. (2010)
MOF-5/Matrimid	2	35	53.8	–	120.0	–	–	Perez et al. (2009)
Matrimid-ZIF-8	–	60	28.39	–	27.3	–	–	Song et al. (2012)
Polyimide/silica	–	–	41	–	–	4.0	1.33	Smaihi et al. (1999)
PAI/TiO$_2$	–	–	66.3	–	51	0.65		Hu et al. (1997)
Matrimid/MgO	–	–	–	4.60	90.1	–	–	Hosseini et al. (2007)
PSf/MgO	–	–	–	–	42.31	–	–	Ahn et al. (2010)

(Continued)

TABLE 8.14 (*Continued*)
MMMs for Hydrogen Separation

MMMs	P (Bar)	T (°C)	P(H_2) (Barrer)	Selectivity				References
				H_2/CO_2	H_2/CH_4	CO_2/H_2	H_2/N_2	
Polyimide Matrimid/ mesoporous silicate spheres	–	–	26.5	79.2	–	–	–	Zornoza, Téllez and Coronas (2011)
NH_2-MIL-53/PI	–	–	0.45	7.7	–	–	–	Perez et al. (2017)
Zeolite FAU/ Matrimid	–	–	6	10.3	–	–	–	Mundstock, Friebe and Caro (2017)
ZIF/Matrimid	–	–	0.6	4.4	–	–	–	Ordoñez et al. (2010)
UiO-66/PI	–	–	2.5	5.1	–	–	–	Friebe et al. (2017)
Cyclic olefin copolymer/GO	–	–	0.05	40	–	–	–	Doğu and Ercan (2016)
Silica/PPO	–	–	23	3.6	–	–	–	Zhuang, Wey and Tseng (2015)
$Cu_3(BTC)_2$/PI	–	–	420	27.8	–	–	–	Hu et al., (2010)

8.5 CONCLUSION

This chapter shows a global overview of the polymer and inorganic membranes used for the separation and purification of the hydrogen current, as well as how we can use the advantages of these membranes to reduce the problems they have when only used for separation. This idea leads to the development of a new structure called MMMs.

The hydrogen current is a renewable, alternative, and above all, environmentally friendly energy source thanks to the separation/purification process adopted by MMMs. Presently, various technologies are employed for hydrogen production and separation, including existing methods such as PSA (Pressure Swing Adsorption) and cryogenic distillation. However, membrane technology stands out as an appealing approach to contribute to the future of the hydrogen economy. Currently, there are several membranes on a commercial scale, which have the ability to separate and purify the current of hydrogen; these membranes are organic and inorganic, and their structures have several advantages but also some disadvantages that limit their use. The main objective of the researchers is to be able to develop membranes that show solutions in a structure with different characteristics (advantages) of polymer membranes (ease of preparation, low cost, and wide availability of materials) together with inorganic membranes (great separation capacity of hydrogen, high mechanical and thermal resistance); this idea of being able to unify these two membranes in a single structure is the so-called MMM.

The chapter describes in a simplified and clear manner the different membranes (organic, inorganic, and MMM) used for the separation of hydrogen; indeed, it explains different production methods, the transport mechanisms involved in passing the permeant flow through the membrane, the different advantages and disadvantages that the membranes offer, and finally, it shows for each membrane a table with different structures applied to the field of gas separation.

As we have described, MMM structures contain a dispersed phase (filler-inorganic) and a continuous phase (organic/polymeric). The inorganic fillers available to be used for the preparation of MMM are hybrid, zeolite, MOF, COF, and graphene, which are materials that can be applied to the polymer matrix. But there are several characteristics that must possess membranes to be used in the manufacture of MMM. For example, inorganic fillers must show their benefits such as physicochemical properties, morphology, porosity, pore size, and ability to adhere to the matrix (polymer). During the manufacturing process of membranes, they present different challenges related to the aggregation of charges in the polymer matrix during the sizing of the hierarchical structure and non-ideal interfacial effects. These challenges are addressed in the chapter by indicating its consequences and possible solutions (such as mixing, polymerization in situ, sol–gel, and use of additives), in order to obtain an MMM with the desired characteristics.

It also provides a critical analysis of the hydrogen separation performance of hybrid membranes present in the specialized literature, evaluating in particular the permeability and selectivity properties of various materials.

Finally, different mechanisms that can explain the process of hydrogen permeation/separation using hybrid membranes have also been considered, describing in detail the Maxwell, Bruggeman, and Lewis–Nielsen models, as an ideal model, but in practice, the interaction of polymer–inorganic particle is not perfect, which leads to the formation of gaps between them, but it also describes several mechanisms that explain the transport of gas in the non-ideal case.

In conclusion, the addition of inorganic charges dispersed in the polymer matrix has become a growing field of research, but further efforts are still needed to develop MMMs as an established technology to be enhanced industrially.

ACRONYMS

Tg	Transition temperature
PC	Polycarbonates
CA	Cellulose acetate
PI	Polyimides
PS	Polysulfone
PPO	Poly(methylene oxide)
PDMS	Poly(dimethyl siloxane)
PIM	Intrinsically microporous polymers
IP	Interfacial polymerization
TFC	Thin film compounds

OR	Reverse osmosis
NF	Nanofiltration
PES	Polyether sulfone
PSA	Pressure swing adsorption
OEA	Oxygen-enriched air
PEBAX	Poly (ether block amide)
PMMs	Poly (methyl methacrylate)
PTMSP	Poly (1-trimethylsilyl-1-propyne)
CVD	Chemical vapor deposition
SEM	Scanning electron microscopy
MMM	Mixed matrix membrane
MOF	Metal organic framework
CNT	Carbon nanotubes
SWCNT	Single-wall nanotubes
MWCNT	Multiple-wall nanotubes

REFERENCES

Aberg, C.M. (2006) 'Membranes for hydrogen purification: An important step toward a hydrogen-based economy', *MRS Bulletin*, 31(10), pp. 735–741. Available at: https://doi.org/10.1557/mrs2006.186.

Adams, B.D. and Chen, A. (2011) 'The role of palladium in a hydrogen economy', *Materials Today*, 14(6), pp. 282–289.

Ahmad, J. and Hägg, M.B. (2013) 'Preparation and characterization of polyvinyl acetate/zeolite 4A mixed matrix membrane for gas separation', *Journal of Membrane Science*, 427, pp. 73–84. Available at: https://doi.org/10.1016/j.memsci.2012.09.036.

Ahmadd Fauzii Ismaill, K.C.K.T.M. (2015) *Gas Separation Membranes Polymeric and Inorganic*. New York, Dordrecht, London: Springer.

Ahn, J. et al. (2010) 'Gas transport behavior of mixed-matrix membranes composed of silica nanoparticles in a polymer of intrinsic microporosity (PIM-1)', *Journal of Membrane Science*, 346(2), pp. 280–287. Available at: https://doi.org/10.1016/j.memsci.2009.09.047.

Amin, M. et al. (2023) 'Issues and challenges in hydrogen separation technologies', *Energy Reports*, 9, pp. 894–911. Available at: https://doi.org/10.1016/j.egyr.2022.12.014.

Anson, M. et al. (2004) 'ABS copolymer-activated carbon mixed matrix membranes for CO_2/CH_4 separation', *Journal of Membrane Science*, 243(1–2), pp. 19–28. Available at: https://doi.org/10.1016/j.memsci.2004.05.008.

Baker, R.W. (2002) 'Future directions of membrane gas separation technology', *Industrial & Engineering Chemistry Research*, 41(6), pp. 1393–1411. Available at; https://doi.org/10.1021/ie0108088.

Baker, R.W. and Lokhandwala, K. (2008) 'Natural gas processing with membranes: An overview', *Industrial and Engineering Chemistry Research*, 47(7), pp. 2109–2121. Available at: https://doi.org/10.1021/ie071083w.

Balachandran, U. et al. (2006) 'Hydrogen separation by dense cermet membranes', *Fuel*, 85(2), pp. 150–155. Available at: https://doi.org/10.1016/j.fuel.2005.05.027.

Bastani, D., Esmaeili, N. and Asadollahi, M. (2013) 'Polymeric mixed matrix membranes containing zeolites as a filler for gas separation applications: A review', *Journal of Industrial and Engineering Chemistry*, 19(2), pp. 375–393. Available at: https://doi.org/10.1016/j.jiec.2012.09.019.

Brisotto, M. et al. (2013) 'Evoluzione della microstruttura e dei microstrain dello strato funzionale di Pd in membrane dense per la separazione dell'idrogeno Parole chiave: Rivestimenti-Trattamenti termici-Diffrattometria-Energia-Idrogeno', *La Metallurgia Italiana*.

Budd, P.M. and McKeown, N.B. (2010) 'Highly permeable polymers for gas separation membranes', *Polymer Chemistry*, 1(1), pp. 63–68. Available at: https://doi.org/10.1039/b9py00319c.

Cardoso, S.P. et al. (2018) 'Inorganic membranes for hydrogen separation', *Separation & Purification Reviews*, 47(3), pp. 229–266. Available at: https://doi.org/10.1080/15422 119.2017.1383917.

Caro, J. et al. (2000) 'Zeolite membranes—State of their development and perspective', *Microporous and Mesoporous Materials*, 38(1), pp. 3–24. Available at: https://doi. org/10.1016/S1387-1811(99)00295-4.

Carreon, M. et al. (2017) *Mixed Matrix Membranes for Gas Separation Applications.* USA. Available at: www.worldscientific.com.

Castarlenas, S., Téllez, C. and Coronas, J. (2017) 'Gas separation with mixed matrix membranes obtained from MOF UiO-66-graphite oxide hybrids', *Journal of Membrane Science*, 526, pp. 205–211. Available at: https://doi.org/10.1016/j.memsci.2016.12.041.

Cheng, L. et al. (2020) 'Cysteamine-crosslinked graphene oxide membrane with enhanced hydrogen separation property', *Journal of Membrane Science*, 595, 117568. Available at: https://doi.org/10.1016/j.memsci.2019.117568.

Cheng, Y. et al. (2018) 'Advanced porous materials in mixed matrix membranes', *Advanced Materials*, 30(47), 1802401. Available at: https://doi.org/10.1002/adma.201802401.

Chiappetta, G., Clarizia, G. and Drioli, E. (2006) 'Design of an integrated membrane system for a high level hydrogen purification', *Chemical Engineering Journal*, 124(1–3), pp. 29–40. Available at: https://doi.org/10.1016/j.cej.2006.08.009.

Choi, S. et al. (2008) 'Fabrication and gas separation properties of polybenzimidazole (PBI)/ nanoporous silicates hybrid membranes', *Journal of Membrane Science*, 316(1–2), pp. 145–152. Available at: https://doi.org/10.1016/j.memsci.2007.09.026.

Chuah, C.Y. et al. (2021) 'Recent progress in mixed-matrix membranes for hydrogen separation', *Membranes*, 11(9), 666. Available at: https://doi.org/10.3390/membranes11090666.

Chung, T.-S. et al. (2007) 'Mixed matrix membranes (MMMs) comprising organic polymers with dispersed inorganic fillers for gas separation', *Progress in Polymer Science*, 32(4), pp. 483–507. Available at: https://doi.org/10.1016/j.progpolymsci.2007.01.008.

Conde, J.J., Maroño, M. and Sánchez-Hervás, J.M. (2017) 'Pd-based membranes for hydrogen separation: Review of alloying elements and their influence on membrane properties', *Separation and Purification Reviews*, 46(2), 152–177. Available at: https://doi.org/10.1 080/15422119.2016.1212379.

De, D. (2017) *Overview on Porous Inorganic Membranes for Gas Separation.* Italy: ENEA. Available at: https://www.enea.it/it/produzione-scientifica/rapporti-tecnici.

Diestel, L. et al. (2015) 'MOF based MMMs with enhanced selectivity due to hindered linker distortion', *Journal of Membrane Science*, 492, pp. 181–186. Available at: https://doi. org/10.1016/j.memsci.2015.04.069.

Doğu, M. and Ercan, N. (2016) 'High performance cyclic olefin copolymer (COC) membranes prepared with melt processing method and using of surface modified graphitic nano-sheets for H2/CH4 and H_2/CO_2 separation', *Chemical Engineering Research and Design*, 109, pp. 455–463. Available at: https://doi.org/10.1016/j.cherd.2016.02.021.

Dong, G., Li, H. and Chen, V. (2013) 'Challenges and opportunities for mixed-matrix membranes for gas separation', *Journal of Materials Chemistry A*, 1(15), pp. 4610–4630. Available at: https://doi.org/10.1039/c3ta00927k.

Dunn, S. (2002) 'Hydrogen futures: Toward a sustainable energy system', *International Journal of Hydrogen Energy*, 27(3), pp. 235–264. Available at: https://doi.org/10.1016/S0360-3199(01)00131-8.

Eden, C.L. and Daramola, M.O. (2021) 'Evaluation of silica sodalite infused polysulfone mixed matrix membranes during H_2/CO_2 separation', In Materials Today: Proceedings. Elsevier Ltd, pp. 522–527. Available at: https://doi.org/10.1016/j.matpr.2020.02.393.

Felske, J.D. (2004) 'Effective thermal conductivity of composite spheres in a continuous medium with contact resistance', *International Journal of Heat and Mass Transfer*, 47(14–16), pp. 3453–3461. Available at: https://doi.org/10.1016/j.ijheatmasstransfer.2004.01.013.

Friebe, S. et al. (2017) 'Metal-organic framework UiO-66 layer: A highly oriented membrane with good selectivity and hydrogen permeance', *ACS Applied Materials and Interfaces*, 9(14), pp. 12878–12885. Available at: https://doi.org/10.1021/acsami.7b02105.

Fryčová, M. et al. (2012) 'Mixed matrix membranes based on 3-aminopropyltriethoxysilane endcapped polyimides and silicalite-1', *Journal of Applied Polymer Science*, 124(Suppl. 1), pp. E233–E240. Available at: https://doi.org/10.1002/app.36466.

Funk, C. V. and Lloyd, D.R. (2008) 'Zeolite-filled microporous mixed matrix (ZeoTIPS) membranes: Prediction of gas separation performance', *Journal of Membrane Science*, 313(1–2), pp. 224–231. Available at: https://doi.org/10.1016/j.memsci.2008.01.002.

Ghasemzadeh, K. et al. (2020) 'Membranes for hydrogen separation', In *Current Trends and Future Developments on (Bio-) Membranes: Membrane Systems for Hydrogen Production*. UK: Elsevier, pp. 91–134. Available at: https://doi.org/10.1016/B978-0-12-817110-3.00004-7.

Ghomshani, A.D. et al. (2016) 'Improvement of H_2/CH_4 separation performance of PES hollow fiber membranes by addition of MWCNTs into polymeric matrix', *Polymer: Plastics Technology and Engineering*, 55(11), pp. 1155–1166. Available at: https://doi.org/10.1080/03602559.2015.1132460.

Goh, P.S. et al. (2011) 'Recent advances of inorganic fillers in mixed matrix membrane for gas separation', *Separation and Purification Technology*, 81(3), pp. 243–264. Available at: https://doi.org/10.1016/j.seppur.2011.07.042.

Hamid, M.R.A. and Jeong, H.K. (2018) 'Recent advances on mixed-matrix membranes for gas separation: Opportunities and engineering challenges', *Korean Journal of Chemical Engineering*, 35, pp. 1577–1600. Available at: https://doi.org/10.1007/s11814-018-0081-1.

Hosseini, S.S. et al. (2007) 'Enhanced gas separation performance of nanocomposite membranes using MgO nanoparticles', *Journal of Membrane Science*, 302(1–2), pp. 207–217. Available at: https://doi.org/10.1016/j.memsci.2007.06.062.

Hsieh, H.P. (1996) *Inorganic Membranes for Separation and Reaction*, Vol. 3, pp. 1–487, USA: Elsevier Science.

Hu, J. et al. (2010) 'Mixed-matrix membrane hollow fibers of $Cu_3(BTC)_2$ MOF and polyimide for gas separation and adsorption', *Industrial and Engineering Chemistry Research*, 49(24), pp. 12605–12612. Available at: https://doi.org/10.1021/ie1014958.

Hu, L. et al. (2017) 'Composites of ionic liquid and amine-modified SAPO 34 improve CO 2 separation of CO_2-selective polymer membranes', *Applied Surface Science*, 410, pp. 249–258. Available at: https://doi.org/10.1016/j.apsusc.2017.03.045.

Hu, Q. et al. (1997) 'Poly(amide-imide)/TiO2 nano-composite gas separation membranes: Fabrication and characterization', *Journal of Membrane Science*, 135(1), pp. 67–79.

Husain, S. and Koros, W.J. (2007) 'Mixed matrix hollow fiber membranes made with modified HSSZ-13 zeolite in polyetherimide polymer matrix for gas separation', *Journal of Membrane Science*, 288(1–2), pp. 195–207. Available at: https://doi.org/10.1016/j.memsci.2006.11.016.

Ismail, A. and David, L. (2001) 'A review on the latest development of carbon membranes for gas separation', *Journal of Membrane Science*, 193(1), pp. 1–18. Available at: https://doi.org/10.1016/S0376-7388(01)00510-5.

Ismail, A.F., Kusworo, T.D. and Mustafa, A. (2008) 'Enhanced gas permeation performance of polyethersulfone mixed matrix hollow fiber membranes using novel Dynasylan Ameo silane agent', *Journal of Membrane Science*, 319(1–2), pp. 306–312. Available at: https://doi.org/10.1016/j.memsci.2008.03.067.

Iulianelli, A., and Drioli, E. (2020) 'Membrane engineering: latest advancements in gas separation and pre-treatment processes, petrochemical industry and refinery, and future perspectives in emerging applications', *Fuel Processing Technology*, 206, pp. 106464–106497. Available at: https://doi.org/10.1016/j.fuproc.2020.106464.

Iwata, M. et al. (2003) 'Hybrid sol-gel membranes of polyacrylonitrile-tetraethoxysilane composites for gas permselectivity', *Journal of Applied Polymer Science*, 88(7), pp. 1752–1759. Available at: https://doi.org/10.1002/app.11895.

Javaid, A. (2005) 'Membranes for solubility-based gas separation applications', *Chemical Engineering Journal*, 112(1–3), pp. 219–226. Available at: https://doi.org/10.1016/j.cej.2005.07.010.

Jeong, H.K. et al. (2004) 'Fabrication of polymer/selective-flake nanocomposite membranes and their use in gas separation', *Chemistry of Materials*, 16(20), pp. 3838–3845. Available at: https://doi.org/10.1021/cm049154u.

Jiang, L.Y. et al. (2005) 'Fundamental understanding of nano-sized zeolite distribution in the formation of the mixed matrix single- and dual-layer asymmetric hollow fiber membranes', *Journal of Membrane Science*, 252(1–2), pp. 89–100. Available at: https://doi.org/10.1016/j.memsci.2004.12.004.

Jiang, L.Y., Chung, T.S. and Kulprathipanja, S. (2006) 'An investigation to revitalize the separation performance of hollow fibers with a thin mixed matrix composite skin for gas separation', *Journal of Membrane Science*, 276(1–2), pp. 113–125. Available at: https://doi.org/10.1016/j.memsci.2005.09.041.

Kajiwara, M. et al. (2000) 'Hydrogen permeation properties through composite membranes of platinum supported on porous alumina', *Catalysis Today*, 56(1–3), pp. 65–73. Available at: https://doi.org/10.1016/S0920-5861(99)00263-1.

Kamble, A.R., Patel, C.M. and Murthy, Z.V.P. (2021) 'A review on the recent advances in mixed matrix membranes for gas separation processes', *Renewable and Sustainable Energy Reviews*. 145, p. 111062. Available at: https://doi.org/10.1016/j.rser.2021.111062.

Keizer, K. and Verweij, H. (1996) 'Progress in inorganic membranes'. *Chemtech*, 26(1), pp. 37–41.

Kim, J.H. and Lee, Y.M. (2001) 'Gas permeation properties of poly(amide-6-b-ethylene oxide)-silica hybrid membranes', *Journal of Membrane Science*, 192(2), pp. 209–225.

Lin, H. et al. (2018) 'Permselective H_2/CO_2 separation and desalination of hybrid GO/rGO membranes with controlled pre-cross-linking', *ACS Applied Materials and Interfaces*, 10(33), pp. 28166–28175. Available at: https://doi.org/10.1021/acsami.8b05296.

Mahajan, R. et al. (2002) 'Challenges in forming successful mixed matrix membranes with rigid polymeric materials', *Journal of Applied Polymer Science*, 86(4), pp. 881–890. Available at: https://doi.org/10.1002/app.10998.

Mahajan, R. and Koros, W.J. (2002) 'Mixed matrix membrane materials with glassy polymers. Part 2, *Polymer Engineering & Science*, 42(7), pp. 1432–1441.

Merkel, T.C. et al. (2002) 'Ultrapermeable, reverse-selective nanocomposite membranes', *Science*, 296(5567), pp. 519–522. Available at: https://science.sciencemag.org/.

Møller, K.T. et al. (2017) 'Hydrogen—A sustainable energy carrier', *Progress in Natural Science: Materials International*, 27(1), pp. 34–40. Available at: https://doi.org/10.1016/j.pnsc.2016.12.014.

Momirlan, M. and Veziroglu, T.N. (2005) 'The properties of hydrogen as fuel tomorrow in sustainable energy system for a cleaner planet', *International Journal of Hydrogen Energy*, 30(7), pp. 795–802. Available at: https://doi.org/10.1016/j.ijhydene.2004.10.011.

Moore, T.T. and Koros, W.J. (2005) 'Non-ideal effects in organic-inorganic materials for gas separation membranes', *Journal of Molecular Structure*, 739(1–3), pp. 87–98. Available at: https://doi.org/10.1016/j.molstruc.2004.05.043.

Morooka, S. and Kusakabe, K. (1999) 'Microporous inorganic membranes for gas separation', *MRS Bulletin*, 24(3), pp. 25–29.

Mulder, M. (1997) *Basic Principles Membrane Technology*, second edition. Netherlands: Kluwer Academic Publishers.

Mundstock, A., Friebe, S. and Caro, J. (2017) 'On comparing permeation through Matrimid(r)-based mixed matrix and multilayer sandwich FAU membranes: H_2/CO_2 separation, support functionalization and ion exchange', *International Journal of Hydrogen Energy*, 42(1), pp. 279–288. Available at: https://doi.org/10.1016/j.ijhydene.2016.10.161.

Ockwig, N.W. and Nenoff, T.M. (2007) 'Membranes for hydrogen separation', *Chemical Reviews*, 107(10), pp. 4078–4110. Available at: https://doi.org/10.1021/cr0501792.

Ordoñez, M.J.C. et al. (2010) 'Molecular sieving realized with ZIF-8/Matrimid(r) mixed-matrix membranes', *Journal of Membrane Science*, 361(1–2), pp. 28–37. Available at: https://doi.org/10.1016/j.memsci.2010.06.017.

Pal, N. et al. (2020) 'A review on types, fabrication and support material of hydrogen separation membrane', *Materials Today: Proceedings*, 28, pp. 1386–1391. Available at: https://doi.org/10.1016/j.matpr.2020.04.806.

Pal, N. and Agarwal, M. (2021) 'Advances in materials process and separation mechanism of the membrane towards hydrogen separation', *International Journal of Hydrogen Energy*, 46(53), pp. 27062–27087. Available at: https://doi.org/10.1016/j.ijhydene.2021.05.175.

Pandey, P. and Chauhan, R.S. (2001) 'Membranes for gas separation', *Progress in Polymer Science*, 26(6), pp. 853–893. Available at: www.elsevier.com/locate/ppolysci.

Pechar, T.W. et al. (2006) 'Fabrication and characterization of polyimide-zeolite L mixed matrix membranes for gas separations', *Journal of Membrane Science*, 277(1–2), pp. 195–202. Available at: https://doi.org/10.1016/j.memsci.2005.10.029.

Perez, E.V. et al. (2009) 'Mixed-matrix membranes containing MOF-5 for gas separations', *Journal of Membrane Science*, 328(1–2), pp. 165–173. Available at: https://doi.org/10.1016/j.memsci.2008.12.006.

Perez, E.V. et al. (2017) 'Amine-functionalized (Al) MIL-53/VTECTM mixed-matrix membranes for H2/CO2 mixture separations at high pressure and high temperature', *Journal of Membrane Science*, 530, pp. 201–212. Available at: https://doi.org/10.1016/j.memsci.2017.02.003.

Peydayesh, M., Mohammadi, T. and Bakhtiari, O. (2017) 'Effective hydrogen purification from methane via polyimide Matrimid(r) 5218- Deca-dodecasil 3R type zeolite mixed matrix membrane', *Energy*, 141, pp. 2100–2107. Available at: https://doi.org/10.1016/j.energy.2017.11.101.

Ricci, E. and De Angelis, M.G. (2019) 'Modelling mixed-gas sorption in glassy polymers for CO_2 removal: A sensitivity analysis of the dual mode sorption model', *Membranes*, 9(1), p. 8. Available at: https://doi.org/10.3390/membranes9010008.

Safak Boroglu, M. and Yumru, A.B. (2017) 'Gas separation performance of 6FDA-DAM-ZIF-11 mixed-matrix membranes for H_2/CH_4 and CO_2/CH_4 separation', *Separation and Purification Technology*, 173, pp. 269–279. Available at: https://doi.org/10.1016/j.seppur.2016.09.037.

Sánchez-Laínez, J. et al. (2015) 'Beyond the H2/CO2 upper bound: One-step crystallization and separation of nano-sized ZIF-11 by centrifugation and its application in mixed matrix membranes', *Journal of Materials Chemistry A*, 3(12), pp. 6549–6556. Available at: https://doi.org/10.1039/c4ta06820c.

Sánchez-Laínez, J. et al. (2016) 'Influence of ZIF-8 particle size in the performance of polybenzimidazole mixed matrix membranes for pre-combustion CO_2 capture and its validation through interlaboratory test', *Journal of Membrane Science*, 515, pp. 45–53. Available at: https://doi.org/10.1016/j.memsci.2016.05.039.

Sanders, D.F. et al. (2013) 'Energy-efficient polymeric gas separation membranes for a sustainable future: A review', *Polymer*, 54(18), pp. 4729–4761. Available at: https://doi.org/10.1016/j.polymer.2013.05.075.

Shao, L. et al. (2009) 'Polymeric membranes for the hydrogen economy: Contemporary approaches and prospects for the future', *Journal of Membrane Science*, 327(1–2), pp. 18–31. Available at: https://doi.org/10.1016/j.memsci.2008.11.019.

Shen, J. et al. (2016) 'Subnanometer two-dimensional graphene oxide channels for ultrafast gas sieving', *ACS Nano*, 10(3), pp. 3398–3409. Available at: https://doi.org/10.1021/acsnano.5b07304.

Shiraz, H. G. and Shiraz, M.G. (2017) 'Palladium nanoparticle and decorated carbon nanotube for electrochemical hydrogen storage', *International Journal of Hydrogen Energy*, 42(16), pp. 11528–11533. Available at: https://doi.org/10.1016/j.ijhydene.2017.03.129.

Valappil, R.S., Ghasem, N. and Al-Marzouqi, M. (2021) 'Current and future trends in polymer membrane-based gas separation technology: A comprehensive review', *Journal of Industrial and Engineering Chemistry. Korean Society of Industrial Engineering Chemistry*, 98, pp. 103–129. Available at: https://doi.org/10.1016/j.jiec.2021.03.030.

Smaihi, M. et al. (1999) 'Gas separation properties of hybrid imide–siloxane copolymers with various silica contents', *Journal of Membrane Science*, 161(1–2), pp. 95–102.

Song, Q. et al. (2012) 'Zeolitic imidazolate framework (ZIF-8) based polymer nanocomposite membranes for gas separation', *Energy and Environmental Science*, 5(8), pp. 8359–8369. Available at: https://doi.org/10.1039/c2ee21996d.

Swaidan, R. et al. (2014) 'Pure- and mixed-gas CO2/CH4 separation properties of PIM-1 and an amidoxime-functionalized PIM-1', *Journal of Membrane Science*, 457, pp. 95–102. Available at: https://doi.org/10.1016/j.memsci.2014.01.055.

Tanh Jeazet, H.B., Staudt, C. and Janiak, C. (2012) 'Metal-organic frameworks in mixed-matrix membranes for gas separation', *Dalton Trans.*, 41(46), pp. 14003–14027. Available at: https://doi.org/10.1039/c2dt31550e.

U.S. Department of Energy (2005) 'Small business innovation research program and small business technology transfer program FY 2005 solicitations, technical topic descriptions', US DOE, Materials Research, Office of Fossil Energy [Preprint].

Vu, D.Q., Koros, W.J. and Miller, S.J. (2003) 'Mixed matrix membranes using carbon molecular sieves I. Preparation and experimental results', *Journal of Membrane Science*, 211(2), pp. 311–334.

Weber, M. et al. (2020) 'Hydrogen selective palladium-alumina composite membranes prepared by atomic layer deposition', *Journal of Membrane Science*, 596, p. 117701. Available at: https://doi.org/10.1016/j.memsci.2019.117701.

Weng, T.H., Tseng, H.H. and Wey, M.Y. (2009) 'Preparation and characterization of multi-walled carbon nanotube/PBNPI nanocomposite membrane for H2/CH4 separation', *International Journal of Hydrogen Energy*, 34(20), pp. 8707–8715. Available at: https://doi.org/10.1016/j.ijhydene.2009.08.027.

Wijenayake, S.N. et al. (2013) 'Surface cross-linking of ZIF-8/polyimide mixed matrix membranes (MMMs) for gas separation', *Industrial and Engineering Chemistry Research*, 52(21), pp. 6991–7001. Available at: https://doi.org/10.1021/ie400149e.

Yampolskii, Y. (2012) 'Polymeric gas separation membranes', *Macromolecules*, 45(8), pp. 3298–3311. Available at: https://doi.org/10.1021/ma300213b.

Yang, T. and Chung, T.-S. (2013) 'Room-temperature synthesis of ZIF-90 nanocrystals and the derived nano-composite membranes for hydrogen separation', *Journal of Materials Chemistry A*, 1(19), p. 6081. Available at: https://doi.org/10.1039/c3ta10928c.

Yang, T., Shi, G.M. and Chung, T.S.C. (2012) 'Symmetric and asymmetric zeolitic imidazolate frameworks (ZIFs)/polybenzimidazole (PBI) nanocomposite membranes for hydrogen purification at high temperatures', *Advanced Energy Materials*, 2(11), pp. 1358–1367. Available at: https://doi.org/10.1002/aenm.201200200.

Yang, T., Xiao, Y. and Chung, T.-S. (2011) 'Poly-/metal-benzimidazole nano-composite membranes for hydrogen purification', *Energy & Environmental Science*, 4(10), p. 4171. Available at: https://doi.org/10.1039/c1ee01324f.

Yun, S. and Ted Oyama, S. (2011) 'Correlations in palladium membranes for hydrogen separation: A review', *Journal of Membrane Science*, 375(1–2), pp. 28–45. Available at: https://doi.org/10.1016/j.memsci.2011.03.057.

Zhang, Y., Balkus, K.J., et al. (2008) 'Mixed-matrix membranes composed of Matrimid(r) and mesoporous ZSM-5 nanoparticles', *Journal of Membrane Science*, 325(1), pp. 28–39. Available at: https://doi.org/10.1016/j.memsci.2008.04.063.

Zhang, Y., Musselman, I.H., et al. (2008) 'Gas permeability properties of Matrimid(r) membranes containing the metal-organic framework Cu-BPY-HFS', *Journal of Membrane Science*, 313(1–2), pp. 170–181. Available at: https://doi.org/10.1016/j.memsci.2008.01.005.

Zhao, D. et al. (2015) 'Effect of graphene oxide on the behavior of poly(amide-6-b-ethylene oxide)/graphene oxide mixed-matrix membranes in the permeation process', *Journal of Applied Polymer Science*, 132(41). Available at: https://doi.org/10.1002/app.42624.

Zhuang, G.L., Wey, M.Y. and Tseng, H.H. (2015) 'The density and crystallinity properties of PPO-silica mixed-matrix membranes produced via the in situ sol-gel method for H_2/CO_2 separation. II: Effect of thermal annealing treatment', *Chemical Engineering Research and Design*, 104, pp. 319–332. Available at: https://doi.org/10.1016/j.cherd.2015.08.020.

Zornoza, B. et al. (2015) 'Mixed matrix membranes based on 6FDA polyimide with silica and zeolite microsphere dispersed phases', *AIChE Journal*, 61(12), pp. 4481–4490. Available at: https://doi.org/10.1002/aic.15011.

Zornoza, B., Téllez, C. and Coronas, J. (2011) 'Mixed matrix membranes comprising glassy polymers and dispersed mesoporous silica spheres for gas separation', *Journal of Membrane Science*, 368(1–2), pp. 100–109. Available at: https://doi.org/10.1016/j.memsci.2010.11.027.

9 Hollow Fiber Membrane for Hydrogen Separation

Nayef Ghasem

9.1 INTRODUCTION

Hydrogen separation is essential for various industrial processes that require high-purity hydrogen gas. One of the most important uses of hydrogen is producing ammonia for fertilizer production. In this process, hydrogen reacts with nitrogen to produce ammonia. Any impurities in the hydrogen gas can harm the reaction, leading to lower yields and increased costs (Estes et al., 2020). It discusses various impurities that can be present in the feed stream, such as carbon monoxide, carbon dioxide, water, and sulfur, and their impact on the catalysts used in the reaction. The chapter also highlights different methods for removing impurities from the feed stream, such as adsorption and membrane separation.

Hydrogen separation is also essential to produce clean fuels, such as hydrogen fuel cells, which are being developed as an alternative to fossil fuels. Hydrogen fuel cells produce electricity by combining hydrogen and oxygen, with water as the only by-product (Qazi, 2022). For fuel cell applications, hydrogen gas must be of high purity to prevent contamination of the fuel cell membrane and ensure optimal performance (Wang et al., 2021). Other industrial processes that require high-purity hydrogen gas include electronics manufacturing, metal processing, and food processing (Dolci, 2018). In these industries, hydrogen gas is used as a reducing agent and for annealing, brazing, and other high-temperature processes. The presence of impurities in the hydrogen gas can lead to defects in the final product, resulting in increased waste and lower yields (Mahajan et al., 2022). Overall, the importance of hydrogen separation lies in its ability to provide high-purity hydrogen gas for various industrial processes, which can improve yields, reduce waste, and increase efficiency (Sazali, 2020). This review chapter overviews hydrogen separation technologies, including commercial processes and emerging technologies. It discusses the importance of high-purity hydrogen gas in various industrial processes, such as ammonia synthesis, petroleum refining, and fuel cell technology. The chapter also covers various hydrogen separation methods, including membrane separation, pressure swing adsorption (PSA), cryogenic distillation, and hybrid systems, and discusses their advantages, limitations, and applications. Additionally, the chapter highlights the importance of developing efficient and cost-effective hydrogen separation technologies to meet the increasing demand for high-purity hydrogen gas.

This chapter overviews hydrogen separation technologies, including membrane separation, PSA, cryogenic distillation, and hybrid systems. It discusses each technology's advantages, limitations, and applications, as well as their commercial

DOI: 10.1201/9781003382522-11

readiness and potential for further development. Additionally, the chapter highlights the importance of developing efficient and cost-effective hydrogen separation technologies to meet the increasing demand for clean energy and fuel cell applications. The chapter concludes by discussing the prospects and challenges of hydrogen separation (Vermaak, Neomagus and Bessarabov, 2021).

9.2 HYDROGEN SEPARATION TECHNIQUES

Hydrogen separation techniques are methods used to isolate and extract hydrogen gas from various sources, such as natural gas, biogas, and water. These techniques typically involve the use of membranes or chemical reactions to separate hydrogen from other gases or compounds. Some common hydrogen separation techniques include PSA, membrane separation, and electrolysis. These techniques have numerous applications in industries such as energy, transportation, and electronics, as hydrogen is a versatile and clean-burning fuel source with many potential uses.

9.2.1 PRESSURE SWING ADSORPTION

This PSA involves passing a mixture of hydrogen and other gases through a bed of adsorbent material, such as activated carbon, which selectively absorbs the impurities while allowing the hydrogen to pass through. The adsorbent bed is then regenerated by reducing the pressure, which desorbs the impurities and allows the hydrogen to be collected (Luberti and Ahn, 2022). It provides an overview of PSA for hydrogen purification, recovery, and production. It discusses the principles of PSA, including adsorption, desorption, and regeneration processes, and covers several types of adsorbent materials used in PSA, including activated carbon, zeolites, and metal–organic frameworks (MOFs). Additionally, the article discusses the advantages and limitations of PSA and its applications in various industries, such as petroleum refining, ammonia synthesis, and fuel cell technology (Rand and Dell, 2007).

9.2.2 MEMBRANE SEPARATION

This technology involves the use of a selectively permeable membrane that allows hydrogen to pass through while blocking other gases. Membranes can be made from various materials, including polymers, ceramics, and metals (Balachandran, Lee and Dorris, 2007; Lu et al., 2007; David and Kopac, 2011; Thursfield et al., 2012). The separation is driven by the pressure difference across the membrane (Huang, Yao and Cheng, 2017). Hydrogen separation by the membranc is a process that uses membranes to selectively separate hydrogen from other gases, such as nitrogen, carbon dioxide, and methane. The membranes used in this process are typically made of polymers, ceramics, or metals and have varying permeability and selectivity properties. The principle behind hydrogen separation by a membrane is that hydrogen molecules are small and can pass through the membrane, while other larger molecules are blocked (Yin and Yip, 2017). This technology is considered promising for various applications, such as hydrogen production, fuel cell technology, and natural gas purification. The efficiency and cost-effectiveness of hydrogen separation by membrane depend on several factors, including membrane type, operating conditions, and feed gas composition (Yue et al., 2021).

9.2.3 Cryogenic Distillation

This process involves cooling the gas mixture to shallow temperatures, which causes the hydrogen to condense into a liquid. The liquid hydrogen can then be separated from the impurities in the gas phase (Fernandes et al., 2013). The process of hydrogen separation by cryogenic distillation involves cooling the gas mixture to very low temperatures, typically below −150°C, using a cooling system such as a refrigeration unit or a Joule–Thomson expansion system. At these temperatures, hydrogen gas molecules lose their kinetic energy and become more closely packed, causing them to condense into a liquid phase. Other gases, such as nitrogen, oxygen, and carbon dioxide, do not condense at these low temperatures and remain in the gas phase (Aasadnia, Mehrpooya and Ghorbani, 2021). The mixture of liquid hydrogen and impurities is then passed through a series of distillation columns, which separate the liquid hydrogen from the impurities based on their different boiling points. The impurities, which have higher boiling points than hydrogen, condense on the trays or packing in the column, while the liquid hydrogen flows to the bottom of the column (Font-Palma, Cann and Udemu, 2021). The liquid hydrogen is collected in a separate container, while the impurities are collected at the bottom of the distillation column. The purity of the separated hydrogen can be controlled by adjusting the operating conditions of the distillation process, such as the pressure and temperature (Agrawal et al., 1988).

In conclusion, cryogenic hydrogen separation is a highly effective method for producing high-purity hydrogen and is widely used in various industries, including chemical production, electronics manufacturing, and fuel cell technology. However, this method is energy-intensive and requires large-scale equipment, making it more suitable for large-scale hydrogen production rather than small-scale applications.

9.2.4 Electrochemical Separation

This technology involves using an electrochemical cell that selectively transports hydrogen ions across a membrane, generating pure hydrogen gas on one side and a stream of impurities on the other (Ockwig and Nenoff, 2007; Pal et al., 2020). Electrochemical separation is a hydrogen separation process involving an electrochemical cell, also known as an electrolyze, to selectively transport hydrogen ions across a membrane, generating pure hydrogen gas on one side and a stream of impurities on the other side. The process is driven by an electric potential difference between the two sides of the membrane. In an electrolyzer, an electrolyte solution is separated into two compartments by a membrane, which allows the passage of hydrogen ions but blocks the passage of other gases. An electric current is passed through the electrolyte, causing hydrogen ions to move across the membrane to the other compartment, where they combine electrons to form pure hydrogen gas. The impurities remain in the original compartment and can be removed separately (Lee et al., 2004).

Electrochemical separation has several advantages over other hydrogen separation processes, including high purity and selectivity, low energy consumption, and the ability to operate at lower pressures and temperatures. It is commonly used to produce high-purity hydrogen for fuel cell applications and hydrogen recovery from industrial waste streams. However, electrochemical separation also has some limitations, including the membranes' high cost and the process's sensitivity to impurities in the electrolyte solution. Research is ongoing to address these challenges and improve the efficiency and cost-effectiveness of electrochemical hydrogen separation (Venugopalan et al., 2022).

In conclusion, each technology has advantages and limitations, depending on the specific application. For example, PSA and membrane separation are suitable for small-scale applications, while cryogenic distillation is more appropriate for large-scale industrial processes. Electrochemical separation is a relatively modern technology that shows promise for high-purity hydrogen production, but it is still in development. The choice of hydrogen separation technology depends on factors such as the required purity level, capacity, energy efficiency, and cost.

9.3 HOLLOW FIBER MEMBRANES

Hollow fiber membranes are a type of membrane technology that can be used for hydrogen separation. The process involves hydrogen diffusion through a selectively permeable membrane, which separates it from other gases in a mixture. The hollow fiber membrane consists of a bundle of long, thin, hollow fibers made of a polymer material. The fibers have a dense skin layer that is permeable to hydrogen but not to other gases such as nitrogen and carbon dioxide. When a hydrogen-rich gas mixture is passed through the membrane, hydrogen molecules diffuse through the skin layer of the fibers and are collected on the other side. The remaining gases are collected separately (Li et al., 2021). This technology is widely used in industrial processes, such as hydrogen production from natural gas and refining petrochemicals. It is also being explored for use in fuel cells, where the purity of the hydrogen fuel is critical for efficient and reliable operation. Hollow fiber membrane technology offers several advantages over traditional separation methods, including high selectivity for hydrogen, compact size, and low energy consumption. However, it also has some limitations, such as susceptibility to fouling and limited durability in harsh environments. The ongoing research is focused on addressing these challenges and improving technology performance (Sewerin et al., 2021). The transport phenomenon for hydrogen separation is different for dense and porous membranes. Dense membranes are made of a dense layer of material that separates the two gases, and the separation occurs due to differences in the permeability and solubility of the gases. In dense membranes, hydrogen permeates through the membrane due to its small size and high solubility, while other gases are unable to penetrate through the membrane. This selective permeation is due to the differences in diffusion rates and solubility between hydrogen and other gases.

Porous membranes, on the other hand, have small holes or channels that allow gases to flow through the membrane. Hydrogen separation in porous membranes occurs due to the differences in the adsorption and diffusion of gases on the surface of the membrane. Hydrogen molecules adsorb onto the surface of the membrane and diffuse through the pores, while other gases are either too large to penetrate or do not adsorb onto the surface of the membrane (Pal and Agarwal, 2021). In summary, dense membranes rely on the solubility and diffusion of hydrogen through the membrane, while porous membranes rely on the adsorption and diffusion of hydrogen through the pores in the membrane.

9.3.1 Membrane Hydrogen Separation Mechanism

Hydrogen separation membranes can be classified into two main types: dense and porous. Dense membranes are made of materials that do not have any porosity or significant void space, while porous membranes have a significant amount of void space that allows gases to flow through (Figure 9.1). Here are some of the mechanisms that are used in both dense and porous hydrogen separation membranes (Golemme et al., 2006; Husain and Koros, 2007; Pal and Agarwal, 2021):

a. Knudsen Diffusion: This mechanism is based on the size of the gas molecules. When the mean free path of gas molecules is greater than the membrane's pore size, gas molecules will pass through the membrane via a mechanism called Knudsen diffusion. In this mechanism, the gas molecules collide with the membrane surface and diffuse through the membrane by random jumps between pore walls without any significant interaction with the membrane material (Huang, Yao and Cheng, 2017).

b. Surface Diffusion: This mechanism involves the absorption of gas molecules on the surface of the membrane and their diffusion across the surface. The surface properties of the membrane, such as surface energy and surface chemistry, influence this mechanism. Surface diffusion is commonly used in dense membranes.

c. Capillary Condensation: This mechanism is based on the interaction between the gas molecules and the pore walls of the membrane. The condensed pore size of the membrane is in the range of a few nanometers; the surface tension between the gas molecules and the pore walls can cause the gas to condense into a liquid-like phase. This mechanism is commonly used in porous membranes.

d. Molecular Sieving: This mechanism involves the selective permeation of gas molecules based on their size and shape. When the membrane's pore size is smaller than the size of the gas molecules, only the smaller gas molecules will pass through, while the larger ones will be blocked. This mechanism is commonly used in porous membranes (Vermaak, Neomagus and Bessarabov, 2021).

e. Solution Diffusion: This mechanism involves the absorption of gas molecules into the membrane material, followed by their diffusion through the material. The diffusion rate is influenced by the concentration gradient of the gas across the membrane and the material's permeability. This mechanism is commonly used in dense membranes.

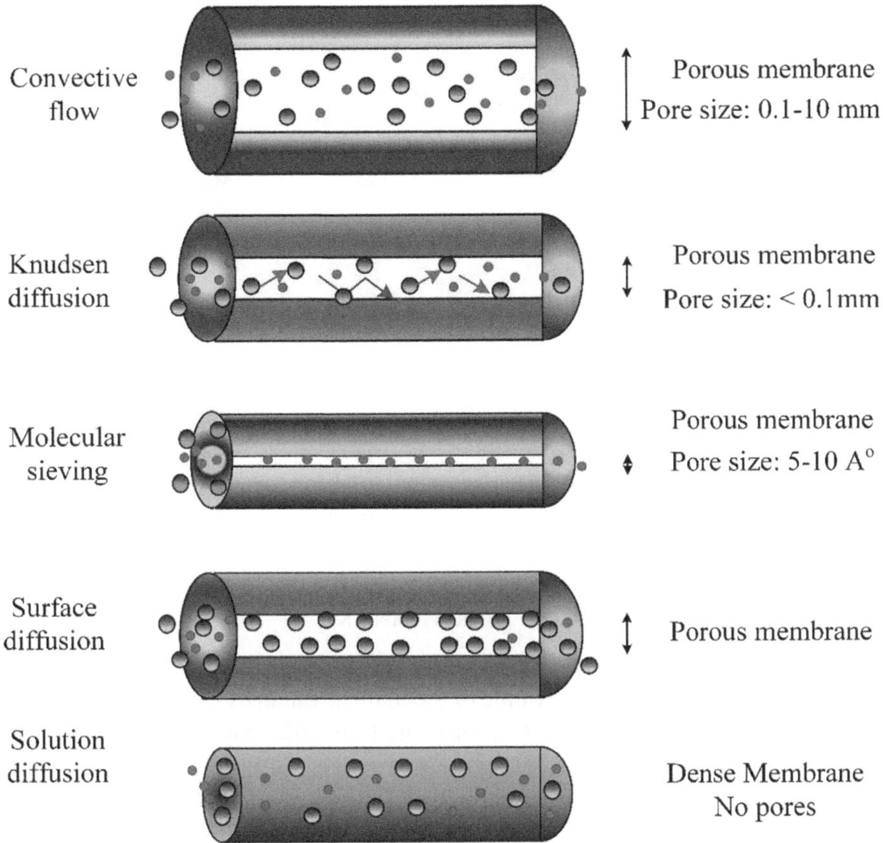

FIGURE 9.1 Gas transport mechanism through microporous and dense membranes (Lasseuguette and Ferrari, 2020).

The specific mechanism used in a hydrogen separation membrane depends on the pore size, surface properties, material properties of the membrane, and the properties of the gas being separated. Different combinations of these mechanisms can be used to achieve high selectivity and high permeability in hydrogen separation membranes.

Glassy polymeric materials, such as polyimide (PI), polycarbonate (PC), polybenzimidazole (PBI), cellulose acetate (CA), polyester (PE), polysulfone (PSF), and polyetherimide (PEI), are commonly studied for gas separation through zeolite-based mixed matrix membranes (MMMs). These materials have strong mechanical properties, are suitable for commercial membrane separations, and exhibit high gas separation performance. Table 9.1 shows the performance of zeolite-based MMMs. Polyimide and polyethersulfone are the most extensively researched polymer materials for hydrogen separation using zeolite-based MMMs.

TABLE 9.1

Performance of Zeolite-Based Mixed Matrix Membranes

Polymer	Zeolite Form	Selectivity H_2/N_2	References
Matrimid/Pd	ZIF-8	95.21	Mirzaei, Navarchian and Tangestaninejad (2020)
Carboxymethyl cellulose	ZIF-L	21.54	Ma et al. (2019)
Polyimide	ZIF-302	62.60	Ghanem et al. (2019)
Polybenzimidazole (PBI)	ZIF-8	26.20	Yang and Chung (2013)
BTDA–MDA polyimide	Sodalite (Sod-N)	281.00	Li et al. (2009)
Poly(phenylene oxide)/ carbon/Al_2O_3	SBA-15	38.90	Weng, Tseng and Wey (2010)
GUS	Zeolite Beta (BEA)	74–267	Kumbar et al. (2010)
Teflon AF 1600	MFI	4.6–4.9	Golemme et al. (2006)
Polyetherimide	HSSZ-13	7.7–8.2	Husain and Koros (2007)
Matrimid®	ZIF-8	78.60	Ordoñez et al. (2010)

9.3.2 Advantages of Hollow Fiber Membranes

Hollow fiber membranes offer several advantages for hydrogen separation compared to other separation technologies. Some of the main advantages include (Nijdam et al., 2005; Turken et al., 2019; Altinbas, Ozturk and İren, 2021; Moattari et al., 2021):

High selectivity: Hollow fiber membranes can selectively permeate hydrogen gas while blocking other gases, making them highly effective for producing high-purity hydrogen.

High permeance: The thin walls of hollow fiber membranes allow for rapid gas transport, resulting in high permeance rates.

Compact size: Hollow fiber membranes can be fabricated in small sizes, making them suitable for applications where space is limited.

Scalability: Hollow fiber membranes can be easily scaled up to increase production capacity, making them suitable for large-scale industrial processes.

Energy efficiency: Hollow fiber membranes require lower energy input than other hydrogen separation technologies such as cryogenic distillation or PSA, making them more energy-efficient.

Low maintenance: Hollow fiber membranes require minimal maintenance compared to other technologies, reducing downtime and operating costs.

Versatility: Hollow fiber membranes can be tailored to specific applications by selecting appropriate materials and modifying the membrane structure.

Overall, the advantages of hollow fiber membranes make them a promising technology for hydrogen separation in various industrial processes, such as ammonia production, refinery hydrogen recovery, and hydrogen purification for fuel cell applications.

9.3.3 Fabrication of Hollow Fiber Membranes

Hollow fiber membranes are commonly used for hydrogen separation due to their high surface area-to-volume ratio, which allows for efficient gas separation. The fabrication process of hollow fiber membranes used in hydrogen separation typically involves the following steps (Singh et al., 2014; Villalobos et al., 2018; Naderi et al., 2019):

- Polymer selection: The first step is to select a suitable polymer for membrane fabrication. Polymers commonly used for hydrogen separation include polyimides, polysulfone, and polyetherimides.
- Polymer synthesis: Once the polymer has been selected, it is synthesized using appropriate chemical reactions. For example, polyimides can be synthesized by reacting dianhydrides with diamines.
- Membrane casting: The polymer is then cast into hollow fiber membranes using a process such as dry/wet spinning or electrospinning. In dry/wet spinning, the polymer is dissolved in a solvent and spun through a spinneret into a coagulation bath, where the solvent is removed and the polymer solidifies into a membrane. In electrospinning, a high voltage is applied to a polymer solution to create a charged jet that is deposited onto a collector to form a membrane.
- Membrane annealing: The membranes are typically annealed after casting to improve their mechanical and chemical stability. This involves heating the membranes to a specific temperature for a set period.
- Membrane surface modification: In order to improve hydrogen separation performance, the surface of the membranes may be modified using techniques such as plasma treatment or chemical functionalization.
- Module assembly: Once the membranes have been fabricated and surface-modified, they are assembled into modules for use in hydrogen separation applications. This typically involves bundling multiple hollow fibers together and sealing them at each end to create a gas-tight seal.

Overall, the fabrication process of hollow fiber membranes used in hydrogen separation is a complex and highly specialized process that requires careful selection of materials and precise control of processing conditions.

9.3.4 Membrane Structure and Properties

Membrane-based separation processes are a promising alternative to traditional separation technologies because of their energy efficiency, environmental friendliness, and low capital cost. Hydrogen separation is one such application where membrane technology has gained significant attention in recent years. The membrane's structure and properties play a crucial role in the separation process, and various types of membranes have been developed to achieve high hydrogen separation performance (Ekiner and Vassilatos, 1990; Kusuki et al., 1997; Wang et al., 2016). Here are some examples of membrane structures and properties that are used in hydrogen separation:

Polymer membranes: These membranes are made of polymers such as polyimides, polyamides, and polyetherimides. They are relatively inexpensive and have good chemical and thermal stability. Polymer membranes work on the basis of selective permeation, where hydrogen molecules pass through the membrane while other gases are rejected. The structure and properties of the polymer membrane can be tailored by changing the type of polymer, its molecular weight, and the crosslinking density. The performance of the polymer membrane depends on its thickness, pore size, and degree of crystallinity (Lin et al., 2006; Perry, Nagai and Koros, 2006; Acharya et al., 2008; Shao et al., 2009; Sharma et al., 2009).

Ceramic membranes: These membranes are made of inorganic materials such as alumina, silica, and zeolites. They have high mechanical strength, good chemical and thermal stability, and high selectivity for hydrogen. Ceramic membranes are typically used in high-temperature applications because of their thermal stability. The structure and properties of the ceramic membrane depend on the type of material used, the porosity, and the size and shape of the pores (Prabhu and Oyama, 2000).

Metal membranes: These membranes are made of metals such as palladium, nickel, and copper. They have high selectivity for hydrogen and can operate at high pressures. Metal membranes work on the basis of hydrogen permeating through the metal lattice, while other gases are blocked. The structure and properties of the metal membrane depend on the type of metal used, the thickness of the membrane, and the degree of porosity (He, 2017).

Composite membranes: These membranes are made of different materials, such as polymers, ceramics, and metals. Composite membranes are designed to take advantage of the unique properties of each material and achieve high selectivity and permeability for hydrogen. The structure and properties of the composite membrane depend on the combination of materials used, the thickness of each layer, and the degree of interlayer adhesion (Athayde, Baker and Nguyen, 1994; Li, Liang and Hughes, 2000; Roa et al., 2003; Gu, Hacarlioglu and Oyama, 2008; Jeon et al., 2011; Strugova et al., 2018; Saini and Awasthi, 2022).

In summary, the structure and properties of membranes used in hydrogen separation depend on the type of material used, the thickness, porosity, and pore size, as well as the degree of selectivity and permeability required for the separation process. Tailoring these properties can help to achieve high hydrogen separation performance and improve the overall efficiency of the separation process.

9.3.5 Performance of Hollow Fiber Membranes

Hollow fiber membranes are widely used in hydrogen separation applications due to their high surface area, compact design, and high selectivity. The performance of these membranes can be affected by several factors, including (Imtiaz et al., 2022):

- Membrane material: The choice of membrane material can significantly affect its performance. For example, membranes made of polymers such as polyimides and polyamides have high hydrogen permeability, while ceramic membranes have high selectivity.
- Membrane thickness: The membrane's thickness can affect selectivity and permeability. Thinner membranes generally have higher permeability but lower selectivity.

- Pore size: The size of the pores in the membrane can also affect selectivity and permeability. Smaller pore sizes generally result in higher selectivity but lower permeability.
- Operating conditions: The operating conditions, such as temperature, pressure, and feed gas composition, can also affect membrane performance. For example, high temperatures can increase membrane permeability and reduce selectivity.
- Fouling: Fouling is the accumulation of contaminants on the surface of the membrane, which can reduce its performance. The type and concentration of contaminants in the feed gas can affect the fouling rate.
- Mechanical stability: The mechanical stability of the membrane is also an important factor, particularly in applications with high pressure or high flow rates. The membrane material, thickness, and support structure can all affect mechanical stability.

Understanding and optimizing these factors can lead to developing hollow fiber membranes with improved performance, enabling more efficient and cost-effective hydrogen separation processes.

9.3.5.1 Effect of Pressure on Membrane Performance

The effect of pressure on the permeance of pure gas permeation for Matrimid® asymmetric hollow fiber membrane depends on the gas being permeated and the operating conditions (Figure 9.2). Matrimid® is a high-performance polymer that is often used

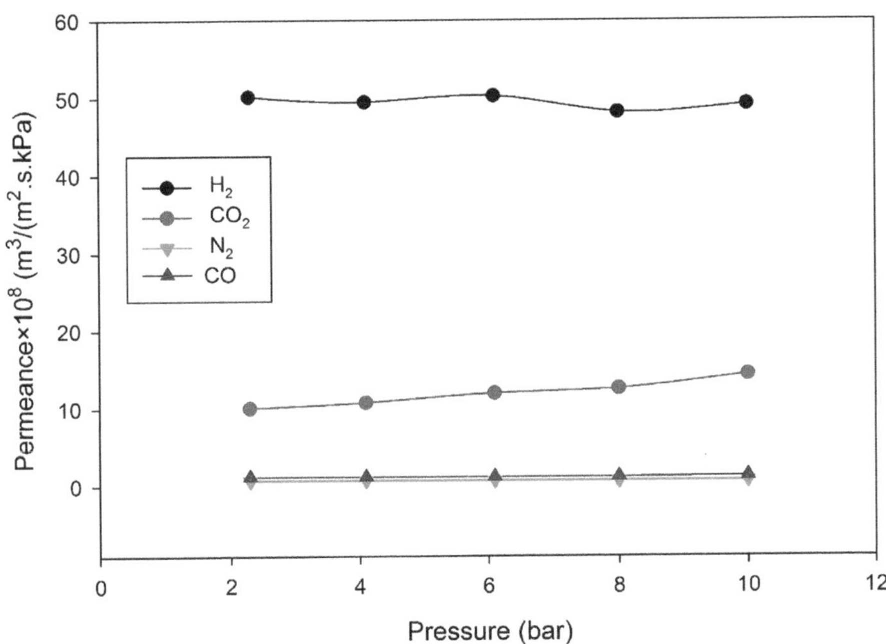

FIGURE 9.2 Effect of pressure on the permeance of pure gas permeation for Matrimid® asymmetric hollow fiber membrane at 30°C (David et al., 2012).

for gas separation membranes due to its high selectivity and permeance for specific gases, such as CO_2 and H_2. However, the permeance of Matrimid® membranes can be affected by pressure, temperature, and other factors. Increasing the pressure can increase the permeance of gases through Matrimid® membranes, as higher pressure gradients can increase the driving force for gas transport. However, this effect may be limited by the structural properties of the membrane, such as its thickness, pore size distribution, and surface morphology. Experimental studies have shown that the permeance of CO_2 and H_2 through Matrimid® membranes increases with increasing pressure up to a certain point, after which the permeance levels decrease. This behavior is often attributed to the formation of concentration polarization at the membrane surface, which can limit the flux of gas through the membrane. Therefore, the effect of pressure on the permeance of pure gas permeation for Matrimid® asymmetric hollow fiber membrane is complex and depends on several factors. In order to optimize the performance of the membrane for a specific gas separation application, experimental studies and modeling simulations should be carried out to determine the optimal operating conditions (David et al., 2012; Mirzaei, Navarchian and Tangestaninejad, 2020).

The effect of pressure on the ideal selectivity of pure gas permeation in an asymmetric hollow fiber membrane depends on the membrane structure and gas permeation properties, which is exhibited in Figure 9.3. Increasing pressure can increase the selectivity of a membrane by reducing the permeation rate of the less selective gas, which may be due to the higher diffusivity of the more selective gas or the more favorable

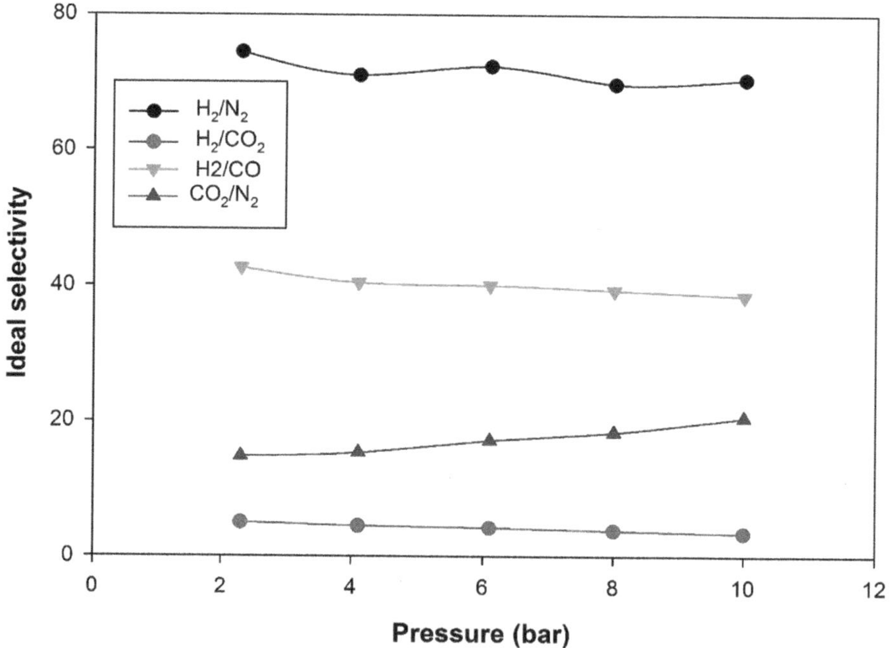

FIGURE 9.3　Effect of pressure on the ideal selectivity of pure gas permeation for Matrimid® asymmetric hollow fiber membrane at 30°C (David et al., 2012).

interaction between the membrane and the more selective gas (CO_2/N_2). However, the effect of pressure on selectivity is not always straightforward and can depend on other factors such as membrane thickness, pore size distribution, and gas solubility. For example, in some cases, increasing pressure can lead to a decrease in selectivity due to the higher concentration polarization at the membrane surface, resulting in a higher permeation rate of the less selective gas (H_2/CO). In addition, defects or imperfections in the membrane structure can also affect the pressure dependence of selectivity.

9.3.6 APPLICATIONS OF HOLLOW FIBER MEMBRANES FOR HYDROGEN SEPARATION

Hollow fiber membranes have a hollow, tubular structure with a selective barrier that allows specific molecules or gases to pass through while rejecting others. These membranes are commonly used in various industrial applications for separating, purifying, and concentrating different gases and liquids. Hydrogen separation is a crucial application of hollow fiber membranes (David et al., 2012). Hydrogen is a process of separating hydrogen gas from a mixture of natural gas, biogas, or syngas. Hydrogen separation is necessary for various applications such as fuel cells, hydrogen production, and refining crude oil. Hollow fiber membranes are an attractive option for hydrogen separation due to their high selectivity, high permeability, and high stability under harsh operating conditions. Some of the applications of hollow fiber membranes for hydrogen separation are:

Hydrogen production: Hollow fiber membranes can be used to separate hydrogen from reformate gas, which is a mixture of hydrogen, carbon dioxide, and carbon monoxide. The separation of hydrogen from reformate gas is necessary to produce high-purity hydrogen for fuel cells, chemical production, and other industrial applications.

Refining of crude oil: Hollow fiber membranes can be used to refine crude oil to remove impurities such as sulfur compounds and other contaminants. This process is known as hydrodesulfurization (HDS), which involves the separation of hydrogen from crude oil to facilitate the removal of sulfur compounds.

Biogas upgrading: Hollow fiber membranes can be used to upgrade biogas, which is a mixture of methane and carbon dioxide produced from organic waste. The separation of hydrogen from biogas is necessary to produce high-purity methane for use as a renewable energy source.

Purification of natural gas: Hollow fiber membranes can be used to purify natural gas, a mixture of methane, ethane, propane, and other gases. The separation of hydrogen from natural gas is necessary to remove impurities such as sulfur compounds and other contaminants that can cause corrosion and damage to pipelines.

Hollow fiber membranes offer an efficient, reliable, and cost-effective solution for hydrogen separation in various industrial applications. With further advancements in membrane technology, hollow fiber membranes are expected to play an increasingly significant role in producing and utilizing hydrogen as a clean and renewable energy source.

9.4 RECENT ADVANCEMENTS

Recent advancements in hollow fiber membrane technology for hydrogen separation have led to improvements in these membranes' selectivity, permeability, and stability. Some of the notable advancements are (Li et al., 2018; Du et al., 2021; Lei et al., 2021):

- Nanocomposite membranes: Nanocomposite membranes are made by incorporating nanoparticles into the membrane matrix, enhancing the membrane's selectivity and permeability. Researchers have developed various nanocomposite membranes for hydrogen separation, such as zeolite/polyimide, carbon nanotube/polyimide, and MOF/polyimide membranes.
- Surface modification: Surface modification of hollow fiber membranes can improve their performance by altering the surface chemistry, morphology, and hydrophobicity/hydrophilicity. Researchers have used various surface modification techniques, such as plasma treatment, chemical grafting, and thin film deposition, to improve the membrane's selectivity and stability.
- High-pressure operation: High-pressure operation of hollow fiber membranes can improve the separation performance by increasing the driving force for gas separation. Researchers have developed high-pressure hollow fiber membrane modules and tested their performance for hydrogen separation at high pressures.
- Hybrid systems: Hybrid systems that combine hollow fiber membranes with other separation technologies, such as PSA and membrane distillation, have been developed to improve the efficiency and selectivity of hydrogen separation.
- Advanced materials: Advanced materials such as graphene, MOFs, and carbon nanotubes are being investigated for their potential use in hollow fiber membranes for hydrogen separation. These materials have unique properties such as high surface area, selectivity, and stability, making them attractive for hydrogen separation.

Overall, these advancements in hollow fiber membrane technology for hydrogen separation have led to improved performance, increased efficiency, and reduced costs. As these advancements continue, it is expected that hollow fiber membranes will play an increasingly important role in producing and utilizing hydrogen as a clean and renewable energy source.

9.5 NEW MATERIALS AND FABRICATION TECHNIQUES

There have been several advancements in materials and fabrication techniques for membranes used for hydrogen separation. Here are some examples (Bitter and Asadi Tashvigh, 2022; Amin et al., 2023):

Nanoporous materials are pores on the nanoscale, allowing for the selective transport of hydrogen molecules while blocking other gases. Examples include zeolites, MOFs, and carbon nanotubes (Prasanth et al., 2011).

Graphene: Graphene is a two-dimensional material with a high surface area and excellent mechanical strength. It has been used to fabricate hydrogen separation membranes with high selectivity and permeability (Zhang et al., 2022).

Thin film membranes: Thin film membranes are made by depositing a thin layer of selective material onto a support layer. They can be made using various techniques, such as physical vapor deposition (PVD), chemical vapor deposition (CVD), and atomic layer deposition (ALD) (Meunier and Manaud, 1992; Ilias et al., 1997; Fernandez et al., 2016).

Asymmetric membranes: Asymmetric membranes have a thin, selective layer on top of a thicker, porous support layer. This design allows for high hydrogen selectivity and permeability while providing mechanical stability and durability (Zhu et al., 2015; Mercadelli et al., 2017; 2020; Chen et al., 2018).

Surface modification: Surface modification techniques can be used to enhance membrane performance by increasing surface hydrophilicity, enhancing surface roughness, and introducing functional groups that can enhance hydrogen–membrane interactions.

Hybrid membranes: Hybrid membranes combine two or more types of membranes with improving performance. For example, combining an asymmetric membrane with a dense, non-porous membrane can improve selectivity and permeability while maintaining mechanical stability.

Overall, these advancements in materials and fabrication techniques have led to improved performance and efficiency of hydrogen separation membranes, enabling more widespread adoption of hydrogen as a clean energy source (Ostwal et al., 2021).

9.6 IMPROVED MEMBRANE PERFORMANCE

Hydrogen separation membranes play a critical role in the efficient and cost-effective production of hydrogen gas, which is used in various industrial applications, such as fuel cells, ammonia production, and refineries (Pulyalina et al., 2018; Cai et al., 2021).

The performance of hydrogen separation membranes is dependent on several factors, including their selectivity, permeability, and stability. In recent years, several advancements in membrane materials and designs have led to improved membrane performance.

One approach to improving membrane performance is using advanced materials, such as nanoporous materials, graphene, and MOFs. These materials have high surface areas and tunable pore sizes, which allows for the selective transport of hydrogen molecules while blocking other gases (Naghdi et al., 2023).

Another approach to improving membrane performance is the development of asymmetric membrane structures. Asymmetric membranes have a thin, selective layer on top of a thicker, porous support layer. This design allows for high hydrogen selectivity and permeability while providing mechanical stability and durability (Pulyalina et al., 2020).

Furthermore, surface modification techniques can improve membrane performance by increasing surface hydrophilicity, enhancing surface roughness, and introducing functional groups to enhance hydrogen–membrane interactions.

Finally, the use of hybrid membrane systems that combine two or more types of membranes can also improve performance. For example, combining an asymmetric membrane with a dense, non-porous membrane can improve selectivity and permeability while maintaining mechanical stability. These advancements in membrane materials, design, and surface modification techniques have led to improved hydrogen separation membrane performance, enabling more efficient and cost-effective production of hydrogen gas.

9.7 CHALLENGES AND PROSPECTS

While hollow fiber membranes offer promising potential for hydrogen separation, some challenges still need to be addressed for broader adoption in commercial applications. Some of the challenges and prospects are (Lau and Yong, 2021):

- Scaling up: The scale-up of hollow fiber membrane modules for industrial applications remains challenging. The fabrication and assembly of large-scale modules require special skills and equipment, and the cost of production can be high.
- Stability under harsh conditions: Hollow fiber membranes must operate under harsh conditions such as high pressure, elevated temperature, and aggressive chemical environments. The stability and durability of the membrane under such conditions remain a challenge.
- Cost-effectiveness: The cost-effectiveness of hollow fiber membranes compared to other separation technologies, such as PSA and cryogenic distillation, needs to be improved for broader adoption.

9.8 CONCLUSION

In conclusion, developing improved membranes for hydrogen separation is critical to the efficient and cost-effective production of hydrogen gas. The performance of hydrogen separation membranes is dependent on several factors, including their selectivity, permeability, and stability. Advancements in membrane materials and fabrication techniques have led to the development of nanoporous materials, graphene, thin film membranes, asymmetric membranes, surface modification techniques, and hybrid membranes. These advancements have led to improved membrane performance, enabling more efficient and cost-effective production of hydrogen gas. As hydrogen continues to be seen as a promising clean energy source, further research and development in membrane technology will be crucial to improve hydrogen separation processes and ultimately reduce the cost and environmental impact of producing hydrogen.

ACRONYMS

MMMs Mixed Matrix Membranes
PSA Pressure swing adsorption
MOFs Metal–organic frameworks
PVD Physical vapor deposition
CVD Chemical vapor deposition
ALD Atomic layer deposition
GUS Glycerinedimethacrylateurethanetriethoxysilane

REFERENCES

Aasadnia, M., Mehrpooya, M. and Ghorbani, B., 2021. A novel integrated structure for hydrogen purification using the cryogenic method. *Journal of Cleaner Production*, 278, p.123872.

Acharya, N.K., Kulshrestha, V., Awasthi, K., Jain, A.K., Singh, M. and Vijay, Y.K., 2008. Hydrogen separation in doped and blend polymer membranes. *International Journal of Hydrogen Energy*, 33(1), pp.327–331.

Agrawal, R., Auvil, S.R., DiMartino, S.P., Choe, J.S. and Hopkins, J.A., 1988. Membrane/cryogenic hybrid processes for hydrogen purification. *Gas Separation & Purification*, 2(1), pp.9–15.

Altinbas, M., Ozturk, H. and İren, E., 2021. Full scale sanitary landfill leachate treatment by MBR: Flat sheet vs. hollow fiber membrane. *Journal of Membrane Science and Research*, 7(2), pp.118–124.

Amin, M., Butt, A.S., Ahmad, J., Lee, C., Azam, S.U., Mannan, H.A., Naveed, A.B., Farooqi, Z.U.R., Chung, E. and Iqbal, A., 2023. Issues and challenges in hydrogen separation technologies. *Energy Reports*, 9, pp.894–911. https://doi.org/10.1016/j.egyr.2022.12.014.

Athayde, A.L., Baker, R.W. and Nguyen, P., 1994. Metal composite membranes for hydrogen separation. *Journal of Membrane Science*, 94(1), pp.299–311.

Balachandran, U.B., Lee, T.H. and Dorris, S.E., 2007. Hydrogen production by water dissociation using mixed conducting dense ceramic membranes. *International Journal of Hydrogen Energy*, 32(4), pp.451–456.

Bitter, J.H. and Asadi Tashvigh, A., 2022. Recent advances in polybenzimidazole membranes for hydrogen purification. *Industrial & Engineering Chemistry Research*, 61(18), pp.6125–6134. https://doi.org/10.1021/acs.iecr.2c00645.

Cai, L., Cao, Z., Zhu, X. and Yang, W., 2021. Improved hydrogen separation performance of asymmetric oxygen transport membranes by grooving in the porous support layer. *Green Chemical Engineering*, 2(1), pp.96–103. https://doi.org/10.1016/j.gce.2020.11.003.

Chen, L., Liu, L., Xue, J., Zhuang, L. and Wang, H., 2018. Asymmetric membrane structure: An efficient approach to enhance hydrogen separation performance. *Separation and Purification Technology*, 207, pp.363–369.

David, O.C., Gorri, D., Nijmeijer, K., Ortiz, I. and Urtiaga, A., 2012. Hydrogen separation from multicomponent gas mixtures containing CO, N2 and CO2 using Matrimid(r) asymmetric hollow fiber membranes. *Journal of Membrane Science*, 419–420, pp.49–56. https://doi.org/10.1016/j.memsci.2012.06.038.

David, E. and Kopac, J., 2011. Devlopment of palladium/ceramic membranes for hydrogen separation. *International Journal of Hydrogen Energy*, 36(7), pp.4498–4506.

Dolci, F., 2018. Green hydrogen opportunities in selected industrial processes. JRC Technical Report 2018.

Du, P., Song, J., Wang, X., Zhang, Y., Xie, J., Liu, G., Liu, Y., Wang, Z., Hong, Z. and Gu, X., 2021. Efficient scale-up synthesis and hydrogen separation of hollow fiber DD3R zeolite membranes. *Journal of Membrane Science*, 636, p.119546.

Ekiner, O.M. and Vassilatos, G., 1990. Polyaramide hollow fibers for hydrogen/methane separation-spinning and properties. *Journal of Membrane Science*, 53(3), pp.259–273.

Estes, D.P., Leutzsch, M., Schubert, L., Bordet, A. and Leitner, W., 2020. Effect of ligand electronics on the reversible catalytic hydrogenation of CO2 to formic acid using ruthenium polyhydride complexes: A thermodynamic and kinetic study. *ACS Catalysis*, 10(5), pp.2990–2998. https://doi.org/10.1021/acscatal.0c00404.

Fernandes, T.R., Pimenta, R., Correas, L., García-Camús, J.M., Cabral, A.M., Reyes, F.D., Grano, B., Guerra, R., Couhert, C. and Chacón, E., 2013. Platform for promoting a hydrogen economy in Southwest Europe: The HYRREG project. *International Journal of Hydrogen Energy*, 38(18), pp.7594–7598. https://doi.org/10.1016/j.ijhydene.2013.01.131.

Fernandez, E., Medrano, J.A., Melendez, J., Parco, M., Viviente, J.L., van Sint Annaland, M., Gallucci, F. and Tanaka, D.A.P., 2016. Preparation and characterization of metallic supported thin Pd-Ag membranes for hydrogen separation. *Chemical Engineering Journal*, 305, pp.182–190.

Font-Palma, C., Cann, D. and Udemu, C., 2021. Review of cryogenic carbon capture innovations and their potential applications. *C*, 7(3), p.58.

Ghanem, A.S., Ba-shammakh, M., Usman, M., Khan, M.F., Dafallah, H. and Habib, M.A.M., 2019. High gas permselectivity in ZIF-302/polyimide self-consistent mixed-matrix membrane. *Journal of Applied Polymer Science*, 48513, pp.1–11. Available at: <https://www.scopus.com/inward/record.uri?eid=2-s2.0-85109072433&partnerID=40&md5=aeff8b7755dbef5cebd356a570770880>.

Golemme, G., Bruno, A., Manes, R. and Muoio, D., 2006. Preparation and properties of super-glassy polymers—zeolite mixed matrix membranes. *Desalination*, 200(1), pp.440–442. https://doi.org/10.1016/j.desal.2006.03.396.

Gu, Y., Hacarlioglu, P. and Oyama, S.T., 2008. Hydrothermally stable silica-alumina composite membranes for hydrogen separation. *Journal of Membrane Science*, 310(1–2), pp.28–37.

He, Y.-H., 2017. Chapter 15- Metal membranes. In: L.Y. Jiang and L. Na, eds. *Membrane-Based Separations in Metallurgy: Principles and Applications*. Amsterdam: Elsevier. pp.371–390. https://doi.org/10.1016/B978-0-12-803410-1.00015-3.

Huang, X., Yao, H. and Cheng, Z., 2017. Hydrogen separation membranes of polymeric materials BT - nanostructured materials for next-generation energy storage and conversion: Hydrogen production, storage, and utilization. In: Y.- P. Chen, S. Bashir and J.L. Liu, eds. *Nanostructured Materials for Next-Generation Energy Storage and Conversion*. Berlin, Heidelberg: Springer. pp.85–116. https://doi.org/10.1007/978-3-662-53514-1_3.

Husain, S. and Koros, W.J., 2007. Mixed matrix hollow fiber membranes made with modified HSSZ-13 zeolite in polyetherimide polymer matrix for gas separation. *Journal of Membrane Science*, 288(1), pp.195–207. https://doi.org/10.1016/j.memsci.2006.11.016.

Ilias, S., Su, N., Udo-Aka, U.I. and King, F.G., 1997. Application of electroless deposited thin-film palladium composite membrane in hydrogen separation. *Separation Science and Technology*, 32(1–4), pp.487–504.

Imtiaz, A., Othman, M.H.D., Jilani, A., Khan, I.U., Kamaludin, R., Iqbal, J., Al-Sehemi, A.G., Lau, H.S., Yong, W.F., Hao, A., Wan, X., Liu, X., Yu, R. and Shui, J., 2022. Recent progress and prospects of polymeric hollow fiber membranes for gas application, water vapor separation and particulate matter removal. *Membranes*, 9(7), p.e9120013.

Jeon, S.-Y., Lim, D.-K., Choi, M.-B., Wachsman, E.D. and Song, S.-J., 2011. Hydrogen separation by Pd-CaZr0. 9Y0. 1O3− δ cermet composite membranes. *Separation and Purification Technology*, 79(3), pp.337–341.

Kumbar, S.M., Selvam, T., Gellermann, C., Storch, W., Ballweg, T., Breu, J. and Sextl, G., 2010. ORMOCERs (organic-inorganic hybrid copolymers)-zeolite Beta (BEA) nano-composite membranes for gas separation applications. *Journal of Membrane Science*, 347(1), pp.132–140. https://doi.org/10.1016/j.memsci.2009.10.014.

Kusuki, Y., Shimazaki, H., Tanihara, N., Nakanishi, S. and Yoshinaga, T., 1997. Gas permeation properties and characterization of asymmetric carbon membranes prepared by pyrolyzing asymmetric polyimide hollow fiber membrane. *Journal of Membrane Science*, 134(2), pp.245–253.

Lasseuguette, E. and Ferrari, M.-C., 2020. Chapter 10- Polymer membranes for sustainable gas separation. In: G. Szekely and A. Livingston, eds. *Sustainable Nanoscale Engineering*. Elsevier. pp.265–296. https://doi.org/10.1016/B978-0-12-814681-1.00010-2.

Lau, H.S. and Yong, W.F., 2021. Recent progress and prospects of polymeric hollow fiber membranes for gas application, water vapor separation and particulate matter removal. *Journal of Materials Chemistry A*, 9(47), pp.26454–26497.

Lee, H.K., Choi, H.Y., Choi, K.H., Park, J.H. and Lee, T.H., 2004. Hydrogen separation using electrochemical method. *Journal of Power Sources*, 132(1), pp.92–98. https://doi.org/10.1016/j.jpowsour.2003.12.056.

Lei, L., Pan, F., Lindbråthen, A., Zhang, X., Hillestad, M., Nie, Y., Bai, L., He, X. and Guiver, M.D., 2021. Carbon hollow fiber membranes for a molecular sieve with precise-cutoff ultramicropores for superior hydrogen separation. *Nature Communications*, 12(1), p.268.

Li, G., Kujawski, W., Válek, R. and Koter, S., 2021. A review: The development of hollow fibre membranes for gas separation processes. *International Journal of Greenhouse Gas Control*, 104, p.103195. https://doi.org/10.1016/j.ijggc.2020.103195.

Li, A., Liang, W. and Hughes, R., 2000. Fabrication of dense palladium composite membranes for hydrogen separation. *Catalysis Today*, 56(1–3), pp.45–51.

Li, Y., Zhang, M., Chu, Y., Tan, X., Gao, J., Wang, S. and Liu, S., 2018. Design of metallic nickel hollow fiber membrane modules for pure hydrogen separation. *AIChE Journal*, 64(10), pp.3662–3670.

Li, D., Zhu, H.Y., Ratinac, K.R., Ringer, S.P. and Wang, H., 2009. Synthesis and characterization of sodalite-polyimide nanocomposite membranes. *Microporous and Mesoporous Materials*, 126(1), pp.14–19. https://doi.org/10.1016/j.micromeso.2009.05.014.

Lin, H., Van Wagner, E., Freeman, B.D., Toy, L.G. and Gupta, R.P., 2006. Plasticization-enhanced hydrogen purification using polymeric membranes. *Science*, 311(5761), pp.639–642.

Lu, G.Q., Da Costa, J.C.D., Duke, M., Giessler, S., Socolow, R., Williams, R.H. and Kreutz, T., 2007. Inorganic membranes for hydrogen production and purification: A critical review and perspective. *Journal of Colloid and Interface Science*, 314(2), pp.589–603.

Luberti, M. and Ahn, H., 2022. Review of Polybed pressure swing adsorption for hydrogen purification. *International Journal of Hydrogen Energy*, 47(20), pp.10911–10933. https://doi.org/10.1016/j.ijhydene.2022.01.147.

Ma, X., Wu, X., Caro, J. and Huang, A., 2019. Polymer composite membrane with penetrating ZIF-7 sheets displays high hydrogen permselectivity. *Angewandte Chemie*, 131(45), pp.16302–16306.

Mahajan, D., Tan, K., Venkatesh, T., Kileti, P. and Clayton, C.R., 2022. Hydrogen blending in gas pipeline networks: A review. *Energies*. https://doi.org/10.3390/en15103582.

Mercadelli, E., Gondolini, A., Montaleone, D., Pinasco, P., Escolástico, S., Serra, J.M. and Sanson, A., 2020. Production strategies of asymmetric BaCe0. 65Zr0. 20Y0. 15O3-δ-Ce0. 8Gd0. 2O2-δ membrane for hydrogen separation. *International Journal of Hydrogen Energy*, 45(12), pp.7468–7478.

Mercadelli, E., Montaleone, D., Gondolini, A., Pinasco, P. and Sanson, A., 2017. Tape-cast asymmetric membranes for hydrogen separation. *Ceramics International*, 43(11), pp.8010–8017.

Meunier, G. and Manaud, J.P., 1992. Thin film permeation membranes for hydrogen purification. *International Journal of Hydrogen Energy*, 17(8), pp.599–602.

Mirzaei, A., Navarchian, A.H. and Tangestaninejad, S., 2020. Mixed matrix membranes on the basis of Matrimid and palladium-zeolitic imidazolate framework for hydrogen separation. *Iranian Polymer Journal*, 29, pp.479–491.

Moattari, R.M., Mohammadi, T., Rajabzadeh, S., Dabiryan, H. and Matsuyama, H., 2021. Reinforced hollow fiber membranes: A comprehensive review. *Journal of the Taiwan Institute of Chemical Engineers*, 122, pp.284–310.

Naderi, A., Chung, T.-S., Weber, M. and Maletzko, C., 2019. High performance dual-layer hollow fiber membrane of sulfonated polyphenylsulfone/polybenzimidazole for hydrogen purification. *Journal of Membrane Science*, 591, p.117292.

Naghdi, S., Shahrestani, M.M., Zendehbad, M., Djahaniani, H., Kazemian, H. and Eder, D., 2023. Recent advances in application of metal-organic frameworks (MOFs) as adsorbent and catalyst in removal of persistent organic pollutants (POPs). *Journal of Hazardous Materials*, 442, p.130127. https://doi.org/10.1016/j.jhazmat.2022.130127.

Nijdam, W., De Jong, J., Van Rijn, C.J.M., Visser, T., Versteeg, L., Kapantaidakis, G., Koops, G.-H. and Wessling, M., 2005. High performance micro-engineered hollow fiber membranes by smart spinneret design. *Journal of Membrane Science*, 256(1–2), pp.209–215.

Ockwig, N.W. and Nenoff, T.M., 2007. Membranes for hydrogen separation. *Chemical Reviews*, 107(10), pp.4078–4110.

Ordoñez, M.J.C., Balkus, K.J., Ferraris, J.P. and Musselman, I.H., 2010. Molecular sieving realized with ZIF-8/Matrimid(r) mixed-matrix membranes. *Journal of Membrane Science*, 361(1), pp.28–37. https://doi.org/10.1016/j.memsci.2010.06.017.

Ostwal, M., Shinde, D.B., Wang, X., Gadwal, I., Lai, Z., Jia, M., Zhang, X.-F., Yao, J., Tseng, H.-H., Shiu, P.-T., Lin, Y.-S., Gao, C., Liao, J., Lu, J., Ma, J. and Kianfar, E., 2021. Graphene oxide-molybdenum disulfide hybrid membranes for hydrogen separation. *Reviews in Inorganic Chemistry*, 36(23), pp.1–20.

Pal, N. and Agarwal, M., 2021. Advances in materials process and separation mechanism of the membrane towards hydrogen separation. *International Journal of Hydrogen Energy*, 46(53), pp.27062–27087.

Pal, N., Agarwal, M., Maheshwari, K. and Solanki, Y.S., 2020. A review on types, fabrication and support material of hydrogen separation membrane. *Materials Today: Proceedings*, 28, pp.1386–1391.

Perry, J.D., Nagai, K. and Koros, W.J., 2006. Polymer membranes for hydrogen separations. *MRS Bulletin*, 31(10), pp.745–749.

Prabhu, A.K. and Oyama, S.T., 2000. Highly hydrogen selective ceramic membranes: Application to the transformation of greenhouse gases. *Journal of Membrane Science*, 176(2), pp.233–248.

Prasanth, K.P., Rallapalli, P., Raj, M.C., Bajaj, H.C. and Jasra, R.V., 2011. Enhanced hydrogen sorption in single walled carbon nanotube incorporated MIL-101 composite metal-organic framework. *International Journal of Hydrogen Energy*, 36(13), pp.7594–7601.

Pulyalina, A., Polotskaya, G., Rostovtseva, V., Pientka, Z. and Toikka, A., 2018. Improved hydrogen separation using hybrid membrane composed of nanodiamonds and P84 copolyimide. *Polymers*. https://doi.org/10.3390/polym10080828.

Pulyalina, A., Tataurov, M., Faykov, I., Rostovtseva, V. and Polotskaya, G., 2020. Polyimide asymmetric membrane vs. dense film for purification of MTBE oxygenate by pervaporation. *Symmetry*. https://doi.org/10.3390/sym12030436.

Qazi, U.Y., 2022. Future of hydrogen as an alternative fuel for next-generation industrial applications; challenges and expected opportunities. *Energies*. https://doi.org/10.3390/en15134741.

Rand, D.A.J. and Dell, R.M., 2007. *Hydrogen Energy: Challenges and Prospects*. RSC Publishing, Heidelberg, Germany.

Roa, F., Way, J.D., McCormick, R.L. and Paglieri, S.N., 2003. Preparation and characterization of Pd-Cu composite membranes for hydrogen separation. *Chemical Engineering Journal*, 93(1), pp.11–22.

Saini, N. and Awasthi, K., 2022. Insights into the progress of polymeric nano-composite membranes for hydrogen separation and purification in the direction of sustainable energy resources. *Separation and Purification Technology*, 282, p.120029.

Sazali, N., 2020. Emerging technologies by hydrogen: A review. *International Journal of Hydrogen Energy*, 45(38), pp.18753–18771. https://doi.org/10.1016/j.ijhydene.2020.05.021.

Sewerin, T., Elshof, M.G., Matencio, S., Boerrigter, M., Yu, J. and de Grooth, J., 2021. Advances and applications of hollow fiber nanofiltration membranes: A review. *Membranes*, https://doi.org/10.3390/membranes11110890.

Shao, L., Low, B.T., Chung, T.-S. and Greenberg, A.R., 2009. Polymeric membranes for the hydrogen economy: Contemporary approaches and prospects for the future. *Journal of Membrane Science*, 327(1–2), pp.18–31.

Sharma, A., Kumar, S., Tripathi, B., Singh, M. and Vijay, Y.K., 2009. Aligned CNT/Polymer nanocomposite membranes for hydrogen separation. *International Journal of Hydrogen Energy*, 34(9), pp.3977–3982.

Singh, R.P., Dahe, G.J., Dudeck, K.W., Welch, C.F. and Berchtold, K.A., 2014. High temperature polybenzimidazole hollow fiber membranes for hydrogen separation and carbon dioxide capture from synthesis gas. *Energy Procedia*, 63, pp.153–159.

Strugova, D. V, Zadorozhnyy, M.Y., Berdonosova, E.A., Yablokova, M.Y., Konik, P.A., Zheleznyi, M.V, Semenov, D.V, Milovzorov, G.S., Padaki, M. and Kaloshkin, S.D., 2018. Novel process for preparation of metal-polymer composite membranes for hydrogen separation. *International Journal of Hydrogen Energy*, 43(27), pp.12146–12152.

Thursfield, A., Murugan, A., Franca, R. and Metcalfe, I.S., 2012. Chemical looping and oxygen permeable ceramic membranes for hydrogen production: A review. *Energy & Environmental Science*, 5(6), pp.7421–7459.

Turken, T., Sengur-Tasdemir, R., Ates-Genceli, E., Tarabara, V.V. and Koyuncu, I., 2019. Progress on reinforced braided hollow fiber membranes in separation technologies: A review. *Journal of Water Process Engineering*, 32, p.100938.

Venugopalan, G., Bhattacharya, D., Andrews, E., Briceno-Mena, L., Romagnoli, J., Flake, J. and Arges, C.G., 2022. Electrochemical pumping for challenging hydrogen separations. *ACS Energy Letters*, 7(4), pp.1322–1329. https://doi.org/10.1021/acsenergylett.1c02853.

Vermaak, L., Neomagus, H.W.J.P. and Bessarabov, D.G., 2021. Recent advances in membrane-based electrochemical hydrogen separation: A review. *Membranes*, 11(2), p.127.

Villalobos, L.F., Hilke, R., Akhtar, F.H. and Peinemann, K., 2018. Fabrication of polybenzimidazole/palladium nanoparticles hollow fiber membranes for hydrogen purification. *Advanced Energy Materials*, 8(3), p.1701567.

Wang, M., Huang, M.-L., Cao, Y., Ma, X.-H. and Xu, Z.-L., 2016. Fabrication, characterization and separation properties of three-channel stainless steel hollow fiber membrane. *Journal of Membrane Science*, 515, pp.144–153.

Wang, Y., Yuan, H., Martinez, A., Hong, P., Xu, H. and Bockmiller, F.R., 2021. Polymer electrolyte membrane fuel cell and hydrogen station networks for automobiles: Status, technology, and perspectives. *Advances in Applied Energy*, 2, p.100011. https://doi.org/10.1016/j.adapen.2021.100011.

Weng, T.-H., Tseng, H.-H. and Wey, M.-Y., 2010. Fabrication and characterization of poly(phenylene oxide)/SBA-15/carbon molecule sieve multilayer mixed matrix membrane for gas separation. *International Journal of Hydrogen Energy*, 35(13), pp.6971–6983. https://doi.org/10.1016/j.ijhydene.2010.04.024.

Yang, T. and Chung, T.-S., 2013. High performance ZIF-8/PBI nano-composite membranes for high temperature hydrogen separation consisting of carbon monoxide and water vapor. *International Journal of Hydrogen Energy*, 38(1), pp.229–239. https://doi.org/10.1016/j.ijhydene.2012.10.045.

Yin, H. and Yip, A.C.K., 2017. A review on the production and purification of biomass-derived hydrogen using emerging membrane technologies. *Catalysts*. https://doi.org/10.3390/catal7100297.

Yue, M., Lambert, H., Pahon, E., Roche, R., Jemei, S. and Hissel, D., 2021. Hydrogen energy systems: A critical review of technologies, applications, trends and challenges. *Renewable and Sustainable Energy Reviews*, 146, p.111180. https://doi.org/10.1016/j.rser.2021.111180.

Zhang, X., Zhang, S., Tang, Y., Huang, X. and Pang, H., 2022. Recent advances and challenges of metal-organic framework/graphene-based composites. *Composites Part B: Engineering*, 230, p.109532. https://doi.org/10.1016/j.compositesb.2021.109532.

Zhu, Z., Sun, W., Wang, Z., Cao, J., Dong, Y. and Liu, W., 2015. A high stability Ni-La0. 5Ce0. 5O2– δ asymmetrical metal-ceramic membrane for hydrogen separation and generation. *Journal of Power Sources*, 281, pp.417–424.

10 Porous Inorganic Membranes for Hydrogen Separation

Maryam Takht Ravanchi

10.1 INTRODUCTION

Because of the increase in world population, economy and urbanization, energy usage and demand are increased as well. In the past years, energy was highly depended on hydrocarbon sources such as fossil fuel; but due to their depletion and their negative impact on greenhouse gas and global warming effects, recently developed clean renewable energy technologies are needed.

Naturally, hydrogen is an abundant chemical, its combustion product (H_2O) is non-pollutant, and it is reported as a "secure and clean energy future". But unfortunately, other energy sources (such as solar, nuclear and fossil) must be consumed for its manufacture and purification. Hydrogen is used in ammonia and methanol production, petroleum processing and manufacturing of fertilizers.

In the growing energy crisis, "hydrogen economy" is a valuable long-term remedy for increased global willingness for an efficient clean energy. "Hydrogen economy" is a new solution to this problem. In this methodology, an energy storage system based on hydrogen is needed. In this process, renewable energy is collected and stored in hydrogen, an energy carrier that is stable, easily transportable and simply utilizable (Chehade et al., 2020) (Figure 10.1). It is predicted that global energy consumption will double by 2050, and consequently, in order to have a competitive sustainable hydrogen economy, many technological and scientific issues for hydrogen production, utilization and storage must be resolved (Ockwig and Nenoff, 2007).

In hydrogen energy system, the required energy can be obtained from biomass or solar energies, which will change into hydrogen carrier by gasification, reforming, water splitting, etc. By this system, hydrogen was intermediary for relocating the energy between energy resources to utilization by consumers (Yin and Yip, 2017).

Normally, steam reforming of methane is used for hydrogen production from fossil fuel. This is a two-step process, in which at first, methane and steam react at 820°C over supported Ni catalysts for the production of carbon monoxide and hydrogen (Equation 10.1). In the subsequent step, water gas shift (WGS) reaction

DOI: 10.1201/9781003382522-12

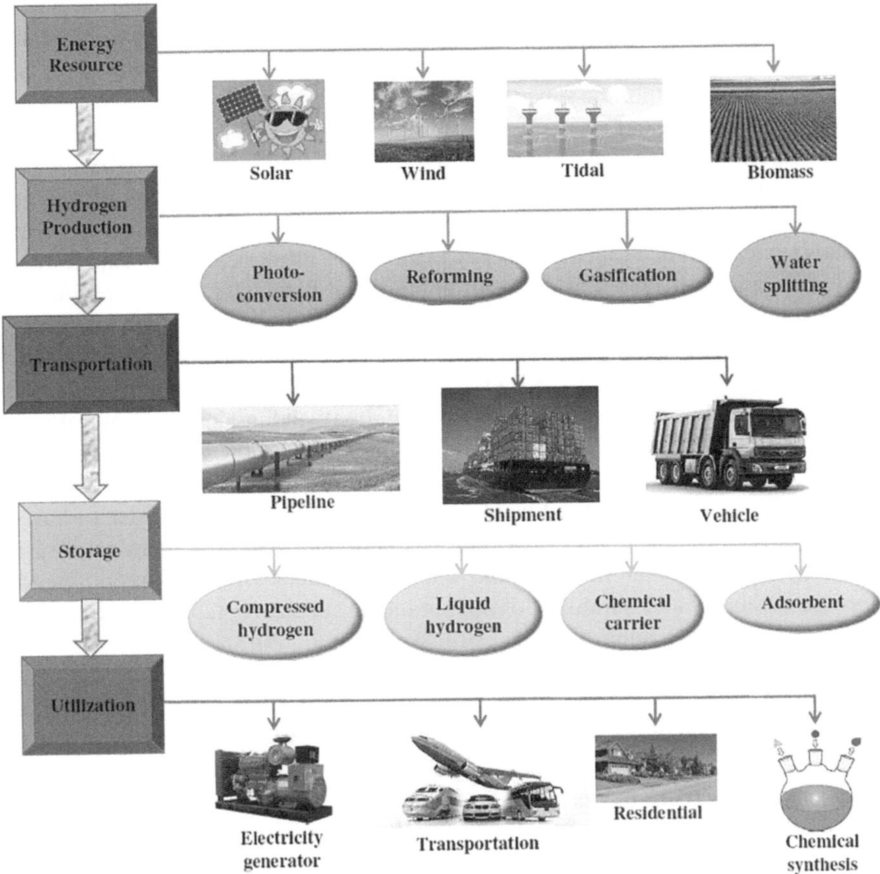

FIGURE 10.1 A summarized scheme for "hydrogen economy" concept.

(Equation 10.2) is used in which carbon monoxide and steam react and extra H_2 is obtained (Sahebdelfar et al., 2022).

$$CH_4 + H_2O \Leftrightarrow CO + 3H_2 \tag{10.1}$$

$$CO + H_2O \Leftrightarrow CO_2 + H_2 \tag{10.2}$$

The nature of the shift process determines the obtained hydrogen purity. For the HTS (high temperature shift) reactor (that operates at 350°C), 74% hydrogen, 7% methane, 18% carbon dioxide and 1% carbon monoxide are present in the product stream. For the LTS (low temperature shift) reactor (that operates at 190°C–210°C), 74% hydrogen, 6.9% methane, 19% carbon dioxide and 0.1% carbon monoxide are observed in the product stream (Sahebdelfar et al., 2022; Takht Ravanchi and Sahebdelfar, 2021). Generally, the main purpose of hydrogen purification is for CO and CO_2 removal.

Selective hydrogen removal from reaction mixture shifted the thermodynamic equilibria to the product side with more CH_4 conversions even at low temperature. Different processes, namely pressure swing adsorption (PSA), fractional or cryogenic distillation, membrane separation and amine absorption, are used for hydrogen separation (Adhikari and Fernando, 2006; Bredesen et al., 2004; Sircar and Golden, 2000). Although they are not cost effective and need high energy utilization, PSA and distillation are commercial processes. The main drawbacks of PSA, amine absorption and cryogenic distillation are being expensive, having negative environmental impacts and needing high energy loadings. In refineries, due to capital investment costs and unit recovery costs, in comparison to membrane separation units, PSA process is less economical (Sołowski et al., 2018). Moreover, in the concept of hydrogen economy, sufficient required hydrogen purity cannot be achieved by these two separation methods. Because of its low energy consumption, capital cost and investment cost, continuous operation, simplicity and modularity, ease of operation, cost-effectiveness, environmental friendliness (due to low waste production and storable precious raw materials) and energy productivity, membrane separation is currently the promising method used for hydrogen separation. In order to have an efficient successful membrane, two crucial parameters, namely permeability and selectivity, must be taken into account. Their high values are highly recommended. Moreover, stability and durability of membrane must be acceptable as well (Li et al., 2015; Ma et al., 2016; Sunarso et al., 2017; Wassie et al., 2018).

For power generation, instead of fossil fuel combustion, proton exchange membrane (PEM) fuel cells are another replacement. In these fuel cells, electrical energy is produced from chemical energy by using hydrogen as fuel (Castel et al., 2018). In comparison to internal combustion engines, fuel cells have some benefits, the most important of which are possessing higher fuel efficiency and producing low carbon dioxide with approximately no CO, NO_x or HC (hydrocarbon). However, a suitable hydrogen supply distribution is needed for these fuel cells (Collier et al., 2006). Being mostly available, hydrogen can completely be extracted from water, natural gas, biomass or coal. Electricity obtained from wind, solar or biomass or nuclear energy can also be used for hydrogen production (Burra et al., 2018).

For a useful valid hydrogen economy, hydrogen production cost must be reduced by a factor of 4. Ockwig and Nenoff (2007) announced that fossil fuel sources such as steam reforming of methane are major hydrogen producers. Water electrolysis, photoconversion and gasification are other producers. Generally, most of this hydrogen is used in textile, petrochemical, metallurgical, pharmaceutical, and chemical industries for the production of steel alloys, chemicals (such as CH_3OH, NH_3 and H_2O_2), semiconductors and vitamins. It is worth mentioning that for the case of large-scale hydrogen production, separation and purification steps are needed, and consequently, the need to large capital investment is inevitable as hydrogen separation from less desirable chemicals is always necessary.

Naturally, hydrogen is present in complex molecules (such as H_2O and CH_4) and high energy is required for its liberation. Subsequent purification, compression and liquefaction processes are also needed that are not environmentally friendly ones (because of the production of unwanted greenhouse gases) and subsequent sequestration processes are also needed to reduce CO_2 emission (Mulder and Mulder, 1996).

This chapter is intended to provide a comprehensive investigation on membranes for hydrogen purification. The main aim of this chapter is providing overview of hydrogen separation by inorganic membranes. Their recent developments, pros and cons, synthesis methods and transport phenomena are provided.

10.2 HYDROGEN SEPARATION MEMBRANE

Generally, membrane is a physical barrier between liquid and/or gas phases by which selective mass transfer occurred by a driving force. They are tailor-made with semi-permeable structures that can control molecules permeation by their pore size and driving force. Consequently, one or more components of a feed stream preferentially transmit the membrane. Permeate is those part of the feed that passes through the membrane, while retentate is the rest that retained above the membrane (Cardoso et al., 2018).

Membrane separation performance is evaluated by selectivity and permeance. Permeance is the flow rate of component A passing through the membrane per unit of transmembrane driving force and per unit of membrane area. The membrane ability for the separation of a specific component from feed stream is called selectivity, which is defined as the ratio of distribution of two components between permeate and retentate stream (Das and Veziroğlu, 2001; Takht Ravanchi et al., 2009).

The application of membrane gas separation technology dated back to 1970s when polymeric membranes were used for separating hydrogen from ammonia purge gas streams and also for adjusting the hydrogen–carbon monoxide ratio in syngas stream (Spillman, 1995). Sazali et al. (2020) reported the application of membrane for different gas separation. They announced that 53% is used for nitrogen production, 20% for natural gas treatment, 13% for hydrogen production, 7% for vapor recovery and the rest is used for other separation processes.

Among different methods available for hydrogen separation, membrane separation has the advantage of being versatile. Membrane formation, material selection, system configuration and modules are supporting areas that must be integrated for designing an effective hydrogen separation scheme (Sazali et al., 2020).

Generally, membranes are classified as natural or synthetic ones (Figure 10.2), with synthetic membranes divided into organic (polymeric), inorganic and hybrid (mixed matrix) membranes.

Organic membranes have a limited operating temperature range of 90°C–100°C. High chemical, thermal and mechanical stability, controllable pore size distribution (that caused better control on permeation and selectivity), and minor plasticization are some advantages of organic membranes (Ismail et al., 2015).

In the field of gas separation, generally, inorganic membranes are classified into porous and non-porous (dense) membranes. Subsequently, non-porous membranes are divided into metallic and ceramic proton conducting membranes. Porous membranes (such as zeolites, silica and carbon molecular sieve) do the separation based on differences in shape, size and/or affinity between permeating molecules and membrane. For this membrane category, membrane morphology (with the emphasis on its pore size and gas molecule size) has strong influence on gas transport (Sazali et al., 2020). Non-porous membranes do the separation based on differences in diffusivity and

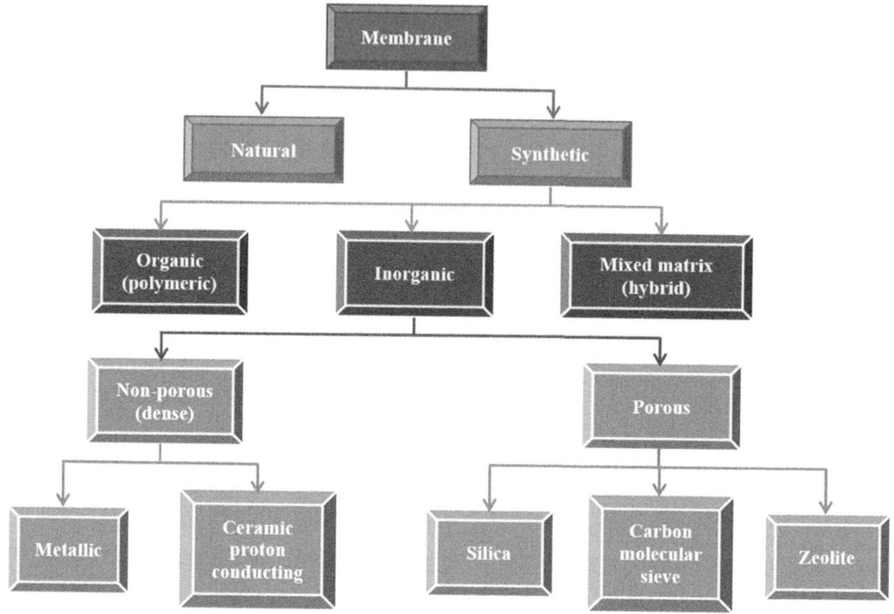

FIGURE 10.2   Membrane's classifications.

solubility (Burggraaf and Cot, 1996; Mulder, 1997). For hydrogen separation, common dense membranes used are palladium, Pd-alloy membranes (such as Pd-Ag and Pd-Ni) and ceramic proton conducting membranes. Among different dense metallic membranes evaluated for hydrogen separation, palladium-based ones are the most studied and industrialized (Cardoso et al., 2018). In Table 10.1, the performance of some of the common inorganic membranes evaluated for H_2 separation from H_2–N_2 mixture is provided.

Recently, researchers focused on developing hybrid membranes by which good capabilities of inorganic membranes and low manufacturing costs of organic membranes are used simultaneously (Chung et al., 2007).

There are different types of transportation mechanism for porous and non-porous membranes, namely Knudsen diffusion, surface diffusion, capillary condensation, molecular sieving, and solution–diffusion mechanism. The first four are valid for porous membranes, while the last is valid for non-porous membranes (Lu et al., 2007; Hao et al., 2022; Perry et al., 2011; Sazali et al., 2020):

- Knudsen diffusion: In this mechanism, for gas molecule, as its mean free path is greater than membrane pore diameter, gas molecule is more likely to collide with pore wall instead of colliding with each other. For Knudsen diffusion, the pore size is in the range of 2–100 nm.
- Surface diffusion: At first, a gas molecule is absorbed on the surface of membrane pore, and then, along the pore wall, it diffused to the membrane's low pressure or concentration side. For surface diffusion, pore size is lower than 10 nm.

TABLE 10.1

Comparison of Different Hydrogen Selective Membranes

Membrane	H_2 Permeance (mol $m^{-2} \cdot s \cdot Pa$)	H_2/N_2 Selectivity	References
Silica (synthesized by sol–gel method)	10×10^{-7}	15	Ghasemzadeh et al. (2015), Tsapatsis et al. (2000)
Silica (synthesized by CVD method)	$0.089 \times 10^{-7} - 0.98 \times 10^{-7}$	>1,000	Yan et al. (1994)
Metallic (Pd)	3.5×10^{-7}	468	Tosto et al. (2020)
HMDSO-derived silica[a]	13×10^{-7}	96	Nishida et al. (2019)
TMOS-derived silica[b]	1.6×10^{-7}	4,230	Nishida et al. (2019)
Silica	0.32×10^{-7}	10	Wei and Kawamoto (2013)
FAO zeolite	4×10^{-7}	6	Hao et al. (2022)
SAPO-34 zeolite	3.7×10^{-7}	32	Hao et al. (2022)
Carbon	1.3×10^{-7}	293	Hao et al. (2022)
ZIF 95 MOF	19×10^{-7}	10.2	Hao et al. (2022)
ZIF 7 MOF	2×10^{-7}	8.2	Hao et al. (2022)

[a] HMDSO: hexamethyldisiloxane.

[b] TMOS: tetramethoxysilane.

- Capillary condensation: Because of having difference in the rate of diffusion between the gas molecule pore size and the membrane pore wall, gas molecule transportation occurred.
- Molecular sieving: When pore size is similar with the diameter of the favorable diffusing molecule, molecular sieving happened. An activation energy is needed in this mechanism, which is highly coordinated by molecular size of the diffusing component. This mechanism is based on the fact that the smallest molecule has higher diffusion rate. This mechanism is applicable to gas molecules with dynamic diameters lower than the membrane pore diameter. Hence, gas molecules with diameters smaller than membrane pore size can pass through membrane and others are blocked.
- Solution–diffusion: There is a chemical potential gradient by which the diffusion of the diffusing molecule into the membrane starts. Solubility and mobility are detrimental factors for separation. Smallest molecule has favorable diffusivity selectivity, and most condensable molecule has favorable solubility selectivity.

10.3 ORGANIC (POLYMER) MEMBRANES

In comparison to conventional separation methods, membranes have many advantages, such as being environmentally friendly, energy productive, cost effective and having simple operation. But due to its low thermal resistance, its application in some industries is limited.

Polymeric membranes are desirable ones as they are cheap and have uncomplicated manufacturing process with commercial applications (Mural et al., 2018).

Polymer-based membranes are cheapest choice and being easily fabricated. Their synthetic composition has the highest flexibility, and their pre- or post-modifications need organic chemicals that are generally available. This is a major advantage of polymer membranes. Among various membranes, polymers are the present commercial choices. Generally, hydrogen separation by polymeric membrane is kinetically controlled, because of its small kinetic diameter.

Generally, "degradation" is a(n) (un)desirable change in polymer properties from its initial state (Kourde-Hanafi et al., 2017). Chemical degradation of polymers has negative influence on their performance and can significantly decline the quality of polymeric material.

Although polymeric membranes can be applied in hard operating conditions, due to their limited thermal stability, it is needed to cool down hot feed streams that is not cost-effective, and consequently, it is impractical at large scale (Cheng et al., 2018). Consequently, researchers focused on synthetic polymer membranes that can tolerate high temperatures. Rezakazemi et al. (2018) introduced carbocyclic and heterocyclic polymers as thermally stable ones; however, they are poorly soluble in organic solvents and unfortunately have high phase transition temperature.

Below their glass transition temperature (T_g), polymers have high structure. Generally, polymers containing C–C backbones are very resistant to degradation, while those containing heteroatoms such as anhydrides and esters are susceptible to degradation. Generally, polymeric membranes used for hydrogen separation can tolerate temperatures of 350°C–450°C. As glassy polymers can tolerate high glassy transition temperature, polymeric membranes made of these polymers, that are hydrogen-selective, can be used at high temperature (above 100°C). Laboratory-scale research announced that thermally rearranged (TR) polymers and metal–organic framework (MOF) membranes (in comparison to conventional membranes) have high H_2 permeability. In case of applying these findings at large scale, the commercial production of TR and MOFs polymers must be evaluated.

Generally, solution–diffusion is the main mechanism of transportation in polymer membranes. Hydrogen has the smallest gas molecule size with high diffusivity and low critical temperature. Hydrogen selective membrane must manipulate high diffusivity and restrict its reduced solubility.

Generally, sorption–diffusion mechanism consists of the following steps (Ockwig and Nenoff, 2007; Wang et al., 2022):

- Movement and diffusion of hydrogen molecule from the bulk phase to the feed stream surface of the membrane.
- Adsorption and dissociation of chemisorbed hydrogen on the surface into H^+ ions and electrons (e^{-1}).
- Dissolution and adsorption of atomic hydrogen from the surface into the membrane bulk.
- Diffusion of H^+ and electrons into the membrane to the other side.
- Precipitation and desorption of H^+ from Pd membrane to the product stream surface.
- Recombination of H^+ ions and electrons on the surface into H_2 molecule and its desorption.
- Leaving hydrogen from the membrane to the bulk phase.

10.4   INORGANIC MEMBRANES

Generally, inorganic membranes are used for separation and purification of hydrogen. Recently, researchers have been focused on the application of non-polymeric materials such as metal, ceramics, silica, zeolites, and carbon molecular sieve that are compared in Table 10.2 (Sazali et al., 2019a).

For industrial applications, because of their acceptable thermal stability and chemically inertness, they are advantageous to polymer membranes; particularly, in zeolite membranes, pore size and shape selectivity are combined with thermal, chemical and mechanical stability that are vital for long-term continuous separation.

As palladium membranes are highly selective and permeable to hydrogen, they are mostly researched and industrialized. Composite palladium-alloy membranes, ceramic proton conducting membrane and non-palladium metallic-alloy membranes are other alternatives used for H_2 separation that are cheaper with acceptable thermal/chemical stability and high permeation rates. Among different porous membranes, amorphous silica is the cheapest, but it is susceptible to moisture presence, which is inevitable in H_2 separation; hence, functionalization with precursors or transition metals is needed to overcome hydro-instability.

Each inorganic membrane has its pros and cons. The feed impurities and their temperature have impact on the performance of dense metallic membranes. On the other

---

TABLE 10.2

The Characteristics of Different Inorganic Membrane Evaluated for H_2 Separation

	Metallic	Ceramic Proton Conducting	Silica	Zeolite	Carbon Molecular Sieve
Transport mechanism	Solution–diffusion	Grotthuss, vehicle	Viscous flow, Knudsen, surface diffusion, activated gaseous diffusion, and bulk or molecular diffusion	Viscous flow, Knudsen, surface diffusion, activated gaseous diffusion, and bulk or molecular diffusion	Viscous flow, Knudsen, surface diffusion, activated gaseous diffusion, and bulk or molecular diffusion
Cost	Expensive	Moderate	Low	Moderate	Moderate
Studied temperature (°C)	250–550	650–1,000	200–600	25–450	25–150
Hydrogen selectivity	1,000	1,000	>100	<50	<100
Hydrogen permeance (mol m^{-2}·s·Pa)	10^{-6}	10^{-9}	10^{-8}	10^{-8}	10^{-8}

Source: Data provided from Cardoso et al. (2018) and Sazali et al. (2019).

hand, expensive fabrication costs are the major drawback of these membranes, and consequently, they are suitable for small-scale applications for hydrogen production with very high purity. Carbon molecular sieves, zeolite and silica membranes (as inorganic ones with microporous structure) have acceptable chemical and mechanical stability (Yang et al., 2017). But in the presence of water vapor in the feed stream, their performance dropped sharply. Moreover, for the scale-up of the synthesis procedure of inorganic membranes, obtaining its surface area at high level is still challenging.

For inorganic porous membranes (such as silica, carbon molecular sieve membrane (CMSM) and zeolite), transport mechanisms are associated with transmembrane pressure difference, pore diameter, permeating species size and molecular–pore wall interaction (Burggraaf, 1996; Krishna, 1993). Generally, for macropores ($d_p > 50\,nm$), the viscous regime wins; for mesopores ($2 < d_p < 50\,nm$), the Knudsen regime dominates; and for micropores ($d_p < 2\,nm$), the activated gaseous or surface diffusion contributes. Among all inorganic membranes, metal and zeolite ones have highest selectivity and flux. Due to their high operating temperature, they have high energy costs. Moreover, they are susceptible to phase transition and hybrid embrittlement. Lito et al. (2015) presented a comprehensive review for different transport mechanisms in microporous membranes.

10.4.1 Non-Porous (Dense) Membrane

Dense membranes have been investigated for many years (Hashim et al., 2018; Iulianelli et al., 2019; Zhuang et al., 2019). Generally, dense membranes are classified as metallic and proton conducting. For the specific case of hydrogen purification, as metals of groups III–V are being permeable to hydrogen, dense metallic membranes are popular ones (Sazali et al., 2019b) among which Pd-based are studied well.

For dense metallic membranes, solution–diffusion is the prevalent mechanism for hydrogen permeation, with the following steps:

- Chemical dissolution of molecular hydrogen on the feed side for the production of atomic hydrogen.
- Atomic hydrogen diffusion over the bulk metal.
- Association of atomic hydrogen for the reproduction of molecular hydrogen in the permeate side.

For the specific case of solution–diffusion mechanism, in non-porous membranes, there is a need to free electrons condensation and also specific catalytic surface for hydrogen dissociation on the feed side and re-association of electrons and protons on the product side (Ockwig and Nenoff, 2007).

High hydrogen selectivity is the main advantage of dense metallic membranes, but because of their low gas permeance, their efficient application is retarded.

Generally, composite metal membranes or metal alloys are used for hydrogen purification (Rahimpour et al., 2017).

Electroless plating, physical and chemical vapor depositions and electroplating are different synthesis methods used for the deposition of dense metallic layer. Based upon its efficiency and capability in the covering of supports with hard geometries, electroless plating is the common applicable technique. In comparison to dense metallic membranes, membranes containing metallic composite are more selective to hydrogen (Barsema et al., 2002). This is because of the fact that thin supported metallic layer is hardly dense causing lower permeation of gas molecule. Palladium membranes have some operational limitations; they encounter H_2-induced embrittlement phenomena, which are phase transformation at temperatures lower than 300°C and pressures lower than 20 bar. Moreover, hydrogen permeation rate can be decreased by surface poisoning with sulfur compounds. Palladium membranes have limited life time of 2–3 years that caused an increase in operating costs at commercial scale. Different metallic compounds such as Cu, Ni, Pt, Au and Ag can be used for the production of Pd alloys (Levin et al., 2004).

Different researchers evaluated dense membranes for hydrogen separation. Zhao et al. (2016) used a Pd-Cu-Ag membrane for hydrogen separation at 500°C and reported hydrogen permeability flux of 0.15–0.45 mol m^{-2}·s. Wang et al. (2018) used Ni membrane in hollow fiber configuration at 1,000°C and reported hydrogen permeability flux of 7.66 mol m^{-2}·s. Song et al. (2016) used perovskite $BaCe_{0.85}Tb_{0.05}Co_{0.1}O_{3-\sigma}$ membrane without coating and with 150 μm thickness in hollow fiber configuration and reported H_2 permeability flux of 0.063 mol m^{-2}·s.

10.4.1.1 Metallic Membranes

For one-step hydrogen production with high purity, dense phase metallic and metallic alloy membranes are highly attractive. As Pd is selectively permeable to H_2, palladium-based membranes are highly preferred. Based upon their high selectivity, Pd-based dense membranes are mostly researched and commercialized for pure hydrogen production (Basile et al., 2008). The utilization of Pd membranes for hydrogen purification dated back to 19th century and its commercialization started in 1964 by Johnson Matthey. Besides palladium (Pd), nickel (Ni) and vanadium (V) can also be used for hydrogen purification. Having good sulfur tolerance, high hydrogen permeability and low costs are recent challenges of metallic membranes.

10.4.1.2 Ceramic Proton Conducting Membranes

Dense ceramic materials that are permeable to hydrogen are oxides with high electron and proton conductivity (Damle, 2008). Perovskite-type oxides, pyrochlores, niobates and tantalates are examples of proton conducting materials. For feed streams containing CO, CO_2 and H_2S, in comparison to Pd-based membranes, pyrochlores and perovskites (as ceramic proton conducting materials) have higher stability, lower cost and higher durability (Tao et al., 2015). Usually, sol–gel or hydrothermal methods are used for the synthesis of ceramic membranes by which the products have high stability and durability in hydrothermal and high-temperature surroundings. Low hydrogen selectivity and high flux are main characteristics of inorganic ceramic membranes.

10.4.2 Porous Membranes

Basically, porous membranes that are based upon inorganic materials consist of ceramic or porous metallic supports under porous layer with different morphology. Kayvani Fard et al. (2018) reported that due to the presence of electrolyte layer, hydrogen can permeate through the pores.

Normally, zeolites, MOFs and silicas are used for the synthesis of porous membranes. As Pera-Titus (2014) reported, inorganic microporous stabilized silica membranes are selective to hydrogen molecules with high permeation at high temperature. As silica membranes are chemically, thermally and mechanically stable in different operating conditions, their application is advantageous (Van Gestel et al., 2014; Zhang et al., 2016a,b). Besides silica, crystal zeolites are also applied for hydrogen separation. Cardoso et al. (2017) performed through investigation on H_2 separation by zeolite membranes. Due to their high purity and acceptable performance, synthetic zeolites are preferable for commercial applications. Currently, lanthanide silicate, silicoaluminophosphate (SAPO) and aluminophosphate (AlPO) as zeotype materials are evaluated for H_2 separation (Caro and Noack, 2008). Zhou et al. (2014) reported a high hydrogen permeability of 6.96×10^{-6} mol m^{-2}·s·Pa for SAPO-34 that is layered on α-Al$_2$O$_3$ tubular support.

10.4.2.1 Silica Porous Membranes

Silica membranes are inorganic ones with amorphous microporous structure in which tetrahedral SiO$_4$ is linked by covalent bonding, and consequently, a three-dimensional network with 0.5 nm pore size is formed. Based upon their kinetic diameters and adsorption behavior, small molecules of hydrogen, oxygen or helium can be separated by silica membranes. Moreover, for a wide range of operating conditions, these membranes are chemically, thermally and mechanically stable. On the other hand, their synthesis method is easy and scalable with low fabrication cost. Morooka and Kusakabe (1999), Prabhu and Oyama (2000), Tsapatsis and Gavalas (1999) and Kluiters (2004) published research with the subject of silica membrane preparation.

Because of their ease of fabrication, scalability and low production cost, silica membranes are good choices for hydrogen separation. Moreover, as silica membranes have no precious metals, they are cheaper than metal ones and they are not susceptible to hydrogen embrittlement.

Regarding porous silica membranes, although they are easily fabricated with low costs and have high hydrogen permeances and moderate to high selectivity, at high temperatures and in the presence of steam, they are instable (Ayral et al., 2008; Tsai et al., 2000). By using microporous and mesoporous membrane and intermediate layers structures, a very high surface of membrane is obtained and any increase in temperature or pressure can cause further phase transformation, densification and structural disintegration. On the other hand, any rapid fluctuations in pressure can cause disruptive tensile stress at the interface of layers (Benes et al., 1999).

Amorphous silica is a suitable choice to be used for hydrogen production membrane, as its operational stability can be optimized. If membrane thickness can be kept below 1 μm, the membrane thermochemical stability and its intermediate layers are the best. On the other hand, production cost of membrane must be reduced. Sometimes, structural defects occurred during synthesis can cause irreproducibility and poor performance (Ockwig and Nenoff, 2007).

Generally, silica membranes have optimum performance as their permeation flow rate is maximized. Normally, they are deposited on a thick porous support such as Vycor glass or α-alumina (Lee et al., 2008; Yoshino et al., 2005). Due to their large pores (>110 nm), alumina supports are more attractive. Moreover, in comparison to

Vycor glass, they are stronger, economical and high temperature resistant. On the other hand, a 2–10 nm intermediate layer is used between silica layer and alumina support to improve selectivity and permeance.

Microporous silica membranes are not completely selective for one component. For a mixture separation, one molecule moves by site-hopping diffusion in the micropore network, and separation occurred by a competitive process (Lito et al., 2010). Generally, silica membranes are composed of membrane layer, intermediate layer and support.

Sol–gel and chemical vapor deposition (CVD) are two techniques used for the synthesis of membrane layer in silica membranes (de Lange et al., 1995; Ha et al., 1993; Kurungot et al., 2003; Nomura et al., 2005). Good membrane permeability and selectivity can be obtained by sol–gel method, but the lack of reproducibility is its major drawback. In contrast, the membrane synthesized by CVD method has enhanced selectivity with loss of permeability. Moreover, high capital investment is needed for CVD method and deposition condition must be carefully controlled.

By dip coating of nanoparticles and subsequent drying and calcination, intermediate layers are synthesized. Transition alumina, zirconia and silica are common nanoparticle materials used. Generally, precipitation from salt solution or organo-metal reagent hydrolysis is used for the preparation of precursor particles. During synthesis, particle agglomeration must be controlled for obtaining final layers with little shrinkage. Any agglomeration after synthesis must be removed. Sonochemical and modified emulsion precipitation techniques can be used to prohibit agglomerate formation (Shi and Verweij, 2005).

Different techniques can be used for support fabrication in silica membranes. "Dip coating" is a common method in this regard in which submicron particles were first obtained by using commercial α-Al_2O_3 powders and then dip coated. In this method, colloidal stabilization is used in which coherent dense packed layers are formed after sintering at 1,000°C (Brinker et al., 1993). By using "colloidal filtration" and "centrifugal casting", a thicker support layer can be obtained with high strength (Nijmeijer et al., 1998).

10.4.2.2 Zeolite Membranes

Zeolites are natural minerals or produced synthetically; the latter is preferred because of higher purity (Feng et al., 2015). Zeolites are inorganic crystalline structures with uniform, molecular size pores. Generally, they are used as adsorbents and catalysts and are made up of TO_2 units in which T is a tetrahedral framework atom such as B, Al, Si, and Ge. Besides neutral silica zeolite frameworks, the net overall charge of the framework is negative and is balanced by the charge of organic or inorganic cations.

Normally, cations reside in the framework pores and the number of T atoms in the ring determines pore size category; zeolite structures with 8-membered oxygen rings are called small-pore zeolites, those with 10-membered rings are called medium-pore zeolites and the ones with 12-membered rings are called large-pore zeolites (Zito et al., 2017). Recently, polycrystalline zeolite membrane is deposited on porous support.

Natural zeolites were discovered in 1756, while synthetic zeolites were obtained in 1962 and their commercialization dated back to 1954 (Szostak, 1989). Since then, researchers did their best to improve zeolite applications and its separation performance. Recently, more than fourteen zeolite structures such as MFI, FAU, LTA (Linde Type A) and MOR are used as hydrogen-selective separation membranes. Due to its pore size and easy preparation procedure, MFI structure is used in zeolite membrane. Silicate-1 and ZSM-5 are examples of MFI structures, which are, respectively, made of pure silica and Al substitution for Si atoms.

Zeolites are hydrated aluminosilicates with crystalline microporous structure composed of AlO_4^{5-} and SiO_4^{4-} building blocks that are connected to each other by oxygen atoms. There are well-defined cavities and channels in which water molecules and alkali or alkali-earth cations are present. These channels are in the range of 0.3–1.0 nm, and by modifying Si/Al ratio, this range can be changed. Moreover, thermal, chemical and mechanical stabilities of zeolites are outstanding (Gascon et al., 2012).

Non-aluminosilicate composites are called zeotypes, which are crystalline materials with a three-dimensional framework, and their tetrahedral sites are occupied by Fe, Al, V, Ti, B, Ta, P, Ge and Sn. AlPO, SAPO, MeAPO (metalloaluminophosphate) and lanthanide silicate are some of these zeotype materials. For zeotype membranes, different researchers reported gas permeation kinetics (Khatib and Oyama, 2013; Lito et al., 2011; Smart et al., 2013; Zito et al., 2017).

Gas molecule transportation through a zeolite membrane, which is a combination of adsorption and diffusion phenomena, occurred according to the model below (Bakker et al., 1996):

- Gas molecule adsorption from bulk phase to the exterior zeolite surface.
- Gas molecule diffusion from zeolite surface to the inner part of zeolite channels.
- Interior diffusion inside zeolite channels.
- Gas molecule back diffusion from zeolite channel to the exterior zeolite surface.
- Gas molecule desorption from exterior zeolite surface to the gas phase.

Adsorption affinity is depended upon zeolite and gas specifications. The interaction strength between molecule and zeolite surface is highly depended upon polarizability and dipole or quadrupole moments of gas as its important characteristics. Polarity, pore size, zeolite framework topology and its counter ion are detrimental parameters of adsorbent. Generally, there is a direct relationship between Al amount of zeolite and its polarity; aluminum-rich zeolites with $Si/Al \approx 1$ are highly selective for polar molecules and silica-rich zeolites with $Si/Al \geq 10$ are highly selective for nonpolar molecules (Calleja et al., 1998; Coterillo et al., 2008).

Liquid and gas separation by zeolite membranes occurred by competitive adsorption and diffusion mechanisms. When molecular size of feed components is in the range of zeolite pore size distribution, size exclusion mechanism prevails (Cui et al., 2004; Lai et al., 2003).

Zeolite phase(s) determine the effective pore size distribution and its subsequent separation performance. A compromise between transmembrane flux and separation performance will determine the optimum zeolite thickness and is tailored according to the envisioned application.

For MFI-type zeolite membranes, the actual gas permeation mechanism depends on gas adsorption properties of zeolites. Non-adsorbing gas molecules can directly enter zeolite pores. Gas molecule mobility inside zeolite pores and its probable entrance into zeolite pores will determine its separation factor (Lovallo et al., 1998). Large gas molecules with low mobility are not prone to permeate through zeolite membrane, while small ones with high mobility will permeate. Based upon operating pressure and temperature, adsorption or activated diffusion is the dominant controlling mechanism for the separation of strongly adsorbing gases by a MFI membrane.

Different researchers have developed surface modification methods for improving zeolite membranes. Generally, post-treatment techniques are used in this regard such as inorganic silylation for reducing pore size and increasing hydrophobicity (Masuda et al., 2001; Park et al., 2001).

Zeolite membranes with small size pores are synthesized for small gas molecules separation. For zeolite A membrane, in the temperature range of 35°C–25°C, Aoki et al. (1998) reported H_2 permeance of 10^{-11}–10^{-10} mol $m^{-2} \cdot s \cdot kPa$ and H_2/N_2 selectivity of 4.8. Guan et al. (2001) announced that any cation change in zeolite structure has influence on hydrogen permeance. Guan et al. (2003) reported that for $AlPO_4$_5 membrane with 0.73 nm pore size, at 35°C, hydrogen permeance was 2×10^{-10} mol $m^{-2} \cdot s \cdot kPa$.

Ockwig and Nenoff (2007) announced the synthesis of silicalite-1 membrane and its application for hydrogen separation at temperatures ranging from 70°C to 300°C. At lower temperature, membrane pores are blocked by adsorbing components, and consequently, hydrogen had low permeation (approximately 3×10^{-11} mol $m^{-2} \cdot s \cdot kPa$) and separation factor varied in the range of 0.13–0.4.

10.4.2.3 Carbon Molecular Sieve Membranes

By carbonization or pyrolysis of a polymeric precursor, in vacuum and under controlled atmosphere, carbon molecular sieve membrane is obtained. An amorphous carbon skeleton containing interconnected pores is obtained by decomposing polymer chain. Having exact control on synthesis conditions, structure and distribution of zeolite pores can be tuned. Because of having high chemical and thermal stability, high permeability and selectivity for H_2, O_2 and N_2, and controllable PSD (pore size distribution), CMSMs are advantageous for gas mixture separation. For hydrogen separation, CMSMs (in comparison to zeolites) are more selective, but in comparison to dense membranes, it is less selective.

In 1973, Ash et al. (1973) announced the concept of carbon membrane for the first time. Later, in 1983, Koresh and Sofer (2006) announced the application of CMSM with micropores in the range of 0.3–0.5 nm for gas separation. Rao and Sircar (1993a,b) used carbon membrane with micropores in the range of 0.5–0.7 nm.

Molecular sieving is the dominant separation mechanism in CMSM. For micropores in the range of 0.5–0.7 nm that are larger than the diffusing molecules, adsorption occurred and non-adsorbable or weakly adsorbable components (such as oxygen, nitrogen and methane) from adsorbable molecules (such as ammonia, sulfur dioxide and hydrogen sulfide) are separated (Ismail et al., 2011).

10.5 CONCLUSION

Among different available separation methods, membranes and specifically inorganic ones have higher separation of hydrogen gas. At high operating temperature and pressure, due to their tunable nature and acceptable stability, inorganic silica and zeolite membranes can be used at industrial scale. Based upon their easy fabrication, scalability and low production costs, silica membranes are valuable candidates for hydrogen separation. Moreover, in comparison to metal membranes, they are cheap and not susceptible to hydrogen embrittlement. From the viewpoint of thermal, chemical and mechanical stability, zeolite membranes are inherently stable. As these membranes are stable at high temperature, they are good candidates to be used for H_2 production via reforming of natural gas. As a concluding remark, lack of chemical stability is the major problem for all membrane classes. Their major contaminants are water, CO, sulfur-containing species and acidic vapors. The cost and viability of a specific material to be used as membrane for hydrogen separation is determined based upon their thermal and chemical performance.

ACRONYMS

AlPO	Aluminophosphate
CMSM	Carbon molecular sieve membrane
CVD	Chemical vapor deposition
HC	Hydrocarbon
HMDSO	Hexamethyldisiloxane
HTS	High temperature shift
LTA	Linde-Type A
LTS	Low temperature shift
MeAPO	Metalloaluminophosphate
MOF	Metal–organic frameworks
PEM	Proton exchange membrane
PSA	Pressure swing adsorption
PSD	Pore size distribution
SAPO	Silicoaluminophosphate
TMOS	Tetramethoxysilane
TR	Thermally rearranged

REFERENCES

Adhikari, S., Fernando, S., 2006. Hydrogen membrane separation techniques. *Ind. Eng. Chem. Res.*, 45, 875–881.

Aoki, K., Kusakabe, K., Morooka, S., 1998. Gas permeation properties of A-type zeolite membrane formed on porous substrate by hydrothermal synthesis. *J. Membr. Sci.*, 141, 197–205.

Ash, R., Barrer, R.M., Lowson, R.T., 1973. Transport of single gases and of binary gas mixtures in a microporous carbon membrane. *J. Chem. Soc. Faraday Trans. 1 Phys. Chem. Condens. Phases*, 69, 2166–2178.

Ayral, A., Julbe, A., Rouessac, V., Roualdes, S., Durand, J., 2008. Microporous silica membrane: Basic principles and recent advances. In: Mallada, R., Menéndez, M., (Eds.) *Inorganic Membranes: Synthesis, Characterization and Applications*, Elsevier: Oxford, 33–79.

Bakker, W.J.W., Kapteijn, F., Poppe, J., Moulijn, J.A., 1996. Permeation characteristics of a metal-supported silicalite-1 zeolite membrane. *J. Membr. Sci.*, 117, 57–78.

Barsema, J.N., van der Vegt, N.F.A., Koops, G.H., Wessling, M., 2002. Carbon molecular sieve membranes prepared from porous fiber precursor. *J. Membr. Sci.*, 205, 239–246.

Basile, A., Gallucci, F., Tosti, S., 2008. Synthesis, characterization, and applications of palladium membranes. In: Mallada, R., Menéndez, M. (Eds.) *Inorganic Membranes: Synthesis, Characterization and Applications*, Elsevier: Oxford, 255–323.

Benes, N.E., Biesheuvel, P.M., Verweij, H., 1999. Tensile stress in a porous medium due to gas expansion. *AIChE J.*, 45, 1322–1328.

Bredesen, R., Jordal, K., Bollard, O., 2004. High-temperature membranes in power generation with CO2 capture. *Chem. Eng. Process.*, 43, 1129–1158.

Brinker, C.J., Ward, T.L., Sehgal, R., Raman, N.K., Hietala, S.L., Smith, D.M., Hua, D.-W., Headley, T.J., 1993. Ultramicroporous silica-based supported inorganic membranes. *J. Membr. Sci.*, 77, 165–179.

Burggraaf, A.J., 1996. Important characteristics of inorganic membranes. In: Burggraaf, A.J., Cot, L., (Eds.) *Fundamentals of Inorganic Membrane Science and Technology*, Elsevier: Amsterdam, Netherlands, 22–28.

Burggraaf, A.J., Cot, L., 1996. *Fundamentals of Inorganic Membranes Science and Technology*, Elsevier: Amsterdam, Netherlands.

Burra, K.R.G., Bassioni, G., Gupta, A.K., 2018. Catalytic transformation of H2S for H2 production. *Int. J. Hydrog. Energy*, 43, 22852–22860.

Calleja, G., Pau, J., Calles, J.A., 1998. Pure and multicomponent adsorption equilibrium of carbon dioxide, ethylene, and propane on ZSM-5 zeolites with different Si/Al ratios. *J. Chem. Eng. Data*, 43, 994–1003.

Cardoso, S.P., Azenha, I.S., Lin, Z., Portugal, I., Rodrigues, A.E., Silva, C.M., 2018. Inorganic membranes for hydrogen separation. *Sep. Purif. Rev.*, 47, 229–266.

Cardoso, S.P., Azenha, I.S., Portugal, I., Lin, Z., Rodrigues, A.E., Silva, C.M., 2017. Single and binary surface diffusion permeation through zeolite membranes using new Maxwell-Stefan factors for Dubinin-type isotherms and occupancy-dependent kinetics. *Sep. Purif. Technol.*, 182, 207–218.

Caro, J., Noack, M., 2008. Zeolite membranes: Recent developments and progress. *Micro. Meso. Mat.*, 115, 215–233.

Castel, C., Wang, L., Corriou, J.P., Favre, E., 2018. Steady vs unsteady membrane gas separation processes. *Chem. Eng. Sci.*, 183, 136–147.

Chehade, G., Lytle, S., Ishaq, H., Dincer, I., 2020. Hydrogen production by microwave based plasma dissociation of water. *Fuel*, 264, 116831.

Cheng, X.Q., Wang, Z.X., Jiang, X., Li, T., Lau, C.H., Guo, Z., Ma, J., Shao, L., 2018. Towards sustainable ultrafast molecular-separation membranes: From conventional polymers to emerging materials. *Prog. Mater. Sci.*, 92, 258–283.

Chung, T.S., Jiang, L.Y., Li, Y., Kulprathipanja, S., 2007. Mixed matrix membranes (MMMs) comprising organic polymers with dispersed inorganic fillers for gas separation. *Prog. Polym. Sci.*, 32, 483–507.

Collier, A., Wang, H., Yuan, X.Z., Zhang, J., Wilkinson, D.P., 2006. Degradation of polymer electrolyte membranes. *Int. J. Hydrog. Energy*, 31, 1838–1854.

Coterillo, C.C., Mendia, A.M.U., Uribe, I.O., 2008. Pervaporation and gas separation using microporous membranes. In: Mallada, R., Menéndez, M., (Eds.) *Inorganic Membranes: Synthesis, Characterization and Applications*, Elsevier: Oxford, 217–253.

Cui, Y., Kita, H., Okamoto, K., 2004. Preparation and gas separation performance of zeolite T membrane. *J. Mater. Chem.*, 14, 924–932.

Damle, A., 2008. Hydrogen separation and purification. In: Gupta, R.B., (Ed.) *Hydrogen Fuel: Production, Transport, and Storage*, CRC Press: Boca Raton, FL, 283–324.

Das, D., Veziroğlu, T.N., 2001. Hydrogen production by biological processes: A survey of literature. *Int. J. Hydrog. Energy*, 26, 13–28.

de Lange, R.S.A., Keizer, K., Burggraaf, A.J., 1995. Analysis and theory of gas transport in microporous sol-gel derived ceramic membranes. *J. Membr. Sci.*, 104, 81–100.

Feng, C., Khulbe, K.C., Matsuura, T., Farnood, R., Ismail, A., 2015. Recent progress in zeolite/zeotype membranes. *J. Membr. Sci. Res.*, 1, 49–72.

Gascon, J., Kapteijn, F., Zornoza, B., Sebastián, V., Casado, C., Coronas, J., 2012. Practical approach to zeolitic membranes and coatings: State of the art, opportunities, barriers, and future perspectives. *Chem. Mater.*, 24, 2829–2844.

Ghasemzadeh, K., Aghaeinejad-Meybodi, A., Vaezi, M.J., Gholizadeh, A., Abdi, M.A., Babaluo, A.A., Haghighi, M., Basile, A., 2015. Hydrogen production *via* silica membrane reactor during the methanol steam reforming process: Experimental study. *RSC Adv.*, 5, 95823–95832.

Guan, G., Kusakabe, K., Morooka, S., 2001. Gas permeation properties of ion-exchanged LTA-type zeolite membranes. *Sep. Sci. Technol.*, 36, 2233–2245.

Guan, G.Q., Tanaka, T., Kusakabe, K., Sotowa, K.I., Morooka, S., 2003. Characterization of AlPO4-type molecular sieving membranes formed on a porous α-alumina tube. *J. Membr. Sci.*, 214, 191–198.

Ha, H.Y., Nam, S.W., Hong, S.A.; Lee, W.K., 1993. Chemical vapor deposition of hydrogen-permselective silica films on porous glass supports from tetraethylorthosilicate. *J. Membr. Sci.*, 85, 279–290.

Hao, A., Wan, X., Liu, X., Yu, R., Shui, J., 2022. Inorganic microporous membranes for hydrogen separation: Challenges and solutions. *Nano Res. Energy*, 1, e9120013.

Hashim, S.S., Somalu, M.R., Loh, K.S., Liu, S., Zhou, W., Sunarso, J., 2018. Perovskite-based proton conducting membranes for hydrogen separation: A review. *Int. J. Hydrog. Energy*, 43, 15281–15305.

Ismail, A.F., Khulbe, K.C., Matsuura, T., 2015. *Gas Separation Membranes*, Springer: Switzerland.

Ismail, A., Rana, D., Matsuura, T., Foley, H., 2011. Transport mechanism of carbon membranes. In: Ismail, A.F., Rana, D., Matsuura, T., Foley, H., (Eds.) *Carbon-Based Membranes for Separation Processes*, Springer: New York, 5–16.

Iulianelli, A., Jansen, J.C., Esposito, E., Longo, M., Dalena, F., Basile, A., 2019. Hydrogen permeation and separation characteristics of a thin Pd-Au/Al$_2$O$_3$ membrane: The effect of the intermediate layer absence. *Catal. Today*, 330, 32–38.

Kayvani Fard, A., McKay, G., Buekenhoudt, A., Al Sulaiti, H., Motmans, F., Khraisheh, M., Atieh, M., 2018. Inorganic membranes: Preparation and application for water treatment and desalination. *Materials*, 11, 74.

Khatib, S.J., Oyama, S.T., 2013. Silica membranes for hydrogen separation prepared by chemical vapor deposition (CVD). *Sep. Purif. Technol.*, 111, 20–42.

Kluiters, S.C.A., 2004. *Status Review on Membrane Systems for Hydrogen Separation*, Energy Center of Netherlands: Petten, The Netherlands.

Koresh, J.E., Sofer, A., 2006. Molecular sieve carbon permselective membrane. Part I. Presentation of a new device for gas mixture separation. *Sep. Sci. Technol.*, 18, 723–734.

Kourde-Hanafi, Y., Loulergue, P., Szymczyk, A., Van der Bruggen, B., Nachtnebel, M., Rabiller-Baudry, M., Audic, J.L., Pölt, P., Baddari, K., 2017. Influence of PVP content on degradation of PES/PVP membranes: Insights from characterization of membranes with controlled composition. *J. Membr. Sci.*, 533, 261–269.

Krishna, R., 1993. Problems and pitfalls in the use of the fick formulation for intraparticle diffusion. *Chem. Eng. Sci.*, 48, 845–861.

Kurungot, S., Yamaguchi, T., Nakao, S.I., 2003. Rh/γ-Al$_2$O$_3$ catalytic layer integrated with sol-gel synthesized microporous silica membrane for compact membrane reactor applications. *Catal. Lett.*, 86, 273–278.

Lai, Z.P., Bonilla, G., Diaz, I., Nery, J.G., Sujaoti, K., Amat, A.M., Kokkoli, E., Terasaki, O., Thompson, R.W., Tsapatsis, M., Vlachos, D.G., 2003. Microstructural optimization of a zeolite membrane for organic vapor separation. *Science*, 300, 456–460.

Lee, D.W., Yu, C.Y., Lee, K.H., 2008. Synthesis of Pd particle deposited microporous silica membranes via a vacuum-impregnation method and their gas permeation behavior. *J. Colloid Interface Sci.*, 325, 447–452.

Levin, D.B., Pitt, L., Love, M., 2004. Biohydrogen production: Prospects and limitations to practical application. *Int. J. Hydro. Energy*, 29, 173–185.

Li, P., Wang, Z., Qiao, Z., Liu, Y., Cao, X., Li, W., Wang, J., Wang, S., 2015. Recent developments in membranes for efficient hydrogen purification. *J. Membr. Sci.*, 495, 130–168.

Lito, P.F., Cardoso, S.P., Rodrigues, A.E., Silva, C.M., 2015. Kinetic modeling of pure and multicomponent gas permeation through microporous membranes: Diffusion mechanisms and influence of isotherm type. *Sep. Purif. Rev.*, 44, 283–307.

Lito, P.F., Santiago, A.S., Cardoso, S.P., Figueiredo, B.R., Silva, C.M., 2011. New expressions for single and binary permeation through zeolite membranes for different isotherm models. *J. Membr. Sci.*, 367, 21–32.

Lito, P.F., Zhou, C.F., Santiago, A.S., Rodrigues, A.E., Rocha, J., Lin, Z., Silva, C.M., 2010. Modelling gas permeation through new microporous titanosilicate AM-3 membranes. *Chem. Eng. J.*, 165, 395–404.

Lovallo, M.C., Gouzinis, A., Tsapatsis, M., 1998. Synthesis and characterization of oriented MFI membranes prepared by secondary growth. *AIChE J.*, 44, 1903–1913.

Lu, G.Q., Diniz da Costa, J.C., Duke, M., Giessler, S., Socolow, R., Williams, R.H., Kreutz, T., 2007. Inorganic membranes for hydrogen production and purification: A critical review and perspective. *J. Colloid Interface Sci.*, 314, 589–603.

Ma, C., Yu, J., Wang, B., Song, Z., Xiang, J., Hu, S., Su, S., Sun, L., 2016. Chemical recycling of brominated flame retarded plastics from waste for clean fuels production: A review. *Renew. Sust. Energ. Rev.*, 61, 433–450.

Masuda, T., Fukumoto, N., Kitamura, M., Mukai, S.R., Hashimoto, K., Tanaka, T., Funabiki, T., 2001. Modification of pore size of MFI-type zeolite by catalytic cracking of silane and application to preparation of H2-separating zeolite membrane. *Micro. Meso. Mater.*, 48, 239–245.

Morooka, S., Kusakabe, K., 1999. Microporous inorganic membranes for gas separation. *MRS Bull.*, 24, 25–29.

Mulder, M. (1997) *Basic Principles of Membrane Technology*, second edition, Kluwer Academic Publishers: Dordrecht.

Mulder, M., Mulder, J., 1996. *Basic Principles of Membrane Technology*, Kluwer Academic Publishers: Dordretch.

Mural, P.K.S., Madras, G., Bose, S., 2018. Polymeric membranes derived from immiscible blends with hierarchical porous structures, tailored bio-interfaces and enhanced flux: Potential and key challenges. *Nano-Struct. Nano-Objects*, 14, 149–165.

Nijmeijer, A., Huiskes, C., Sibelt, N.G.M., Kruidhof, H., Verweij, H., 1998. Centrifugal casting of tubular membrane supports. *Am. Ceram. Soc. Bull.*, 77, 95–98.

Nishida, R., Tago, T., Saitoh, T., Seshimo, M., Nakao, S., 2019. Development of CVD silica membranes having high hydrogen permeance and steam durability and a membrane reactor for a water gas shift reaction. *Membranes*, 9, 140.

Nomura, M., Ono, K., Gopalakrishnan, S., Sugawara, T., Nakao, S.I., 2005. Preparation of a stable silica membrane by a counter diffusion chemical vapor deposition method. *J. Membr. Sci.*, 251, 151–158.

Ockwig, N.W., Nenoff, T.M., 2007. Membranes for hydrogen separation. *Chem. Rev.*, 107, 4078–4110.

Park, D.H., Nishiyama, N., Egashira, Y., Ueyama, K., 2001. Enhancement of hydrothermal stability and hydrophobicity of a silica MCM-48 membrane by silylation. *Ind. Eng. Chem. Res.*, 40, 6105–6110.

Pera-Titus M., 2014. Porous inorganic membranes for CO_2 capture: Present and prospects. *Chem. Rev.*, 114, 1413–1492.

Perry, J.D., Nagai, K., Koros, W.J., 2011. Polymer membranes for hydrogen separations. *MRS Bull.*, 31, 745–749.

Prabhu, A.K., Oyama, S.T., 2000. Highly hydrogen selective ceramic membranes: Application to the transformation of greenhouse gases. *J. Membr. Sci.*, 176, 233–248.

Rahimpour, M.R., Samimi, F., Babapoor, A., Tohidian, T., Mohebi, S., 2017. Palladium membranes applications in reaction systems for hydrogen separation and purification: A review. *Chem. Eng. Proc. Proc. Intens.*, 121, 24–49.

Rao, M.B., Sircar, S., 1993a. Nanoporous carbon membranes for separation of gas mixtures by selective surface flow. *J. Membr. Sci.*, 85, 253–264.

Rao, M.B., Sircar, S., 1993b, Nanoporous carbon membrane for gas separation. *Gas Sep. Purif.*, 7, 279–284.

Rezakazemi, M., Sadrzadeh, M., Matsuura, T., 2018. Thermally stable polymers for advanced high-performance gas separation membranes. *Prog. Energ. Comb. Sci.*, 66, 1–41.

Sahebdelfar, S., Takht Ravanchi, M., Nadda, A.K., 2022. *C1 Chemistry Principles and Processes*, first edition, Taylor and Francis: Boca Raton.

Sazali, N., Ghazali, M.F., Ibrahim, H., Paiman, S.H., Moslan, M.S., 2019a. A review on hydrogen separation through inorganic membranes. *J. Adv. Res. Fluid Mech. Therm. Sci.*, 63, 238–248.

Sazali, N., Salleh, W.N.W., Ismail, A.F., Wong, K.C., Iwamoto, Y., 2019b. Exploiting pyrolysis protocols on BTDA-TDI/MDI (P84) polyimide/nanocrystalline cellulose carbon membrane for gas separations. *J. Appl. Polym. Sci.*, 136, 46901.

Sazali, N., Mohamed, M.A., Salleh, W.N.W., 2020. Membranes for hydrogen separation: A significant review. *Int. J. Adv. Man. Tech.*, 107, 1859–1881.

Shi, J.Y., Verweij, H., 2005. Synthesis and purification of oxide nanoparticle dispersions by modified emulsion precipitation. *Langmuir*, 21, 5570–5575.

Sircar, S., Golden, T.C., 2000. Purification of hydrogen by pressure swing adsorption. *Sep. Sci. Tech.*, 35, 667–687.

Smart, S., Beltramini, J., Da Costa, J.C.D., Katikaneni, S.P., Pham, T., 2013. Microporous silica membranes: Fundamentals and applications in membrane reactors for hydrogen separation. In: Basile, A., (Ed.) *Handbook of Membrane Reactors*, Elsevier: Cambridge, 337–369.

Sołowski, G., Shalaby, M.S., Abdallah, H., Shaban, A.M., Cenian, A., 2018. Production of hydrogen from biomass and its separation using membrane technology. *Renew. Sust. Energy Rev.*, 82, 3152–3167.

Song, J., Kang, J., Tan, X., Meng, B., Liu, S., 2016. Proton conducting perovskite hollow fiber membranes with surface catalytic modification for enhanced hydrogen separation. *J. Euro. Ceram. Soc.*, 36, 1669–1677.

Spillman, R., 1995. Economics of gas separation membrane processes. In: Noble, R.D., Stern, S.A. (Eds.) *Membrane Science and Technology*, Elsevier: England, 589–667.

Sunarso, J., Hashim, S.S., Lin, Y.S., Liu, S.M., 2017. Membranes for helium recovery: An overview on the context, materials and future directions. *Sep. Purif. Technol.*, 176, 335–383.

Szostak, R., 1989. *Molecular Sieves Principles of Synthesis and Identification*, Van Nostrand Reinhold: New York.

Takht Ravanchi, M., Kaghazchi, T., Kargari, A., 2009. Application of membrane separation processes in petrochemical industry: A review, *Desalination*, 235, 199–244.

Takht Ravanchi, M., Sahebdelfar, S., 2021. Catalytic conversions of CO_2 to help mitigate climate change: Recent process developments. *Process. Safety Environ. Prot.*, 145, 172–194.

Tao, Z., Yan, L., Qiao, J., Wang, B., Zhang, L., Zhang, J., 2015. A review of advanced proton-conducting materials for hydrogen separation. *Prog. Mater. Sci.*, 74, 1–50.

Tosto, E., Alique, D., Martinez-Diaz, D., Sanz, R., Calles, J.A., Caravella, A., Maderan, J.A., Gallucci, F., 2020. Stability of pore-plated membranes for hydrogen production in fluidized-bed membrane reactors. *Int. J. Hydrogen Energy*, 45, 7374–7385.

Tsai, C.Y., Tam, S.Y., Lu, Y.F., Brinker, C.J., 2000. Dual-layer asymmetric microporous silica membranes. *J. Membr. Sci.*, 169, 255–268.

Tsapatsis, M., Gavalas, G.R., 1999. Synthesis of porous inorganic membranes. *MRS Bull.*, 24, 30–35.

Tsapatsis, M., Gavalas, G.R., Xomeritakis, G., 2000. Chemical vapor deposition membranes. In: Kanellopoulos, N.K. (Ed.) *Recent Advances in Gas Separation by Microporous Ceramic Membranes*, Elsevier: Amsterdam.

Van Gestel, T., Hauler, F., Bram, M., Meulenberg, W.A., Buchkremer, H.P., 2014. Synthesis and characterization of hydrogen selective sol-gel SiO_2 membranes supported on ceramic and stainless-steel supports. *Sep. Purif. Technol.*, 121, 20–29.

Wang, H., Wang, X., Meng, B., Tan, X., Loh, K.S., Sunarso, J., Liu, S., 2018. Perovskite-based mixed protonic-electronic conducting membranes for hydrogen separation: Recent status and advances. *J. Ind. Eng. Chem.*, 60, 297–306.

Wang, W., Olguin, G., Hotza, D., Seelro, M.A., Fu, W., Gao, Y., Ji, G., 2022, Inorganic membranes for in-situ separation of hydrogen and enhancement of hydrogen production from thermochemical reactions. *Renew. Sustain. Energy Rev.*, 160, 112124.

Wassie, S.A., Cloete, S., Spallina, V., Gallucci, F., Amini, S., van Sint Annaland, M., 2018. Techno-economic assessment of membrane assisted gas switching reforming for pure H2 production with CO_2 capture. *Int. J. Greenhouse Gas Cont.*, 72, 163–174.

Wei, L., Kawamoto, K., 2013. Upgrading of simulated syngas by using a nanoporous silica membrane reactor. *Chem. Eng. Technol.* 36, 650–656.

Yan, S., Maeda, H., Kusakabe, K., Morooka, S., Akiyama, Y., 1994. Hydrogen-permselective SiO_2 membrane formed in pores of alumina support tube by chemical vapor deposition with tetraethylorthosilicate. *Ind. Eng. Chem. Res.*, 33, 2096–2101.

Yang, Y., Le, T., Kang, F., Inagaki, M., 2017. Polymer blend techniques for designing carbon materials. *Carbon*, 111, 546–568.

Yin, H., Yip, A., 2017. A review on the production and purification of biomass-derived hydrogen using emerging membrane technologies. *Catalysts*, 7, 297.

Yoshino, Y., Suzuki, T., Nair, B., Taguchi, H., Itoh, N., 2005. Development of tubular substrates, silica-based membranes and membrane modules for hydrogen separation at high temperature. *J. Membr. Sci.*, 267, 8–17.

Zhang, X.L., Akamatsu, K., Nakao, S., 2016b. Hydrogen separation in hydrogen-methylcyclohexane-toluene gaseous mixtures through triphenylmethoxysilane-derived silica membranes prepared by chemical vapor deposition. *Ind. Eng. Chem. Res.*, 55, 5395–5402.

Zhang, X.L., Yamada, H., Saito, T., Kai, T., Murakami, K., Nakashima, M., Ohshita, J., Akamatsu, K., Nakao, S.I., 2016a. Development of hydrogen-selective triphenylmethoxysilane-derived silica membranes with tailored pore size by chemical vapor deposition. *J. Membr. Sci.*, 499, 28–35.

Zhao, L., Goldbach, A., Xu, H., 2016. Tailoring palladium alloy membranes for hydrogen separation from sulfur contaminated gas streams. *J. Membr. Sci.*, 507, 55–62.

Zhou, L., Yang, J., Li, G., Wang, J., Zhang, Y., Lu, J., Yin, D., 2014. Highly H2 permeable SAPO-34 membranes by steam-assisted conversion seeding. *Int. J. Hydrog. Energy*, 39, 14949–14954.

Zhuang, L., Li, J., Xue, J., Jiang, Z., Wang, H., 2019. Evaluation of hydrogen separation performance of Ni-$BaCe_{0.85}Fe_{0.15}O_{3-\delta}$ cermet membranes. *Ceram. Int.*, 45, 10120–10125.

Zito, P.F., Caravella, A., Brunetti, A., Drioli, E., Barbieri, G., 2017. Knudsen and surface diffusion competing for gas permeation inside silicalite membranes. *J. Membr. Sci.*, 523, 456–469.

Section III

Absorption and Absorption Methods for Hydrogen Purification

11 Hydrogen Purification by Ionic Liquids and Deep Eutectic Solvents

Bilal Kazmi and Syed Ali Ammar Taqvi

11.1 INTRODUCTION

The rise of the world's population, combined with the effects of urbanization and industrialization, is one of the primary contributors to the ever-rising need for energy (U.S. Energy Information Administration, 2019). As a result, it is essential to research and develop renewable sources of energy that may satisfy this ever-demand while simultaneously minimizing the negative effects of energy production on the natural environment (Yadav and Mishra, 2013). The urgency of lowering emissions of greenhouse gases and mitigating the effects of climate change is the primary underlying reason behind the exploration of renewable sources of energy (Cownden, Mullen, and Lucquiaud, 2023). Burning fossil fuels like coal, oil, and natural gas are the principal source of greenhouse gas emissions, which are contributing to both global warming and climate change (U.S. Energy Information Administration, 2019).

Hydrogen (H_2) is regarded as a clean energy source since it can be produced using processes that are both renewable and low-carbon, and when it is burned, it does not result in the release of any greenhouse gases (Dawood, Anda, and Shafiullah, 2020). The primary benefits that H_2 offers as a fuel for clean energy are: because the sole by-product of using H_2 as fuel is water vapor (H_2O), this makes H_2 an eco-sustainable and emission-free substitute for conventional fuels such as fossil fuels (Kazmi et al., 2023). H_2 is attributed to its high energy density, which indicates that it stores a significant quantity of energy relative to its volume. Because of this, it is a fuel that has very high efficiency and can be readily stored and transported. H_2 is a versatile energy carrier since it may be put to use in a variety of contexts, including the transportation sector, the production of electricity, and many industrial operations (Yang and Li, 2023). H_2 can also be created by low-carbon technologies such as steam methane reforming with technology that captures and stores carbon, making it a viable alternative to fossil fuels. Hence, it can be attributed that H_2 can be produced and stored locally, which lessens our reliance on fossil fuel supplies sourced from other countries and increases our energy independence. Moreover, enhanced scale development of H_2 production and utilization results in costs decreasing in the future as a result of technological advancements and economies of scale, which will make H_2 an increasingly cost-effective alternative to fossil fuels. Unlike fossil fuels, H_2 does not

DOI: 10.1201/9781003382522-14

produce any harmful pollutants such as particulate matter, nitrogen oxides (NO), or sulfur dioxide (SO_2), which helps improve air and H_2O quality (Dash, Chakraborty and Elangovan, 2023).

H_2 is a source of energy that is both clean and effective, and it may be put to use in a multitude of different ways, including the production of electricity and the operation of manufacturing processes (Khojasteh Salkuyeh, Saville and MacLean, 2017). However, for H_2 to be useful, it must first be sequestered from other gases and then purified. This seems to be critical to achieving the required levels of purity (Muhammad Mustafa Rizvi et al., 2023). There are a few different approaches for separating and purifying H_2; some of which include pressure swing adsorption (PSA) (Kakiuchi et al., 2023), Membrane-based process (Bernardo et al., 2020), cryogenic process (Agrawal et al., 1988), and physical/chemical absorption (Hren et al., 2023). The PSA method makes use of a bed of adsorbent material, such as zeolite, to selectively adsorb pollutants like CO_2 and H_2O, while still permitting H_2 to permeate through the system (Santos and Hanak, 2022). After that, the pressure is lowered to renew the bed, which in turn releases the impurities and makes it possible to reuse the adsorbent. The cryogenic distillation process separates H_2 and other gases by taking advantage of the disparity in their respective boiling points (Naquash et al., 2023). After being brought down to a low temperature, the combination of gases is cooled, which causes the H_2 to condense and separate from the other gases. The Membrane Separation technique separates H_2 from other gases by using a membrane that allows certain gases to pass through but not others. H_2 molecules are of a size that allows them to pass through the membrane, whereas contaminants such as carbon dioxide (CO_2) and H_2O molecules are too large and are therefore prevented (Salim and Ho, 2018).

In recent times, the ionic liquid (IL) separation method has gained much attention as a solvent that selectively binds to impurities such as CO_2 and H_2O while allowing pure H_2 to pass through. After heating, the IL is free of impurities and can be reused. Distinctive features, such as low vapor pressure, good thermal stability, and high solubility, separate the class of substances known as "ILs." Because of these characteristics, they can be used as solvents in a wide variety of chemical procedures, including the purification of H_2.

ILs are used to purify H_2 by passing it through a column that also contains the polluted gas (Alves and Cserj, 2009). Pure H_2 is allowed to pass through, whereas contaminants like CO_2 and H_2O are selectively absorbed by the IL (Tu et al., 2022). Once the IL has been heated to remove the contaminants, it can be reused.

ILs offer several benefits over more conventional H_2 purification processes, including reduced energy use and increased selectivity (Alves and Cserj, 2009). ILs are more environmentally friendly than other methods of H_2 purification since they are nonflammable, non-toxic, and recyclable (Kazmi, Taqvi, and Ali, 2022). Low vapor pressure, high thermal stability, and high solubility are just a few of the peculiar properties exhibited by a class of substances known as ILs (Haider et al., 2022). Their stated properties make them appropriate for implementation as solvents in distinct chemical procedures, encompassing the purification of H_2. The process of purifying H_2 gas through the use of ILs involves the passage of impure H_2 gas through a column that is filled with an IL. The IL exhibits preferential affinity toward impurities, namely CO_2 and H_2O, thereby enabling the unhindered passage of pure H_2 (Gouveia et al., 2021). Lower energy usage and more selectivity are merely two of the many benefits of using ILs to purify H_2 instead of more conventional approaches. In addition to being a more environmentally friendly solution for H_2 purification, ILs are nonflammable, non-toxic, and recyclable. It's important to remember that there are still obstacles to H_2 production and use, including the high cost of manufacturing and storage and the requirement of targeted technology. However, research and development efforts are making progress toward solutions to many of these issues.

11.2 PRINCIPLES AND METHODS

The most common approaches to H_2 purification and separation include pressure swing adsorption (PSA), cryogenic separation, and membrane separation. After separation, the H_2 is purified using distillation, adsorption, and catalytic oxidation. There are benefits and drawbacks to each of these approaches (Figure 11.1). The description for the conventionally most used technologies are listed below.

11.2.1 MEMBRANE-BASED PURIFICATION OF H_2

To purify H_2 using a membrane, only H_2 molecules are allowed to pass while all other gases are hindered (Peschel, 2020). Since H_2 molecules are very small, they can flow through the membrane, but molecules of other gases are too large and are thus blocked. The membrane is constructed from a very thin layer of material (usually polymeric, inorganic, or metal) that prevents gas from passing through. H_2 molecules can pass across the membrane, while other gases including nitrogen (N_2), O_2, and CO_2 are kept out.

Membrane-based H_2 purification usually involves this scenario, the feed gas is introduced to the membrane module. Hydrogen molecules in the input gas will pass freely across the membrane (He, 2017). The concentration difference between the input gas and the permeate side of the membrane is the factor that initiates this process. The feed gas's other components, because of their greater size or different

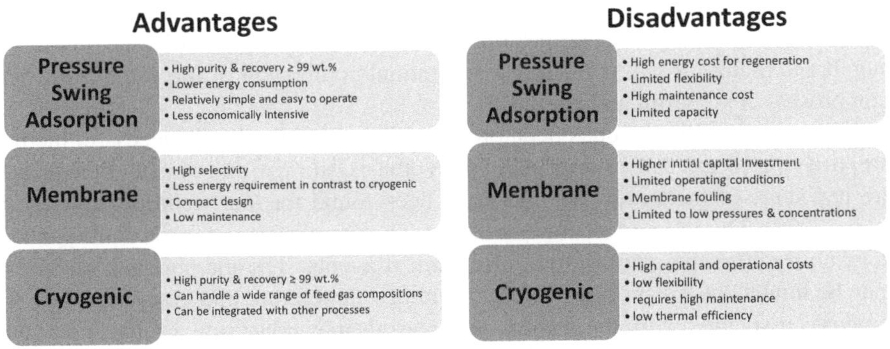

FIGURE 11.1 Advantages and disadvantages of the conventional technologies used for the separation and purification of H_2 (Bernardo et al., 2020).

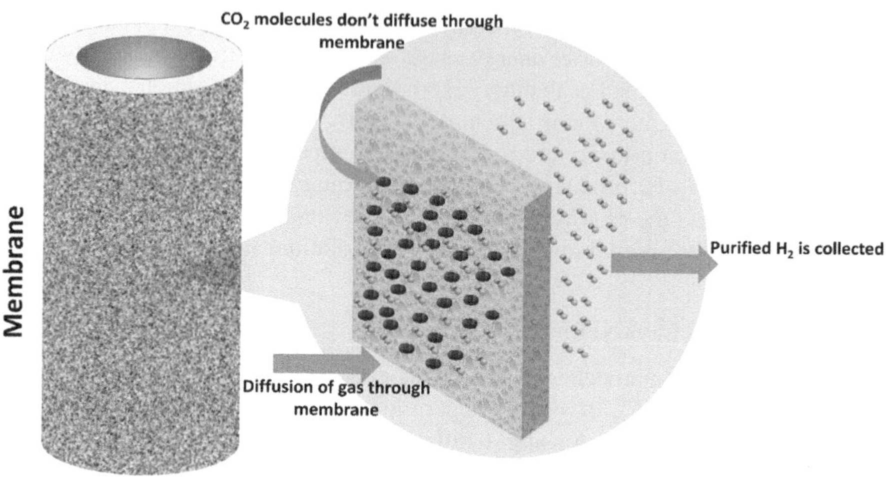

FIGURE 11.2 Schematic diagram of membrane-based process for separation and purification of H$_2$.

chemical properties, will be unlikely to diffuse across the membrane (Lei et al., 2021). Purified H$_2$ is retained on the other side of the membrane, while the non-permeated gases are expelled. The way a polymeric, inorganic, or metallic membrane accomplishes its function is membrane-specific. Physical separation is the method by which polymeric membranes function, while chemical contact is the process by which inorganic and metallic membranes function. The schematic diagram of the process is shown in Figure 11.2.

Polymeric membranes exhibit selective permeability, allowing for the passage of H$_2$ molecules owing to their relatively small size, while impeding larger molecules such as O$_2$ and N$_2$ (Shao et al., 2009). In the context of inorganic and metal membranes, the interaction between H$_2$ molecules and the membrane surface results in the absorption of H$_2$ and subsequent permeation, whereas other gases are unable to interact with the membrane surface and are consequently impeded. It is noteworthy that the membrane's selectivity can be influenced by various factors, including temperature, pressure, and the composition of the feed gas. Additionally, the cost of the membrane materials can constitute a substantial component of the overall purification process cost (Salim and Ho, 2018).

Membrane purification has been proven to be an efficient and successful method for purifying H$_2$ gas from impurities. The synthesis, storage, and transfer of hydrogen are just some of the many uses that have been found for this technique (Bernardo et al., 2020). The process is characterized by its simplicity and affordability, and it does not require the consumption of chemical agents. The membrane's selectivity may be influenced by various factors, including temperature and pressure, while the expenses associated with the membrane materials may constitute a substantial component of the purification process's total cost.

11.2.2 Pressure Swing Adsorption-Based H_2 Purification

PSA is a method of H_2 purification that uses adsorption to separate H_2 from other gases (Li et al., 2016a). The procedure involves the passage of a feed gas mixture, such as natural gas, through a high-pressure bed of adsorbent material, such as activated carbon or zeolite. The adsorbent substance exhibits a selective adsorption behavior toward impurities, namely CO_2, methane (CH_4), and H_2O, while simultaneously permitting the passage of H_2 gas (Silva et al., 2013).

In the PSA process, the feed gas is compressed to a high pressure, typically between 100 and 150 bar (Zhang et al., 2021). The process involves the passage of compressed feed gas through a bed of adsorbent material, which selectively adsorbs impurities, such as CH_4 and H_2O vapor (Zhang et al., 2023). Subsequently, the H_2 is permitted to traverse the bed. Following a specific duration, the pressure within the adsorption bed is reduced, leading to the desorption of impurities and the regeneration of the adsorbent material. Subsequently, the elimination of impurities from the adsorbent substance is carried out through diverse techniques, including but not limited to, heating or purging with a clean gas. Subsequently, the feed gas is subjected to compression and subsequently directed through the adsorption bed once more, thereby initiating a recurring cycle. The schematic diagram of the process is illustrated in Figure 11.3.

PSA offers a significant benefit in that it is a constant procedure that can be employed for the substantial-scale purification of hydrogen gas. Furthermore, the PSA method can attain higher levels of hydrogen purities, generally above 99.99% (Luberti and Ahn, 2022). Nevertheless, the process of PSA is characterized by a considerable consumption of energy and may entail significant operational

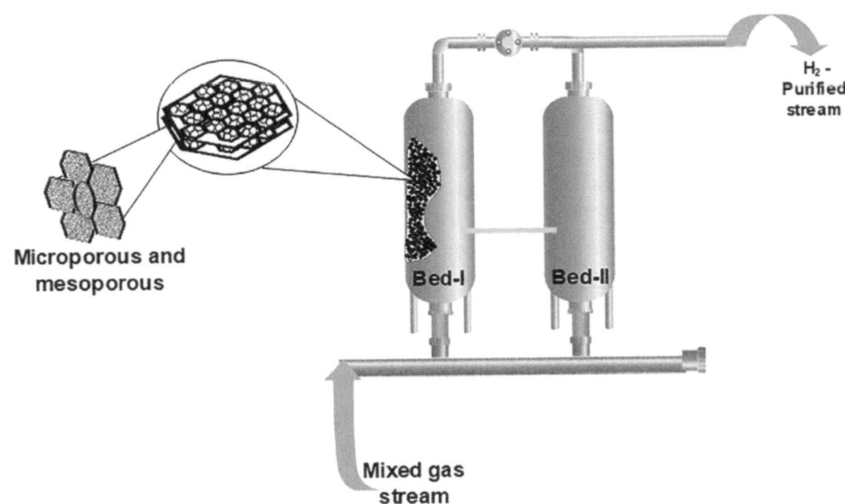

FIGURE 11.3 Schematic diagram of the pressure swing adsorption technique for the purification and separation of H_2.

expenses. The utilization of this technology is primarily observed in industrial settings, specifically in the production of H_2 from natural gas, the purification of hydrogen for fuel cell systems, and the purification of H_2 for various industrial processes (Wu, Lan, and Yao, 2023).

11.2.3 Cryogenic-Based Purification of H_2

The technology utilized for the separation and purification of H_2 through cryogenic means is particularly efficient, resulting in the attainment of exceptionally high levels of purity, typically reaching 99% (Aasadnia, Mehrpooya, and Ghorbani, 2021). Nevertheless, it is a rather complex and energy-demanding process, requiring substantial energy inputs for the cooling of the feed gas and the preservation of the low temperatures (Mehrpooya et al., 2021). Cryogenic H_2 separation and purification can be costly due to the equipment and infrastructure needed. Its primary applications are in H_2 production from natural gas and other hydrocarbons, as well as other industrial settings requiring H_2 of high purity (Genovese et al., 2023). Cryogenic H_2 separation is also used for H_2 purification in H_2 fuel cell systems.

The cryogenic technology utilized for H_2 separation and purification operates on the fundamental principle of gas liquefaction. The process of cooling gases to temperatures below their boiling point can result in their liquefaction. At a temperature of $-253°C$, H_2 undergoes a phase transition from a gaseous state to a liquid state, whereas other gases maintain their gaseous state. This temperature is commonly referred to as the boiling point of H_2 (Agrawal et al., 1988). Initially, the process mechanism involves cooling the feed gas, which is a mixture of gases, to exceptionally low temperatures. There exist several techniques to accomplish this task, including expansion cooling, adiabatic expansion, and Joule–Thomson expansion. Under these conditions of reduced temperature, the H_2 present in the feed gas will undergo a phase transition to a liquid state, while the other constituent gases will maintain their gaseous state. Subsequently, the liquid H_2 is isolated from the remaining gases through the utilization of a separator, which may take the form of a centrifugal or gravity separator. Subsequently, the liquid H_2 is subjected to additional purification procedures to eliminate any impurities that may be present, including but not limited to H_2O, oxygen (O_2), and CH_4. Several methodologies can be utilized to achieve this objective, such as distillation, adsorption, or catalytic oxidation. Following that, the purified aqueous substance H_2 is converted into a gaseous form to facilitate its preservation and transportation.

The process of separating H_2 from other gases can be achieved through the use of a separator after undergoing liquefaction. Separation of H_2 gas from other gases can be accomplished by employing a centrifugal separator that utilizes centrifugal force, or a gravity separator that takes benefit of the difference in density between the gaseous H_2 and other gases.

After undergoing the separation process, it becomes feasible to apply additional purification methods to the liquid H_2, including distillation, adsorption, or catalytic oxidation. Distillation is a method used to purify a liquid substance, such as H_2, by heating it to create vapor and then condensing it back into a liquid state. This process effectively separates impurities from the original substance. The process of adsorption entails the utilization of a substance, such as activated carbon, to adsorb contaminants from the H_2. The process of catalytic oxidation entails the utilization of a catalyst, such as platinum, to facilitate the oxidation of impurities present in the liquid H_2.

To complete the process, the liquid H_2 is compressed back into a gaseous state. A vaporizer can be used to accomplish this task by reverting H_2 from a liquid to a gas by applying heat to it. Cryogenic technology is an effective means of H_2 purification, but it is also a complex and power-intensive process. Its primary application is the purification of H_2 for use in hydrogen fuel cell systems and the manufacture of hydrogen from natural gas and other hydrocarbons.

11.2.4 Factors Affecting the Separation and Purification of H_2

Several factors can affect the purity and separation of H_2 (Figure 11.4). All of these factors contribute to lower process intensification and achieve an optimal design.

The purity of the H_2 produced or separated is highly sensitive to the composition of the feed gas. CO_2, CH_4, and H_2O vapor are instances of impurities that can be challenging to separate from H_2 and reduce the purity of the final product. Feed gas temperature can also have a role in how well H_2 is isolated and purified. Lower temperatures favor cryogenic separation, while higher temperatures favor PSA.

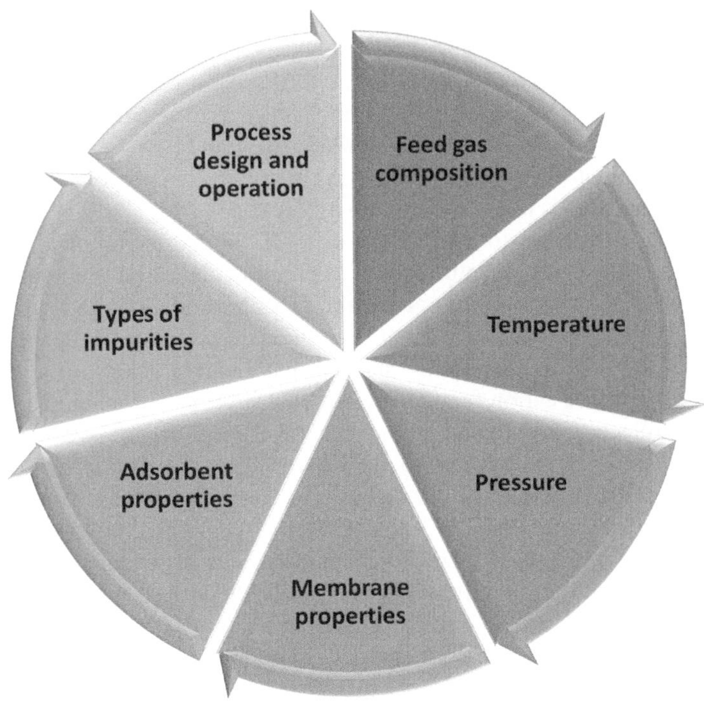

FIGURE 11.4 Schematic diagram of the factors influencing the separation and purification of H_2.

H_2 separation and purity can be impacted by feed gas pressure as well. Some separation techniques, such as PSA, can be optimized at higher pressures. The effectiveness of H_2 separation via membrane separation is very sensitive to membrane parameters such as material, thickness, and porosity. The properties of the adsorbent material, such as the surface area, pore size, and chemical composition, can greatly affect the efficiency of adsorption-based separation methods, such as PSA and adsorption. The amount of impurities within the feed gas may have an impact on the degree of purity and separation achieved for H_2. Some impurities, such as CO_2, pose a greater challenge for separation compared to others, such as H_2O vapor. The purity and separation of H_2 can be significantly influenced by the design and operation of the separation process. Various factors, including the flow rate, number of stages, and regeneration technique, can significantly impact the efficacy of the process.

It is important to consider all of these factors when designing an H_2 separation and purification process to ensure that the final product is of a high enough purity to meet the desired application.

11.3 ADVANCEMENT OF H_2 PURIFICATION AND SEPARATION TECHNOLOGY USING IL/DES

High-pressure H_2 purification procedures can benefit from the use of ILs, a class of solvents comprised of ions and having a very low vapor pressure. They can be used and reused many times over and have a high selectivity for H_2 over other gases like CO_2 and CH_4. However, ILs are still relatively expensive to produce and may not be cost-effective for large-scale H_2 production. DESs are a class of materials that are composed of a mixture of two or more components that form a liquid at low temperatures. They have a low vapor pressure, similar to ILs, and have been shown to have a high selectivity for H_2 over other gases. DESs are often cheaper to produce than ILs, but the selectivity for H_2 is lower than that of ILs.

Shi et al. (2012) have performed a study using IL for the separation of CO_2 and H_2. The results based on the quantum chemical interaction-based methodology of ab initio show that H_2 solubility and permeability in [emim][CH_3COO] are quite low with Henry's law constants about 1×10^4 bar and permeabilities in the range 29–79 barrer at 313–373 K. The results also showed that a higher chain of the alkyl group on imidazolium-based cation functionalized with fluorine on the anion provides six times higher selectivity and permeability for the absorption of H_2. The low H_2 permeability issue can be resolved by using a pyrrolidinium-based cation in situ with acetate as an anion, resulting in the higher selective removal of CO_2 and enhanced separation and purification of H_2. Li et al. (2016b) also studied the separation of CO_2 and H_2 using the amino acid-based IL with polyethylene glycol as the cosolvent. The results indicated that the solubility of H_2 in AAILs-PEG400 is much lower than that of CO_2, which can be attributed to the higher removal of H_2 purified stream during the H_2 purification. The amino acid-based IL in combination with polyethylene glycol 400 provides a higher interface for the efficient removal of CO_2. Hence, from the obtained results, it can be attributed that higher CO_2 capacity and CO_2/H_2 selectivity with reduced viscosity provide higher mass transfer rate and also consequently have high thermal stability and very low volatility, which provides an efficient and practical solution for new H_2 purification solvent. Gouveia et al. (2021) presented the concept of a polymeric IL membrane system for the selective removal of CO_2 from a mixed stream of H_2 and N_2. The results showed that based on the quantum chemical calculation method of COSMO-RS which shows that the mixed H_2, CO_2, and N_2 permeabilities with increasing temperature were shown to be due to the dominant role of gas solubility at low temperature and diffusivity at high temperature. Additionally, the computational approach based on COSMO-RS further explains the experimental mixed gas permeation results. Overall, COSMO-RS predictions revealed that the higher permeability of CO_2 compared to H_2 and N_2 is mainly due to the higher solubility of CO_2 in the IL, while H_2 and N_2 permeabilities seem to be determined by both gas solubility (higher for N_2) and diffusivity (higher for H_2) effects. The small pronounced differences between mixed and ideal CO_2/H_2 permse-lectivities through the prepared PIL–IL composites indicated that membrane separation efficiency can be maintained despite the competition effect between gases in mixed gas experiments. Nagulapati et al. (Mohan et al., 2022) studied the H_2 purification-based system from the process system engineering perspective using Choline chloride-based deep eutectic solvent (DES) on Aspen Plus. The results show that a machine learning approach is being used based on the support vector machine to predict the solubility for mixed gas mixture components resulting in the purification of H_2 with a specific energy consumption of 6.03 kW/kg of H_2. Yusuf et al. (2018) proposed another system based on an adsorption-based technique for the recovery and purity of H_2 as the gas. The proposed process is based on a choline chloride and zinc chloride system for the selective adsorption of CO_2 using activated carbon. The results showed that the activated carbon in combination with proposed IL selectively removes the higher amount of CO_2 and can be a potential system for enhanced purification and separation of H_2 as an eco-friendly approach for effective H_2 recovery. Jomekian et al. (2017) studied the IL liquid in combination with a mixed matrix membrane and studied the effect of selectivity for CO_2 removal. 1,3-Di-n-butyl-2-methylimidazolium chloride (DnBMCl) was selected for sequestration of CO_2, and the incorporation of imidazolium-based IL into Pebax 1657-ZIF-8 further enhanced the separation mechanism with higher interactive interface, which results in better compatibility at particle–polymer interface of mixed matrix membrane and can be an effective solvent and membrane system for higher purification of H_2. Nabais et al. (2019) also studied the combination of polymeric composed IL in conjunction with a suitable mixed matrix membrane for gas separation. In this work, mixed matrix membranes combining a pyrrolidinium-based polymeric IL and metal–organic framework were used resulting in enhancing the CO_2 permeability in the system. Moreover, the results seem to show higher selectivity of CO_2, enhancing the separation and purity of obtained H_2 gas. Thus, it indicates that the proposed polymeric-based membrane system can be a viable approach in the gas separation platform. Ramdin et al. (2014) utilized the imidazolium-based cation functionalized with bis(trifluoromethyl sulfonyl)imide for the separation of the CO_2 and relatively mixed gases. The results show that the proposed solvent is highly selective for CO_2 with almost very lower selectivity for H_2, making it a suitable solvent that can be used in syngas, bio H_2, and precombustion gas purification to separate and recover H_2. Additionally, the obtained solubility data was correlated with the Monte-Carlo-based isotherms to assess the experimental-based solubility of the studied gas system in IL. Jacquemin et al. (2007) studied the solubility o both CO_2 and H_2 by varying the cation while keeping the anion constant. The results obtained show that cation is not the primary function to hinder the solubility of H_2 in IL. The solubility results mark ammonium-based cation to have larger selectivity and less solvation of energy than the imidazolium-based IL. Tu et al. (2022) utilized for the first time a new concept in IL by using protic chlorocuprate-based IL for the selective removal of CO and H_2, which can be effectively used as a raw material for various chemical production. The results indicate that as the CuCl and [TEAH] ratio rises, the permeability for the selected gas increases and enhances the separation process. Alves and Cserj (2009) also studied the effect of using imidazolium-based cation mobilized by varying alkyl groups in combination with fluorine functionalized anion for the selective separation and purification of H_2 from the bio H_2 stream. The result indicated that IL supported membrane provides higher selective separation for the removal of CO_2 and that the less solubility of H_2 is indicative of higher performance in a gas separation process. It was also observed that cation and anion have an effect on the permeability but almost a minor effect on the selectivity in the gas separation process. Zhou et al. (2013) studied the use of 1-n-butyl-3 methylimidazolium heptafluorobutyrate for the separation of CO_2 from the mixture of gases. It was shown that the solubility of CO_2 is higher in the proposed IL than that in [Bmim][CF_3COO] and [Bmim][CF_3SO_3], marking it an effective solvent for gas separation technology. The results also showed that higher fluorination or the addition of a carbonyl group on the anion helps in the ability to capture CO_2 with less solubility of H_2. Barghi, Tsotsis, and Sahimi (2015) studied the solubility and diffusivity of H_2 and CO_2 in the IL 1-butyl-3-methylimidazolium hexafluorophosphate [Bmim][PF_6]. The results indicate that the diffusivity is not the function of the pressure, whereas solubility is the function of both the pressure and temperature. Cserjési et al. (2009) also studied the novel IL-supported membrane-based system for the purification of H_2. The results indicate that the proposed novel IL has a higher solubility for CO_2 than H_2 providing higher selectivity and making it a suitable option for the H_2 purification process. The results also indicate that both the temperature and the transmembrane pressure influence the rate of gas permeation. A higher temperature results in a higher rate of gas permeation, whereas a higher transmembrane pressure results in a lower rate of gas permeation. Table 11.1 lists various forms of IL and DES based on literature finding being used in situ with the separation technologies for the effective separation of H_2.

11.4 CHALLENGES ASSOCIATED WITH THE SEPARATION AND PURIFICATION OF H_2

There are several technological challenges associated with the separation and purification of H_2: One of the main challenges in H_2 purification is achieving high selectivity for H_2 over other gases, such as CO_2, CH_4, and H_2O vapor. This is particularly challenging when using adsorption-based methods, such as PSA, as these impurities can have similar adsorption characteristics to H_2 (Levin and Chahine, 2010). Another challenge is achieving high separation efficiency, particularly when dealing with low concentrations of H_2 in the feed gas. This can be difficult to achieve with

TABLE 11.1

Various Studies for the Purification and Separation of H_2 Using IL/DES

Year	Separation Mechanism	IL Type	IL/DES	CO_2/H_2	References
2012	Absorption	Imidazolium-based IL	[Emim][CH$_3$COO]	37–21	Shi et al. (2012)
2016	Membrane separation with Co-solvent	Amino acid based IL + Polyethylene glycol	[P$_{4444}$][Gly] +PEG-400	120	Li et al. (2016b)
			[P$_{4444}$][Ala] +PEG-400	100–10	
			[P$_{4444}$][Pro] +PEG-400	90–15	
2021	Polymeric membrane	Pyrrolidinium based IL	[Emim][C(CN)$_3$]+ Poly([Pyr$_{11}$][C(CN)$_3$])	8.2–11.4	Gouveia et al. (2021)
			[Bmpyr][NTf$_2$] + Poly([Pyr$_{11}$][NTf$_2$])	5.0–6.6	
			[Emim][NTf$_2$] + Poly([Pyr$_{11}$][NTf$_2$])	6.9	
2018	Adsorption	Choline based ionic liquid	ChCl$_2$ + ZnCl$_2$	–	Yusuf et al. (2018)
2017	Mixed matrix membrane	Imidazolium based IL + Pebax 1657 + ZIF-8	DnBMCl	7–0.8	Jomekian et al. (2017)
2019	Mixed matrix membrane	Polymeric ionic liquid	poly[Pyr$_{11}$][NTf$_2$] + MIL-53(Al)	2.5–12.5	Nabais et al. (2019)
			poly[Pyr$_{11}$][NTf$_2$] + Cu3(BTC)2	~2.5–4	
			poly[Pyr$_{11}$][NTf$_2$] + ZIF-8	~2.5–6	
2014	Absorption	Imidazolium-based IL	[Bmim][NTf$_2$]	30.4	Ramdin et al. (2014)
2007	Absorption	Imidazolium and ammonium-based IL	[Emim][NTf$_2$]	–	Jacquemin et al. (2007)
			[Bmim][NTf$_2$]	–	
			[N$_{4111}$][NTf$_2$]	–	
2022	Membrane	Protic-based IL	[TEAH][CuCl$_2$]	–	Tu et al. (2022)
2009	Membrane	Imidazolium-based IL + PVDF	[BMIM][PF$_6$]	6	Alves and Cserj (2009)
			[HMIM][PF$_6$]	8	
			[OMIM][PF$_6$]	5	
2013	Absorption	Imidazolium-based IL	[Bmim][CF$_3$CF$_2$CF$_2$COO]	–	Zhou et al. (2013)
2015	Absorption	Imidazolium-based IL	[Bmim][PF$_6$]	5.23–9.227	Barghi, Tsotsis and Sahimi (2015)

certain separation methods, such as membrane separation and distillation, which may require multiple stages or large equipment to achieve the desired purity (Dash, Chakraborty, and Elangovan, 2023).

Adsorbents and membranes used in H_2 purification and separation can be fairly costly. This can be an obstacle in the way of H_2 broad use as a renewable energy source. H_2 purification and separation processes are highly demanding on the equipment and materials employed, which must be able to withstand high temperatures and pressures. This can be difficult, especially if you are utilizing a material that degrades over time, like a polymeric membrane. Processes for purifying and separating H_2 can be difficult to scale up from the laboratory to the commercial level due to difficulties including mass transfer restrictions and economic viability. H_2 is a highly combustible gas that requires special handling to prevent accidents. This presents a difficulty in the separation and purification processes, especially when high pressures and temperatures are involved.

Researchers are always working to improve H_2 purification and separation technologies such as upgraded adsorbents, novel membrane materials, and new process designs to meet these problems. It is anticipated that the application of cutting-edge technologies such as AI and ML would aid in the development of better solutions to these problems.

11.4.1 Challenges Associated with the Use of IL/DES for H_2 Separation and Purification

The utilization of IL and DES for the purification of H_2 presents various challenges. The cost of production poses a significant challenge. Both ILs and DESs are comparatively costly to manufacture, which may render them less economically viable to produce H_2 on a large scale (Kazmi, Taqvi, and Ali, 2022). This cost barrier could be reduced by developing more efficient and cost-effective methods for producing these materials.

The selectivity of ILs and DESs poses a significant challenge. Although there is evidence that indicates that they exhibit a considerable degree of selectivity toward H_2 in comparison to other gases, it is possible that their selectivity may not be as pronounced as that of alternative materials employed in the process of H_2 purification. Consequently, a greater amount of ILs or DESs may be required to effectively purify a certain amount of H_2, increasing the overall expenses linked to the entire process. The utilization of ILs and DESs for H_2 separation poses various challenges and opportunities when viewed through the lens of process systems engineering. An obstacle that arises is the integration of ILs and DESs into pre-existing systems for the production and refinement of H_2. The novelty of these materials poses a challenge in terms of comprehensively grasping their properties and behavior, thereby impeding their seamless integration into pre-existing systems. Additional research and development efforts are necessary to enhance the efficiency of these materials within current systems. Furthermore, Mohan et al. (2022) propose the development of proficient and economical process systems for hydrogen separation utilizing ILs and DESs. The distinctive characteristics and functioning of these substances may not be thoroughly recognized, thereby posing a challenge in devising systems that are both efficient and economical. Additional research and development efforts are necessary to enhance the efficacy of these materials in the process of H_2 separation.

Hence, both IL and DES are relatively new materials, and their long-term stability and durability under the conditions used in H_2 purification has not been fully established yet. Therefore, further research is needed to fully understand the stability and durability of these materials over time.

11.4.2 Environmental Perspective of IL/DES for Separation and Purification of H_2

From an environmental perspective, both ILs and DESs are considered to be relatively benign compared to other materials used in H_2 purification and separation. ILs have a low vapor pressure, which means that they are less likely to evaporate and contribute to air pollution (Kazmi, Awan, Hashmi, 2019). In addition, these materials exhibit a reduced level of toxicity and possess the ability to undergo biodegradation, thereby rendering them a more environmentally friendly option in comparison to alternative materials. Nevertheless, the production cost of ILs remains considerably high, and their ecological footprint is contingent upon the particular constituents employed in their fabrication (Kazmi et al., 2023). From an ecological point of view, DESs are likewise thought to be safe. They are biodegradable and have a low vapor pressure, both of which lessen the likelihood of air pollution. They may help maintain the cost of producing H_2 reduced because they are economical for production. Some DESs may be more hazardous or less biodegradable than others, and this is because of the materials employed in their production.

11.5 CONCLUSION

The future of H_2 separation and purification is looking bright with ongoing research and development in the field. ILs and DESs offer a significant advantage in enabling innovative process techniques for H_2 separation. Process simulation software is a proficient approach to enhance and develop H_2 separation and purification procedures that are dependent on ILs and DESs. It allows for the simulation and prediction of the performance of various ILs and DESs under different operational conditions, such as temperature, pressure, and flow rate. It can also aid in the determination of the most favorable temperature and pressure parameters that allow for the effective isolation of H_2 from other gases through the use of ILs or DESs.

Process simulation and life cycle assessment (LCA) can help assess the environmental impacts of H_2 purification and separation methods, including those based on ILs and DESs. LCA can help find sustainable alternatives and reduce side effects, as well as identify process system engineering opportunities for IL/DES-based H_2 separation and purification. Additional investigation and advancement are needed to understand the capabilities of these substances. H_2 separation and purification using ILs and DESs presents a promising opportunity to enhance efficiency and profitability. However, further research is needed to understand the ecological consequences and devise strategies to mitigate any negative impact. Despite this, the prospects of H_2 separation and purification appear promising due to the advent of novel and nascent technologies aimed at enhancing efficacy, affordability, and eco-friendliness.

ACRONYMS

ILs	Ionic liquids
DES	Deep eutectic solvent
LCA	Life cycle assessment
PSA	Pressure swing adsorption
H$_2$	Hydrogen gas
COSMO-RS	Conductor-like screening model for real solvents
CO$_2$	Carbon dioxide
CH$_4$	Methane
N$_2$	Nitrogen gas
O$_2$	Oxygen gas
H$_2$O	Water

REFERENCES

Aasadnia, M., Mehrpooya, M. and Ghorbani, B. (2021) 'A novel integrated structure for hydrogen purification using the cryogenic method', *Journal of Cleaner Production*, 278, pp. 1–15. doi: 10.1016/j.jclepro.2020.123872.

Agrawal, R. et al. (1988) 'Membrane/cryogenic hybrid processes for hydrogen purification', *Gas Separation and Purification*, 2(1), pp. 9–15. doi: 10.1016/0950-4214(88)80036-7.

Alves, V. D. and Cserj, P. (2009) 'Separation of biohydrogen by supported ionic liquid membranes', *International Journal of Hydrogen Energy*, 240(September 2007), pp. 2–6. doi: 10.1016/j.desal.2007.10.095.

Barghi, S. H., Tsotsis, T. T. and Sahimi, M. (2015) 'Solubility and diffusivity of H2 and CO2 in the ionic liquid [bmim][PF6]', *International Journal of Hydrogen Energy*, 40(28), pp. 8713–8720. doi: 10.1016/j.ijhydene.2015.05.037.

Bernardo, G. et al. (2020) 'Recent advances in membrane technologies for hydrogen purification', *International Journal of Hydrogen Energy*, 45(12), pp. 7313–7338. doi: 10.1016/j.ijhydene.2019.06.162.

Cownden, R., Mullen, D. and Lucquiaud, M. (2023) 'Towards net-zero compatible hydrogen from steam reformation: Techno-economic analysis of process design options', *International Journal of Hydrogen Energy*. doi: 10.1016/j.ijhydene.2022.12.349.

Cserjési, P. et al. (2009) 'Study on gas separation by supported liquid membranes applying novel ionic liquids', *Desalination*, 245(1–3), pp. 743–747. doi: 10.1016/j.desal.2009.02.046.

Dash, S. K., Chakraborty, S. and Elangovan, D. (2023) 'A brief review of hydrogen production methods and their challenges', *Energies*, 16(3), p. 1141. doi: 10.3390/en16031141.

Dawood, F., Anda, M. and Shafiullah, G. M. (2020) 'Hydrogen production for energy: An overview', *International Journal of Hydrogen Energy*, 45(7), pp. 3847–3869. doi: 10.1016/j.ijhydene.2019.12.059.

Genovese, M. et al. (2023) 'Current standards and configurations for the permitting and operation of hydrogen refueling stations', *International Journal of Hydrogen Energy*, 48(51), pp. 19357–19371. doi: 10.1016/j.ijhydene.2023.01.324.

Gouveia, A. S. L. et al. (2021) 'CO2/H2 separation through poly (ionic liquid)–ionic liquid membranes : The effect of multicomponent gas mixtures, temperature and gas feed pressure', *Separation and Purification Technology*, 259(November 2020), p. 118113.

Haider, J. et al. (2022) 'State-of-the-art process simulations and techno-economic assessments of ionic liquid-based biogas upgrading techniques: Challenges and prospects', *Fuel*, 314, p. 123064.

He, X. (2017) 'Techno-economic feasibility analysis on carbon membranes for hydrogen purification', *Separation and Purification Technology*, 186, pp. 117–124. doi: 10.1016/j.seppur.2017.05.034.

Hren, R. et al. (2023) 'Hydrogen production, storage and transport for renewable energy and chemicals: An environmental footprint assessment', *Renewable and Sustainable Energy Reviews*. doi: 10.1016/j.rser.2022.113113.

Jacquemin, J. et al. (2007) 'Influence of the cation on the solubility of CO2 and H2 in ionic liquids based on the bis(trifluoromethylsulfonyl)imide anion', *Journal of Solution Chemistry*, 36(8), pp. 967–979. doi: 10.1007/s10953-007-9159-9.

Jomekian, A. et al. (2017) 'Ionic liquid-modified Pebax® 1657 membrane filled by ZIF-8 particles for separation of CO2 from CH4, N2 and H2', *Journal of Membrane Science*, 524(November 2016), pp. 652–662.

Kakiuchi, T. et al. (2023) 'Modeling and optimal design of multicomponent vacuum pressure swing adsorber for simultaneous separation of carbon dioxide and hydrogen from industrial waste gas', *Adsorption*, 29(1), pp. 9–27. doi: 10.1007/s10450-022-00371-x.

Kazmi, B. et al. (2023) 'Exergy-based sustainability analysis of biogas upgrading using a hybrid solvent (imidazolium- based ionic liquid and aqueous monodiethanolamine)', *Biofuel Research Journal*, 10(1), pp. 1774–1785. doi: 10.18331/BRJ2023.10.1.3.

Kazmi, S., Awan, Z. and Hashmi, S. (2019) 'Simulation study of ionic liquid utilization for desulfurization of model gasoline', *Iranian Journal of Chemistry And Chemical Engineering*, 38(4), pp. 209–221.

Kazmi, B., Taqvi, S. A. A. and Ali, S. I. (2022) 'Ionic liquid assessment as suitable solvent for biogas upgrading technology based on the process system engineering perspective', *ChemBioEng Reviews*, 9(2), pp. 190–211. doi: 10.1002/cben.202100036.

Khojasteh Salkuyeh, Y., Saville, B. A. and MacLean, H. L. (2017) 'Techno-economic analysis and life cycle assessment of hydrogen production from natural gas using current and emerging technologies', *International Journal of Hydrogen Energy*, 42(30), pp. 18894–18909. doi: 10.1016/j.ijhydene.2017.05.219.

Lei, L. et al. (2021) 'Carbon molecular sieve membranes for hydrogen purification from a steam methane reforming process', *Journal of Membrane Science*. doi: 10.1016/j.memsci.2021.119241.

Levin, D. B. and Chahine, R. (2010) 'Challenges for renewable hydrogen production from biomass', *International Journal of Hydrogen Energy*, 35(10), pp. 4962–4969. doi: 10.1016/j.ijhydene.2009.08.067.

Li, B. et al. (2016a) 'Pressure swing adsorption / membrane hybrid processes for hydrogen puri fi cation with a high recovery', *Frontiers of Chemical Science and Engineering*, 10(2), pp. 255–264. doi: 10.1007/s11705-016-1567-1.

Li, J. et al. (2016b) 'International journal of greenhouse gas control CO2/H2 separation by amino-acid ionic liquids with polyethylene glycol as co-solvent', *International Journal of Greenhouse Gas Control*, 45, pp. 207–215.

Luberti, M. and Ahn, H. (2022) 'Review of Polybed pressure swing adsorption for hydrogen purification', *International Journal of Hydrogen Energy*, 47, pp. 10911–10933.

Mehrpooya, M. et al. (2021) 'Conceptual design and evaluation of an innovative hydrogen purification process applying diffusion-absorption refrigeration cycle (exergoeconomic and exergy analyses)', *Journal of Cleaner Production*, 316(July). doi: 10.1016/j.jclepro.2021.128271.

Mohan, V. et al. (2022) 'Hybrid machine learning-based model for solubilities prediction of various gases in deep eutectic solvent for rigorous process design of hydrogen purification', *Separation and Purification Technology*, 298(May), p. 121651.

Muhammad Mustafa Rizvi, S. et al. (2023) 'Techno-economic sustainability assessment for bio-hydrogen production based on hybrid blend of biomass: A simulation study', *Fuel*, 347(March), p. 128458. doi: 10.1016/j.fuel.2023.128458.

Nabais, A. R. et al. (2019) 'Poly (ionic liquid) -based engineered mixed matrix membranes for CO2/H2 separation', *Separation and Purification Technology*, 222(November 2018), pp. 168–176.

Nagulapati, V. M. *et al.* (2022) 'Hybrid machine learning-based model for solubilities prediction of various gases in deep eutectic solvent for rigorous process design of hydrogen purification', *Separation and Purification Technology*, 298(June). doi: 10.1016/j.seppur.2022.121651.

Naquash, A. et al. (2023) 'Separation and purification of syngas-derived hydrogen: A comparative evaluation of membrane- and cryogenic-assisted approaches', *Chemosphere*, 313(November 2022), p. 137420. doi: 10.1016/j.chemosphere.2022.137420.

Peschel, A. (2020) 'Industrial perspective on hydrogen purification, compression, storage, and distribution', *Fuel Cells*, 20(4), pp. 385–393. doi: 10.1002/fuce.201900235.

Ramdin, M. et al. (2014) 'Solubility of the precombustion gases CO2, CH4, CO, H2, N2, and H2S in the ionic liquid [bmim][Tf2N] from Monte Carlo simulations', *Journal of Physical Chemistry C*, 118(41), pp. 23599–23604. doi: 10.1021/jp5080434.

Salim, W. and Ho, W. S. W. (2018) 'Hydrogen purification with CO2-selective facilitated transport membranes', *Current Opinion in Chemical Engineering*, 21, pp. 96–102. doi: 10.1016/j.coche.2018.09.004.

Santos, M. P. S. and Hanak, D. P. (2022) 'Techno-economic feasibility assessment of sorption enhanced gasification of municipal solid waste for hydrogen production', *International Journal of Hydrogen Energy*, 47(10), pp. 6586–6604. doi: 10.1016/j.ijhydene.2021.12.037.

Shao, L. et al. (2009) 'Polymeric membranes for the hydrogen economy: Contemporary approaches and prospects for the future', *Journal of Membrane Science*, 327(1–2), pp. 18–31. doi: 10.1016/j.memsci.2008.11.019.

Shi, W. et al. (2012) 'Theoretical and experimental studies of CO2 and H2 separation using the 1-ethyl-3-methylimidazolium acetate ([emim][CH3COO]) ionic liquid', *The Journal of Physical Chemistry B*, pp. 283–295.

Silva, B. et al. (2013) 'H2 purification by pressure swing adsorption using CuBTC', In: *Separations Division 2013- Core Programming Area at the 2013 AIChE Annual Meeting: Global Challenges for Engineering a Sustainable Future*, 118, pp. 744–756. doi: http://dx.doi.org/10.1016/j.seppur.2013.08.024.

Tu, Z. et al. (2022) 'Selective and simultaneous membrane separation of CO and H2 from N2 by protic chlorocuprate ionic liquids', *Renewable Energy*, 196, pp. 912–920. doi: 10.1016/j.renene.2022.06.137.

U.S. Energy Information Administration (2019) 'International Energy Outlook 2019 with projections to 2050', *Choice Reviews Online*. doi: 10.5860/CHOICE.44-3624.

Wu, N., Lan, K. and Yao, Y. (2023) 'An integrated techno-economic and environmental assessment for carbon capture in hydrogen production by biomass gasification', *Resources, Conservation and Recycling*, 188(May 2022). doi: 10.1016/j.resconrec.2022.106693.

Yadav, S. K. and Mishra, G. C. (2013) 'Greenhouse gases emission', nor does it accept any responsibility for the consequences of its use', *International Journal of Engineering Research and Technology*, 6(6), pp. 781–788.

Yang, J. and Li, Y. (2023) 'Study on performance comparison of two hydrogen liquefaction processes based on the Claude cycle and the Brayton refrigeration cycle', *Processes*, 11(3), p. 932.

Yusuf, N. Y. M. et al. (2018) 'Impregnated carbon–ionic liquid as innovative adsorbent for H2/CO2 separation from biohydrogen', *International Journal of Hydrogen Energy*, 44(6), pp. 3414–3424.

Zhang, N. et al. (2021) 'Optimization of pressure swing adsorption for hydrogen purification based on Box-Behnken design method', *International Journal of Hydrogen Energy*, 46(7), pp. 5403–5417. doi: 10.1016/j.ijhydene.2020.11.045.

Zhang, X. et al. (2023) 'Modeling study on a two-stage hydrogen purification process of pressure swing adsorption and carbon monoxide selective methanation for proton exchange membrane fuel cells', *International Journal of Hydrogen Energy*, (48), pp. 1–14. doi: 10.1016/j.ijhydene.2023.01.138.

Zhou, L. et al. (2013) 'Solubilities of CO2, H2, N2 and O2 in ionic liquid 1-n-butyl-3-methylimidazolium heptafluorobutyrate', *Journal of Chemical Thermodynamics*, 59, pp. 28–34.

12 Carbonaceous Sorbents for Hydrogen Purification

*Ahmed Alengebawy, Tanmay Jyoti Deka,
Sagar Ban, Zhonghao Chen, Ahmed I. Osman,
Pow-Seng Yap, and Ping Ai*

12.1 INTRODUCTION

Before using it as a fuel or energy source, hydrogen (H_2) must be purified to remove impurities and achieve a high purity degree (Amin et al., 2023). Hydrogen purification refers to the process of removing impurities from hydrogen gas in order to produce high-purity hydrogen suitable for use in several applications (Du et al., 2021b; Kim et al., 2022). Hydrocarbons, sulfur compounds, halogenated chemicals, and other trace pollutants are the most prevalent impurities in hydrogen. There are several hydrogen purification technologies, such as adsorption, membrane separation, and distillation (Bak et al., 2019). Adsorption is the most preferred approach because it is cost-effective and adaptable and can be utilized for several applications (Gorbounov et al., 2022). Besides, membranes fouling concerns and energy consumption in distillation are obstacles to these technologies (Amin et al., 2023). Therefore, developing cost-effective and environmentally friendly methods to purify hydrogen is a critical issue.

Carbonaceous sorbents are a material used to purify hydrogen gas streams (Wasajja et al., 2020). They are usually formed by activating carbon-rich precursors, preferably renewable feedstocks, such as agricultural biomass, wood, and coconut shells, resulting in a large surface area and permeable structure (Li et al., 2021). Carbonaceous sorbents have a large porosity and surface area, the potential for reusability and regeneration, and stability and durability (Baamran et al., 2023). These characteristics make them extremely effective in capturing hydrogen gas pollutants. There are numerous forms of carbon sorbents that may be used for hydrogen purification. However, we are focusing on the most common five types: activated carbon (AC) (Shamsudin et al., 2019), biochar (BC) (Guo et al., 2022), carbon nanotubes (CNTs) (Bartocci et al., 2020), porous carbon (PC) (Cheng et al., 2022), and carbon molecular sieves (CMS) (Llosa Tanco et al., 2021). Each sorbent has distinct attributes, composition, and benefits, and the choice of sorbent depends on the specific application and the contaminants to be removed.

AC is a sorbent that is commonly utilized in the purification of hydrogen and other gases. Carbon is heated in the presence of an activator, such as steam or carbon dioxide, to produce this substance (AC) (Shamsudin et al., 2019). The resultant AC has a very large surface area and a wide variety of pore diameters, making it particularly

DOI: 10.1201/9781003382522-15

effective against hydrogen gas-expanded contaminants (Far et al., 2021). AC may be used to remove contaminants from hydrogen gas streams, such as carbon dioxide, hydrocarbons, and moisture (Amin et al., 2023). Once AC is used to purify hydrogen, it is typically packed into a bed and a stream of hydrogen gas is fed through it to be purified (Gorbounov et al., 2022). The AC may then be regenerated by heating to a high temperature, which allows the adsorbed impurities to be desorbed and reused (Brea et al., 2019).

BC is a form of AC generated from biomass resources, such as wood, agricultural waste, or grasses. It has been investigated as a possible hydrogen purification sorbent (Manyà & Gascó, 2020). BC can efficiently adsorb different pollutants, such as water vapor and carbon monoxide (Guo et al., 2022). Moreover, BC has a larger surface area and adsorption capacity than traditional AC and may be generated more sustainably and cheaply. It is also a more ecologically friendly alternative because it adsorbs carbon dioxide during manufacturing.

CNTs are promising potential substances for hydrogen cleaning. CNTs are small carbon-atom tubes with a high surface area-to-volume ratio, which makes them valuable sorbents (Cheng et al., 2022). They also demonstrated significant potential as sorbents in various applications, including gas separation, pollutant removal, and energy storage (Bartocci et al., 2020). Besides their high surface areas, the adsorption capability of CNTs can be increased further by the functionalization of the surface with chemical groups (Yang et al., 2020a). They can also be designed to provide selective adsorption capabilities for particular compounds or pollutants, which makes them have several benefits as sorbents in a variety of applications (Kırbıyık & Kuş, 2020).

PC is a material with a densely linked network of pores that is recognized by its highly advanced micro- and mesopore structures, with micropores being crucial for small molecule adsorption (Baamran et al., 2023). Pores are usually classified into three types: macropores, mesopores, and micropores (Manyà & Gascó, 2020). The porous structure of PC allows for the adsorption of many molecules or particles, which is critical for applications, e.g., gas separation, where the selective removal of contaminants is required (Li et al., 2021). To customize the material's interaction with the target molecules, the surface chemistry of PC can be modified using different approaches, such as chemical functionalization or doping (Li et al., 2022). PC reusability and regeneration also contribute to its sustainability and environmental acceptability (Gorbounov et al., 2022).

Besides the common characteristics of the above-mentioned carbonaceous materials, CMS have gained popularity in recent years due to their unique sorption capabilities (Bak et al., 2019; Llosa Tanco et al., 2021). The capacity of CMS as a sorbent to selectively adsorb gas molecules based on their size and shape is its most important benefit (Wahby et al., 2012). This feature is useful in hydrogen purification, which requires the separation of distinct gas molecules depending on their molecular size and shape. CMS may be synthesized in various forms, including powders, pellets, and membranes, making it appropriate for a wide range of applications (Liu et al., 2020a). Moreover, they have excellent thermal and chemical stability, making them long-lasting sorbents (Olabi et al., 2023). It is also worth noting that CMS are energy-efficient sorbents that can help minimize energy usage in gas separation and purification procedures.

One of the important points that must be noted to maximize the benefit of carbonaceous sorbents in hydrogen purification is the origin of these materials. Carbonaceous sorbents should be formed from an environmentally friendly renewable source to reduce greenhouse gases and climate change (Li et al., 2021). Carbonaceous sorbent production from biomass turns the raw material into a carbon-rich solid substance, which is then activated to enhance its porosity and surface area. One of the most frequent ways is pyrolysis, which involves heating the biomass in the absence of oxygen to form a carbon-rich substance (e.g., biochar) (Manyà & Gascó, 2020). This BC can be activated further by chemical or physical techniques. Hydrothermal carbonization is another way to generate carbonaceous sorbents from biomass (Fagnani et al., 2019). By adsorbing carbon dioxide from the environment during biomass development, biomass as a feedstock for carbonaceous sorbents minimizes greenhouse gas emissions. It also encourages the use of renewables, which reduces reliance on fossil fuels (Bos & Broeze, 2020).

Accordingly, the main objectives of this chapter are to give an insight into the use of carbon sorbents for hydrogen purification. Also, discussing the properties and applications of these materials is a more useful point for researchers involved in this field. Moreover, it is an attempt to help select the appropriate purification method based on the main advantages and disadvantages of each technology/material in order to achieve environmental sustainability and preservation of the environment.

12.2 CURRENT HYDROGEN PURIFICATION SORBENTS

To date, scientists have developed a variety of carbon sorbents capable of removing contaminants from hydrogen streams (Baamran et al., 2023). These sorbents will play a growing role in ensuring the cleanliness and dependability of hydrogen streams as the demand for clean hydrogen increases. The following subsections describe the five chosen sorbents in depth, concentrating on their characteristics, production, regeneration, and applications.

12.2.1 ACTIVATED CARBON

12.2.1.1 Physicochemical Properties of Activated Carbon

AC is a carbonaceous material with a heterogeneous crystal structure made by physically or chemically activating a range of carbon-rich compounds (Osman et al., 2019a; Osman et al., 2020a). The microcrystalline structure of AC comprises basic elements such as graphene sheets and fullerene or quasi-graphite pieces that do not spread on a macroscopic scale (Osman et al., 2020b), leading to disordered and highly nanoporous structures (Pallarés et al., 2018). This solid black material is amorphous, has a substantial specific surface area, and has considerable surface reactivity. It resembles powdered or granular charcoal (Yahya et al., 2015). Among different sorbents, AC has a high internal surface area ($500–1500\,m^2g^{-1}$) due to its special sponge-like structure, a well-organized pore size structure ($<1–100\,nm$), and the presence of a wide range of chemical functional groups (carboxylates, carbonyl groups, hydroxyl groups, and amines) on its surface, making it a versatile and multifunctional carbon material (Danish & Ahmad, 2018). Since AC has various surface functional groups, carbon with the same physical attributes may have varying adsorption capabilities for the same sorbents (Zulkefli et al., 2019). The chemical nature of AC indicates the presence of oxygen-containing functional groups that include oxygen, which improves the surface's wettability. Since carbon is amphoteric, it may cohabit with basic and acidic groups on its surface while still demonstrating modest acid–base reactivity. In inert or reducing atmospheres, AC is thermally stable at high temperatures.

12.2.1.2 Activated Carbon Production and Regeneration

By-products or leftovers from industry and agriculture are the primary raw materials that may be cheaply transformed into AC (Nacu et al., 2016). Agricultural wastes are often used for the preparation of AC, such as chestnut shells (Niazi et al., 2018), palm kernel shells (Rashidi & Yusup, 2017), sugarcane bagasse (Osman et al., 2019b), and coconut shells (Tsai & Jiang, 2018). The two-step production process of physically AC consists of pyrolysis of biomass precursors in an inert atmosphere to produce BC, followed by activation of the formed BC in air, CO_2, or steam (Chen et al., 2018). Proper activation temperature and duration are essential to ensure adequate porous development and functional group production, while high temperatures can lead to low carbon yields, pore structure collapse, and ash generation (Zhai et al., 2016). The carbon dioxide and steam activator with heat-absorbing properties facilitate the generation and expansion of microporous pores in AC. Carbonization and activation will occur simultaneously in the presence of the activator at one step of the physical activation process. This method is more straightforward, affordable, and better suited to producing AC.

Physical activation is less flexible than chemical activation, enabling AC to be created with adjustable surface chemistry and porosity (Fuertes et al., 2018). Chemical activators are used to impregnate biomass precursors, and then the impregnated biomass precursors are activated or carbonated (Oginni et al., 2019). In a two-step chemical activation process, AC is created by first carbonizing biomass precursors at temperatures between 300°C and 600°C (Marques et al., 2018). Thermal breakdown causes the gas to produce fumes, tar, and other volatile components while producing pores in the AC. The precursors are soaked in chemicals that activate reactions, such as KOH, H_3PO_4, or $ZnCl_2$ (Pallarés et al., 2018). However, some metal hydroxide activators yield low solids and are environmentally unfriendly, with time-consuming washing procedures and secondary contamination (Rashidi & Yusup, 2017). The production of AC involves the physical and chemical activation of industrial and agricultural waste, as shown in Figure 12.1.

AC regeneration makes possible the stability and recycling of adsorbed AC, the reduction of hazardous waste, and the conservation of natural resources. By absorption or breakdown, this process removes and recovers adsorbents from AC's initial surface area (Nasruddin et al., 2018). Steam, chemical, thermal, microwave-assisted, electrochemical, biological, and moist oxidation are only a few regeneration approaches. Steam regeneration promotes contaminant desorption by rapidly heating the bed and significantly restoring the structural properties of the original AC (Jeong

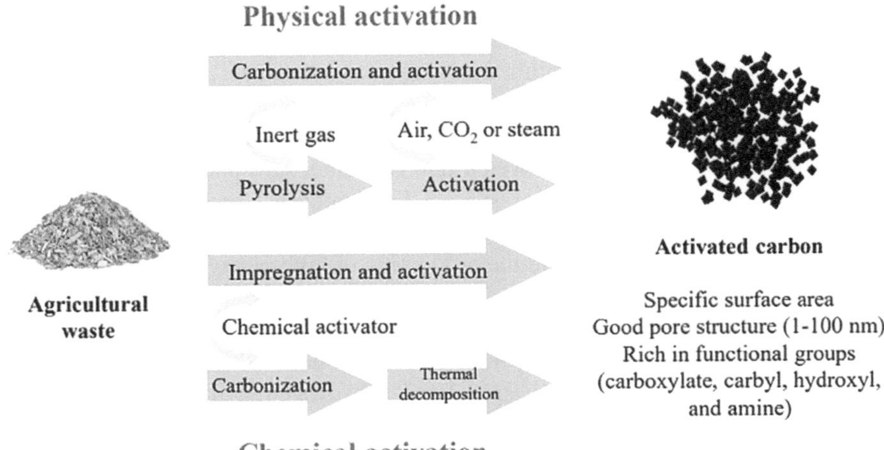

FIGURE 12.1 Activated carbon production by physical and chemical activation.

et al., 2020). Pyrolysis in an inert atmosphere and thermal regeneration of AC using oxidizing gases to gasify organic pollutants are considered the best way to deplete the sorbent (Sun et al., 2020). However, it will affect its mechanical strength; much carbon loss occurs in reducing carbon strength and size (Márquez et al., 2022).

In contrast, chemical regeneration using chemicals to desorb contaminants from the AC does not cause carbon wear (Li et al., 2015). Electrochemical regeneration occurs when an alternating current is placed between two electrodes, and when an electric field is applied, in situ cracking of the contaminant is stimulated (McQuillan et al., 2018). This method is more efficient in terms of energy consumption and has lower CO_2 emissions (Sharif et al., 2017). Dielectric heating at specific frequencies is a microwave regeneration method that penetrates 10–20 mm of AC and allows friction and collisions resulting from the rotation and movement of polar molecules (Ao et al., 2018). The presence of AC creates a liquid–solid surface, which is adapted to the new metabolism of microorganisms and enables biological regeneration on AC surface (El Gamal et al., 2018). Adsorption by AC and the synergistic biodegradation of organic materials by microorganisms result in increased adsorption by AC. Used AC is heated while floating in an aqueous solution, desorbing the impurities. The native solution's oxygen is then permitted to oxidize the desorbed pollutants, with a sizable portion of the desorbed contaminants present in the aqueous phase (El Gamal et al., 2018).

12.2.1.3 Application of Activated Carbon in Hydrogen Purification

Structural properties of AC due to agglomeration or pore-blocking effects are crucial for physical adsorption gas storage and selective gas separation. The industrially produced hydrogen is mixed with a small amount of CO_2 (Osman et al., 2022b). Therefore, AC is widely used in the separation of CO_2 and purification of hydrogen. Proton exchange membrane fuel cells (PEMFCs) employ AC as a sorbent for the first CO adsorption, demonstrating the dependability of this material (Zhang et al., 2023).

Shamsudin et al. (2019) used AC derived from palm kernel shell with a surface area of $698\,m^2\,g^{-1}$ to achieve excellent CO_2 capture by variable pressure adsorption, completing the recovery of hydrogen within 5 min (recovery: 88.43%) and achieving almost 100% purity. Moreover, Shi et al. (2018) employed the extremely adaptable variable pressure adsorption approach to examine AC, silica gel, and 5A zeolite for the purification of hydrogen and discovered that AC had a substantially greater adsorption capacity for various gases than silica gel and zeolite. Combining the two-stage vacuum/vapor adsorption process allows for the capture of CO_2 from steam methane (CH_4) reforming gas mixtures and the production of high purity (99.995%) hydrogen using AC (Liu et al., 2020b).

In this context, Moon et al. (2018) developed an eight-layer bed variable pressure adsorption process by filling AC to the bottom of the bed with zeolites to adsorb CO_2 from syngas, which could produce 99.99+ mol% hydrogen with a recovery of 89.7%. He et al. (2020) used an organic–inorganic binder with exceptional mechanical strength to bond AC from coconut shells with a nickel foam skeleton. The AC's ability to absorb CO_2 was enhanced by applying more pressure. This material was calcined to obtain high bond strength and toughness with higher loading than hydrothermal synthesis. It exhibits better CO_2 adsorption in hydrogen and provides a cost-effective solution for hydrogen purification. Furthermore, Shamsudin et al. (2019) studied the purification of hydrogen from CO_2 using PSA with AC made from palm kernel shells. The adsorption system was able to collect 88.43% of the hydrogen produced while maintaining purity levels close to 100%. The PSA efficiency for hydrogen purification with pressure equalization from syngas was proved by the low energy demand along with the high degree of hydrogen purity and recovery.

12.2.2 BIOCHAR

12.2.2.1 Physicochemical Properties of Biochar

BC, a high carbon-containing solid fuel produced by thermochemical biomass conversion in a limited O_2 environment, has different physicochemical characteristics. These characteristics can be changed based on the feedstock, pyrolysis conditions, and post-treatment procedures (Qiu et al., 2021). Due to BC's high porosity and surface area, its potential to absorb nutrients, water, and contaminants is enhanced. According to previous studies, BC made at higher pyrolysis temperatures tends to have more porosity and surface area (Cha et al., 2016). With the gradual organic decomposition of lignin and cellulose and the development of channel structures or vascular bundles, pyrolysis may increase pore volumes and the surface area of BC (Zhao et al., 2017). The capacity of adsorption and removal mechanism of adsorbate depend on the concentration and type of surface functional groups present in BC (Tomczyk et al., 2020). As per the Fourier Transform Infrared (FTIR) spectra, functional groups of oxygenated hydrocarbons dominate BC, which reflects the carbohydrate structure of hemicellulose and celluloses (Ghani et al., 2013).

BC's pH value is dependent on pyrolysis circumstances and feedstock type, directly impacting the soil pH. A more alkaline pH level in BC is achieved with higher pyrolysis temperatures (Mukherjee et al., 2011). The increase in ash content and oxygen functional groups that occur in the pyrolysis process is the reason for higher pH with increasing temperature (Zhao et al., 2017). Although soil carbon storage for a longer time is achieved through BC, its stability is influenced by its other physicochemical characteristics. Higher pyrolysis temperatures are likely to create more stable BC resistant to microbial deterioration (Cha et al., 2016). The BC constancy is affected by its pH, pore structure, particle size, sorption process, surface area, and minerals (Yaashikaa et al., 2020). BC can impact the electrical conductivity of soil through its mineral content, which directly impacts plant development and nutrient availability. The type of feedstock used directly impacts the BC mineral content. Higher levels of potassium and calcium were found in rice straw BC compared to wheat straw BC (Tian et al., 2021). Moreover, higher levels of calcium, potassium, and magnesium were found in BC produced from hardwood (Mukherjee et al., 2011). Ash concentration and the presence of conductive minerals like calcium, magnesium, and potassium can also determine the electrical conductivity of BC. Also, the BC cation exchange capacity depends on the nature and distribution of O-functional groups on the BC surface (Banik et al., 2018).

12.2.2.2 Biochar Production and Regeneration

Pyrolysis, hydrothermal carbonization, gasification, and torrefaction are common thermochemical techniques used for BC production. Pyrolysis is considered the most popular method used for producing BC. During pyrolysis, under an oxygen-limited environment, the organic materials present in biomass disintegrate at a specific temperature. Process temperature, residence duration, biomass type, and heating rate all impact the pyrolysis product. Temperature is the primary operational process condition that influences product efficiency. With the rise in temperature during pyrolysis, the BC yield decreases while the syngas production increases (Wei et al., 2019; Yaashikaa et al., 2020). The pyrolysis process can be categorized into other subclasses considering different operating conditions, including fast pyrolysis, slow pyrolysis, vacuum pyrolysis, flash pyrolysis, intermediate pyrolysis, and hydro pyrolysis (Tripathi et al., 2016). Without the high energy-consuming pre-drying step, the wet feedstock is converted into hydrochar, a final carbonaceous material, through hydrothermal carbonization (Kumar et al., 2017). However, for the post-treatment of hydrochar, additional thermal energy is needed (Kambo & Dutta, 2015). For gasification-based BC production, different feedstocks, including biomass, are converted into CO and H-rich gases at high temperatures (~700°C) in a controlled environment. Syngas, which primarily contain hydrogen, is the process' primary output; char, along with ash, tar, and some oil, is considered a waste by-product with a relatively low yield (Dasappa & Shivapuji, 2021). Nevertheless, because the BC production from gasification is minimal, commercial usage of gasified BC is not suggested (Yuan et al., 2019). Torrefaction is a different pretreatment technique frequently used to produce BC, and it aids in volatile chemical removal to enhance biomass quality (Chen et al., 2015; Yu et al., 2017). Nevertheless, substantial levels of heavy metals persist in the ash after torrefaction, which is a significant obstacle to using BC as a soil supplement (Kambo and Dutta, 2015).

BC regeneration is considered the inverse of the adsorption process. There are two regeneration principles: (i) adsorbate desorption and (ii) adsorbate decomposition (Omorogie et al., 2016). Regeneration through the adsorbate desorption approach uses heat or non-thermal means to break the connections between the pollutants and BC surface. On the other hand, absorbate decomposition is dependent on entirely mineralizing the adsorbed contaminants or converting them into less harmful by-products, allowing the adsorption capacity of the BC to get restored (Dai et al., 2019; Liao et al., 2022). BC regeneration and cost reduction can be accomplished by using different technologies to treat different types of BC: (i) Thermal regeneration, (ii) microwave irradiation regeneration, (iii) solvent regeneration, and (iv) supercritical fluid regeneration. Thermal regeneration is only employed for waste adsorbate with no recovery requirement, as it is a destructive process. Because oxidation and ultrasonic breakdown methods typically employ considerable thermal energy, they might be classified as a thermal decomposition process. The cost of thermal regeneration is higher because of the high consumption of energy (temperatures up to 800°C), equivalent to around 50% of fresh BC production cost. The sorbents get preserved with non-thermal methods and can be used for many cycles (Dai et al., 2019). BC can also be used effectively for wastewater treatment (Alengebawy et al., 2021), and its operational cost can be significantly reduced because it can be used for multiple process cycles with regeneration. Reusing and recycling sorbents can make wastewater management and gas purification more economically efficient (Liao et al., 2022; Wan et al., 2018).

12.2.2.3 Application of Biochar in Hydrogen Purification

The main impurities produced in conventional hydrogen production through processes like hydrocarbon reforming, biomass gasification, and hydrocarbon pyrolysis are CO_2, CO, N_2, O_2, Cl, and CH_4 (Ihonen et al., 2017). Among these, the presence of CO is considered as most crucial impurity because of its wide presence in hydrogen fuels and its noxious effect on PEMFC in terms of performance and longevity (Pei et al., 2020). Since CO can be produced from CO_2 through reverse water gas shift reaction (WGSR), the removal of CO_2 is also necessary alongside CO for the purification of hydrogen. The superior surface structure of BC is advantageous for efficient CO_2 adsorption and a good replacement for traditional C-based catalysts; both were found to be expensive and unfriendly to the environment (Guo et al., 2022). Cox and Mokaya (2017) studied the CO_2 capture performance of mesoporous carbons (activated BC) with $2,800$–$4,000\,m^2g^{-1}$ surface area and 2.5–$3.6\,cm^3g^{-1}$ pore volume for hydrogen purification. Massive volumetric uptake of up to $480\,g\,L^{-1}$ (at 20 bar), $640\,g\,L^{-1}$ (at 30 bar), $780\,g\,L^{-1}$ (at 40 bar), and $930\,g\,L^{-1}$ (at 50 bar) are achieved by the mesoporous carbons' strong gravimetric CO_2 uptake with 0.25–$0.4\,g\,cm^{-3}$ packing density (at 50 bar). A storage capacity of up to $470\,cm^3$ at 50 bar was found for mesoporous carbon, which is more than ten times the amount of CO_2 that they can hold at 30 bar in a pressurized cylinder. They have found this mesoporous carbon material suitable for hydrogen purification (Cox & Mokaya, 2017).

So far, few studies have evaluated the efficacy of BC-derived carbonaceous sorbents for the adsorption of CO_2 in a CO_2/H_2 mixture in terms of capacity absorption and selectivity. Most studies using AC, such as You et al. (2012), who investigated hydrogen purification using the pressure swing adsorption (PSA) method for two different H_2-rich gas mixtures. The volumetric % of $H_2/CO/CO_2/N_2$ composition of Feed 1 and Feed 2 are 99:0.1:0.05:0.85 and 95:0.3:0.1:0.05:4.55, respectively. Using AC under 6.5 bar adsorption pressure, CO removal of 1.1 ppm with hydrogen purity and recovery of more than 99.99% and 80.0%, respectively, can be achieved for Feed 1. Similarly, for Feed 2, a CO removal of 6.7 ppm with 99.96% hydrogen purity and 78.4% recovery was attained under the same adsorption conditions. They also found that with higher packing density beds in the PSA process, the CO removal and hydrogen purification were improved. He et al. (2020) used laminate-structured AC of a specific surface area of 1,309 m^2/g with high mechanical and binding strength for hydrogen purification. They found a rise in structured sorbent's CO_2 adsorption capacity with pressure. At 0.4 MPa and 200 mL min^{-1}, the static adsorption capacity was found to be 3.94 mmol g^{-1}, while the dynamic capacity was found to be 1.99 mmol g^{-1}.

12.2.3 Carbon Nanotubes

12.2.3.1 Physicochemical Properties of Carbon Nanotubes

CNTs are considered as the most distinctive nanotechnology-based material (Norizan et al., 2020; Sajid et al., 2022). The carbon allotrope known as CNTs has sp^2 hybridization. CNTs are regarded as graphene that has been rolled up and has a nanostructure with a length-to-diameter (L/D) ratio of 1,000,000 or more. CNTs can be classified into single-walled CNTs (SWCNTs), double-walled CNTs (DWCNTs), and multi-walled CNTs (MWCNTs). Figure 12.2 shows the structure of graphene, graphite, SWCNTs, and MWCNTs. There have been numerous studies on the physical properties of CNTs for use in mechanical, electrical, and thermal applications, such as structural composites, heat sinks, health science, and microelectronics. CNTs have some novel characteristics, including their low weight, high aspect ratio, high surface area, compact size, and high mechanical strength properties, making them beneficial for diverse applications (Gupta et al., 2019; Qiu & Yang, 2017). Due to differences in their structural composition, SWCNTs and MWCNTs exhibit different physical characteristics. SWCNTs are significantly thinner than MWCNTs because MWCNTs comprise several concentric cylinders of SWCNTs, while the inner tubes in MWCNTs are protected by the outer tube. DWCNTs are a synthetic combination of both SWCNTs and MWCNTs, and compared to SWCNTs and MWCNTs, it possesses superior thermal and electrical stability and flexibility. DWCNTs have many of the same characteristics and morphologies as SWCNTs, except for their much higher chemical resistance (Bhatt et al., 2016). The physical characteristics of CNTs include electrical, mechanical, vibrational, and thermal properties. Table 12.1 shows a comparison of the physical characteristics of SWCNTs and MWCNTs (Gupta et al., 2019).

The chemical properties of CNTs include solubility, functionalization, and toxicity. Because of their hydrophobic nature, CNTs are insoluble in water, other solvents, and polymeric media. They can be sonicated in various solvents to scatter them, but if the operation is stopped, precipitation will start right away. Nonetheless, they may form suspensions in toluene, DMF, THF, and other solvents despite SWCNT's general insoluble nature (Singh & Gupta, 2016). CNTs can be functionalized to increase their internalization potential into cells, enhance their dispersibility, minimize individual nanotube aggregation, and provide a competent intracellular uptake capability. CNTs may be altered by the surface addition of different functional groups. This is done for the dispersion of the nanocomposites based on carbon by employing grafting peptides, surfactants, nucleic acid, and other substances (Bhatt et al., 2016).

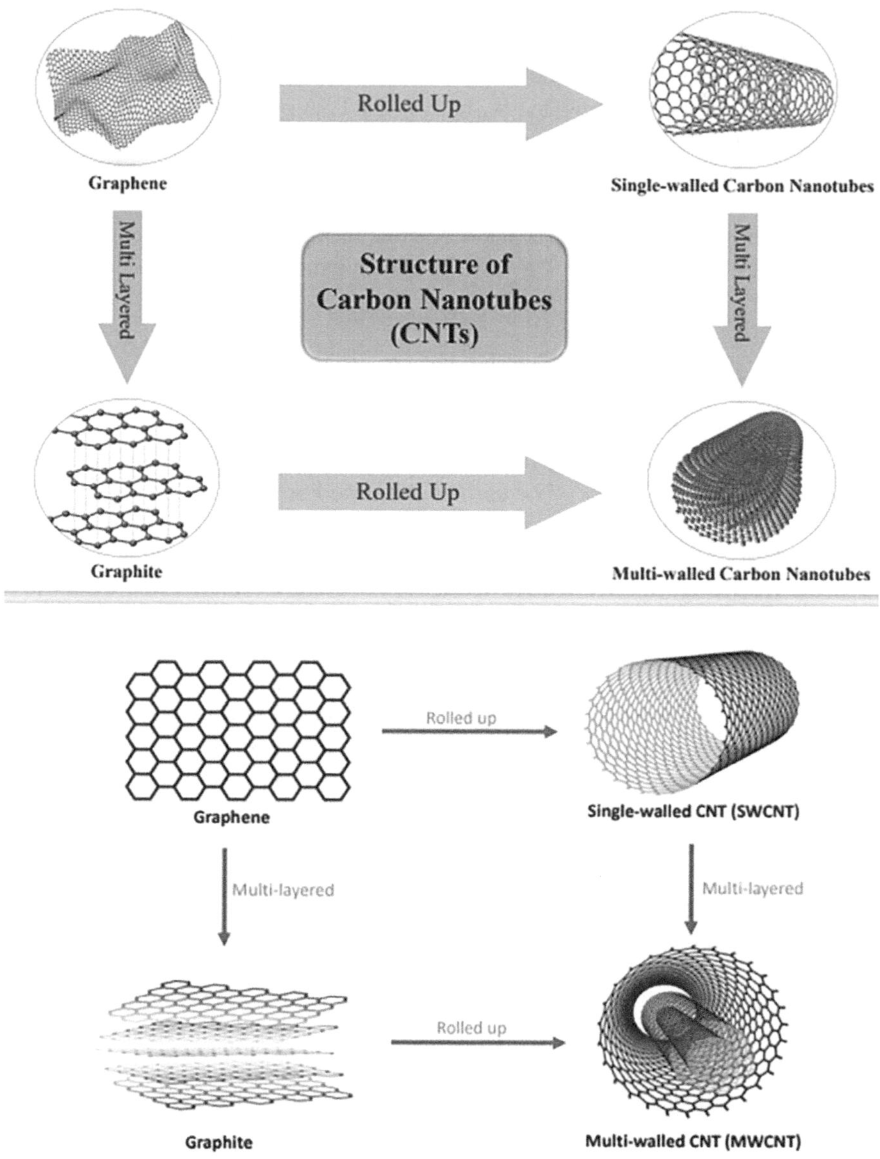

FIGURE 12.2 Structure of graphene, graphite, single-walled carbon nanotubes (SWCNTs), and multi-walled carbon nanotubes (MWCNTs).

12.2.3.2 Carbon Nanotubes Production and Regeneration

High-quality nanotube production is required for both research and extensive industrial uses. The biggest problem with nanotubes is their heterogeneous nature. Many variables, including hydrocarbon, catalyst, pressure, temperature, gas flow rate, reactor design, and deposition duration, are involved in the synthesis of CNTs. The main

TABLE 12.1

Comparison of Physical Properties of Single-Walled Carbon Nanotubes (SWCNTs) and Multi-Walled Carbon Nanotubes (MWCNTs) (Gupta et al., 2019)

Property	SWCNTs	MWCNTs
Diameter	1–2 nm	5–100 nm
Length	100–1,000 nm	15,000 nm
Density	2,600 kg m^{-3}	1,600 kg m^{-3}
Tensile strength	22.2 ± 2.2 GPa	11–63 GPa
Elastic modulus	1,000–3,000 GPa	300–1,000 GPa
Thermal conductivity at 300 K	3,000–6,000 W mK^{-1}	2,000–3,000 W mK^{-1}
Specific heat	425 J kg^{-1} K	796 J kg^{-1} K
Band gap	0–0.5 eV	2.9–3.7 eV
Melting point	3,527°C	3,527°C

SWCNTs, single-walled carbon nanotubes; MWCNTs, multi-walled carbon nanotubes.

methods of CNTs synthesis are: (a) chemical vapor deposition (CVD) method, including thermal-activated CVD, photo-assisted CVD, and plasma-assisted CVD, (b) electric arc-discharge method, and (c) laser ablation method (Gupta et al., 2019; Kumar Jagadeesan et al., 2020). However, the four main obstacles in CNTs synthesis at present are: (i) mass manufacture at a low cost of high-quality CNTs, such as SWCNTs; (ii) CNTs organization, which enables the orientation and placement on a flat substrate; (iii) selective manufacturing, which allows for precise control over the structure and electrical characteristics of the synthesized CNTs; and (iv) being aware of the mechanisms behind the CNTs processes.

There is much interest in environmentally safe and sustainable methods of regenerating used nanomaterials to cost-efficiently utilize resources and materials. Ateia et al. (2018) studied the use of ozone to regenerate magnetic CNTs (MCNTs) after their use for pollutants removal from organic water. They found that in three cycles with only ozone treatment, the capacity of MCNTs to remove atrazine decreased from 57.8 to 27.6 mg g^{-1}. However, the enhancement in atrazine removal capability was seen by ethanol washing of the MNCTs after ozone treatment. Even after ten cycles of regeneration, MCNTs were shown to retain 85%–93% of their initial affinity of adsorption. Toński et al. (2021) investigated the chemical and thermal regeneration of MWCNTs and their potential to be reused for three anticancer medicines, viz., ifosfamide, cyclophosphamide, and 5-fluorouracil. MWCNTs suffered significant losses during thermal regeneration (300°C for 2 hours) with increased temperatures. After 5 contamination– thermal regeneration cycles, no negative effect was seen on the sorption level of any tested medicines. Regarding the degradation, regeneration, and reuse of CNTs and N-doped CNTs in catalytic activity, peroxymonosulfate activation for the degradation of phenol was studied (Hou et al., 2019).

The initial peroxymonosulfate concentration was found to strongly affect the CNTs' catalytic activity. Their study indicates a close correlation of surface functionalities with catalyst regeneration and deactivation of CNT. Bassyouni et al. (2020) found high effectiveness in terms of regeneration for CNTs compared to AC.

12.2.3.3　Application of Carbon Nanotubes in Hydrogen Purification

One of the hydrogen purification technologies is membrane-based separation. Selectivity and permeance are two main parameters that regulate the membrane-based separation activity in gas purification. CNT membranes are useful for gas purification and separation, including hydrogen, as they use an improved Knudson diffusion mechanism. Using CNT-based membranes for hydrogen purification have many advantages over other membranes, like a smooth inner surface, high rigidity because of the graphene plane, and nonpolar nature because of sp^2 hybridization (Bartocci et al., 2020). CO_2 separation is a key issue in hydrogen purification. Yang et al. (2020b) studied the adsorption and diffusion of CO_2 and CH_4 in preabsorbed CNT at 300 K up to 400 bar. They found that the fluid–solid interaction leads to a substantial absorption of CO_2 and CH_4 at low pressures inside very small CNT pores. H_2O promotes CO_2 adsorption in large-diameter CNTs at low pressures. The CO_2 adsorption was found to be more compared to CH_4. Furthermore, Pishnamazi et al. (2020) found about 40% in the CO_2 separation efficiency using CNT with distilled water. Also, Pacheco et al. (2021) studied the effect of CNT mixed matrix membrane in CO_2 separation. They found that CO_2 separation can be improved by CNT functionalization and the use of additives like zeolites, GO, hydrogel, and SiO_2 nanoparticles.

Additionally, Cheng et al. (2022) studied the gas separation efficiency of porous CNTs. They found no chemical reaction between PCNT and the gases (CH_4, CO_2, CO, H_2, O_2, N_2, etc.). Their results showed that porous CNTs effectively worked as a hydrogen purification material. Regmi et al. (2021) investigated the effect of hybrid CNT and titanium dioxide nanotube (TNT)-based hybrid mixed matrix membranes in CO_2, CH_4, and H_2 separation. They found the TNT@CNT hybrid fillers highly selective for H_2 and CO_2 over CH_4. Ashtiani et al. (2021) investigated the effectiveness of a unique composite membrane, CNT-ZIF-8-PDMS, for CO_2 separation. The penetration of CO_2 was found to get increased due to the combined effect of the CNT membrane and zeolitic imidazolate framework-8 (ZIF-8) crystal. The membrane's selective area and stability (mechanical and thermal) were improved due to the spray-coating of the polydimethylsiloxane (PDMS) top layer. High CO_2 permeability and selectivity were observed with the composite membrane (CNT-ZIF-8-PDMS) with high hydrogen purification potential. Also, Nour et al. (2013) investigated the effect of MWCNT concentration (1%, 5%, and 10%) on MWCNT/PDMS composite membranes' selectivity for hydrogen separation from CH_4. With an increase in the MWCNT weight ratio in PDMS composite membrane, the hydrogen selectivity increased through efficient CH_4 blocking. They discussed the membrane as an effective way of cost reduction and industrial application for hydrogen separation from different gas species. Likewise, Boháčová et al. (2019) studied the use of SWCNT that is mildly oxidized as a new hydrogen separation membrane support material. Improved mechanical properties in oxidized SWCNTs were found to be an effective new hydrogen separation membrane.

12.2.4 Porous Carbon

12.2.4.1 Physicochemical Properties of Porous Carbon

PC as a porous solid material has been studied continually for gas purification with a physical adsorptive process. This material has been used extensively in clean energy and for environmental applications, including hydrogen and methane sorption, separation, and storage (Shao et al., 2011). Excellent stability, flexible structure, low preparation cost, and outstanding sustainable and recyclable performance at low regeneration energy have rendered PCs a lightweight advantageous sorbent (Srinivas et al., 2014). Basically, PC consists of defective graphene layers, which are randomly arranged and enclose a number of very high interconnected nanocavities ($>1\,cm^3{\cdot}g^{-1}$) and surface areas that exceed $3,500\,m^2{\cdot}g^{-1}$, achieving pore volumes (Marsh & Rodríguez-Reinoso, 2006). The surface area and volume of pores, alongside pore size, determine their practical application potential. Hence, recent research focuses on enhancing these pore properties of the sorbents. Prior studies have revealed that the creation of hollow or porous frameworks is an effective technique to improve the specific surface area up to $3,000\,m^2 g^{-1}$, especially for certain micro-/mesoporous structures (Xu et al., 2015). The porosity and spread of pore size have an important influence on optimizing the electrochemical process. The increase in porosity increases surface area and thus capacity. Alternatively, micropores store ions and mesopores and macropores may serve as a reservoir or buffers for active ions, all of which influence the delivery and storage of energy (Wang & Hu, 2018).

12.2.4.2 Porous Carbon Production and Regeneration

Several methods are available to prepare PC materials such as: (i) physical (carbonizing followed by activation) as well as chemical activation (mixing the precursors of carbon under inert gas protection), (ii) hard templating method where carbon precursors are filled into template containing pore structure and soft templates using ionic liquids, (iii) direct carbonization method, and (iv) element doping method where several elements (Zn, N, P, and S) are doped into carbon nano-precursors (Li et al., 2020b). A raw material to be used as a precursor for PC preparation must be able to produce large carbon residue after heat treatment under an inert condition. Additionally, they must give non-graphitizable disordered microstructures (Sevilla et al., 2021). These raw materials include biomass and products derived from them, such as biopolymers and saccharides, selective polymers, and coal/petroleum. On the other hand, wood, nutshell, and grass are some of the biomass with a hierarchical structure that is organized naturally, and the hierarchical porosity can be obtained through simple pyrolysis activation. There have been several studies showing the effect of activating conditions and agents on the specific surface area, pore volume, and hierarchical structure of porous materials (Zhou et al., 2021).

Biomass wastes are environmentally-friendly and easily available, such as kenaf stem, corn cobs, dead leaves of *Ficus religiosa* (Sacred Fig), and sunflower heads, and they have attracted widespread attention for use as a carbon precursor. Song et al. (2019) utilized low-cost and sustainable carbon source biowaste, such as husks of mung beans, for preparing a value-added PC material by carbonizing, followed by KOH-based activation. They reported that with a considerable specific surface area of $2,131\,m^2 \cdot g^{-1}$. Similarly, when the hydrothermal treatment method was used for the preparation of BC in an aqueous H_2SO_4 solution followed by KOH-based activation, a specific surface area of $1214\,m^2 \cdot g^{-1}$ was also obtained. Likewise, Zhou et al. (2021) described a novel method for preparing PC material utilizing biomass pyrolysis vapors before these were condensed with a calcium citrate template. The authors reported that at 550°C pyrolysis temperature with 10°C min^{-1} heating rate and preparation of precursor at 200°C, the maximum specific surface area of $1,703\,m^2 \cdot g^{-1}$ was attained with a volume of micropores and microporosity being $0.51\,cm^3 \cdot g^{-1}$ and 24.17%, respectively. Moreover, they presented that the adsorption capacity for hydrogen could reach $170.12\,cm^3 \cdot g^{-1}$ at −196°C and under atmospheric conditions.

Furthermore, Li et al. (2020a) investigated the application of simple activation using KOH for preparing PC sorbents using water-chestnut shells and reported that the sorbents prepared possessed porous structures that were highly developed with good CO_2 uptake at atmospheric pressure of 5.23 and 3.61 mmol·g^{-1} at 0°C and 25°C, respectively. Similarly, in another study, dates were used as a biomass precursor for the preparation of PC materials with varied temperatures for carbonization and chemical activation. It was reported that smaller micropores were produced at high carbonization temperatures suitable for CO_2 storage, with uptake capacities of 6.4 and 4.36 mmol·g^{-1} at 0°C and 25°C, respectively (Li et al., 2019). These studies showed potential for utilizing sustainable and available techniques at low cost for preparing PC materials.

12.2.4.3　Application of Porous Carbon in Hydrogen Purification

The PCs are used as an excellent material for hydrogen purification with high selectivity. Cheng et al. (2022) studied the porous carbon nanotubes (PCNTs) performance for separating hydrogen and reported that PCNTs showed selectivity of up to 1,031 and were far better than porous phosphorene or silicon. Similarly, Baamran et al. (2023) assessed commercial ACs having high pore volume and high surface area for PSA of hydrogen with simulated steam methane reforming off-gas stream. They presented hydrogen purity of 99.6% and recovery of 55.3% with multicomponent selectivity $(CO + CH_4 + CO_2)/H_2$ of 59.86%. In another study, machine learning was applied as a novel method to predict the hydrogen uptake by PC. It was reported that robust predictions were made by two parameters Brunauer–Emmett–Teller (BET) surface area and the pressure, with the smallest effect shown by the pore volume (Maulana Kusdhany and Lyth, 2021). Guo et al. (2020) presented that the process of phase nucleation could stabilize the chemisorbed H on the PC surface and that the specific surface area and sp^2 carbon ratio values must be more than $3500\,m^2 \cdot g^{-1}$ and 0.43, respectively. A theoretical study by Zhang et al. (2016) showed promising potential for hydrogen purification membrane with porous phosphorene showing high selectivity for hydrogen, compared to N_2, H_2O, CO_2, CH_4, and CO, as well as over silicene-based or graphene-based membranes.

12.2.5　CARBON MOLECULAR SIEVES

12.2.5.1　Physicochemical Properties of Carbon Molecular Sieves

CMS are among the most highly researched materials to be used as sorbents and show ever-growing potential with the simple sieving mechanism, as shown in Figure 12.3. With the application of high temperatures and under a selective environment, CMS

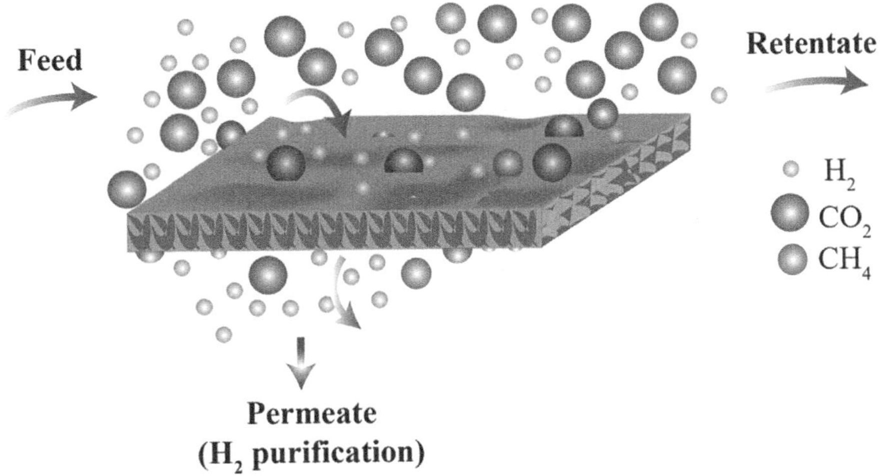

FIGURE 12.3 A schematic diagram of gas purification from feed containing H_2, CO_2, and CH_4 by carbon molecular sieves membranes with high selectivity for H_2.

are formed by carbonizing polymers and cellulosic precursors (Koresh & Soffer, 1980). They show high corrosive resistance and are thermally stable while exhibiting outstanding permeabilities and permselectivity (Steel & Koros, 2003). They possess properties that can also be tuned by altering carbonizing conditions, such as the rate at which the precursors are heated, changing gas atmosphere, soak periods, temperature for thermostabilizing, and end carbonization. Furthermore, pretreating the polymer solution is also an effective way of tuning the properties, and these can be done by polymer blending or by adding nanofillers and other additives. Finally, aging (naturally or in an oxidative atmosphere) and modifying the surface of the substrate are included in the post-treatments of these membranes. The membrane material has structures, which are turbostratic along with pore sizes that show a bimodal distribution with ultramicropores connected by micropores. These micropores (<2 nm) provide sorption, while ultramicropores (<0.7 nm) (called constrictions) have been found responsible for the gas permeation mechanism for molecular sieving (Bernardo et al., 2020; Xu et al., 2011).

12.2.5.2 Carbon Molecular Sieves Production and Regeneration

There have been several studies on the preparation of CMS that were produced utilizing waste materials. CVD method with CH_4 as the carbon source has been used to prepare CMS with soybean straw. The soybean straw was later activated by zinc chloride and under nitrogen atmosphere-carbonization at 500°C for an hour (Xu et al., 2018). It was reported that adsorption selectivity was 20.8, which showed potential to be used in CO_2/CH_4 separation. Kaya et al. (2020) prepared CMS from the spent coffee ground with removed fats, which was also activated with chemical agents (KOH, H_3PO_4, and $ZnCl_2$) along with carbonization at 400°C–900°C. The characterization results showed that there were variations in pore diameters ranging from 2 to 4 Å of the formed CMS and showed potential to be used in removing impurities, such as methane, oxides of nitrogen, CO_2, and other impurities in biogas and natural gas processing. In another study, narrow pore-sized CMS were successfully prepared from chemically activated chestnut shells and their subsequent methane CVD, and they showed a high CO_2 adsorption capacity (Demiral & Demiral, 2018). Carbon sphere molecular nano-sieves were prepared via hydrothermal method at 180°C using cane molasses, and the so-prepared materials were studied for the adsorption process of CO_2/N_2 for potential application in enhanced carbon capture and storage application conditions (Acevedo et al., 2020).

In another study, CMS were prepared using pistachio shells, and it was reported that $895\,m^2{\cdot}g^{-1}$ of the optimum surface was obtained at 800°C with an average pore diameter of 0.27 nm (ultramicropores) (Şahin et al., 2018). Moreover, other CMS have been prepared using agricultural residues utilizing three stages, which are carbonization, activation, and pore modification, and several studies have also been presented showing pore characteristics dependent on temperature and duration for deposition and rate of depositing agent (Kumar et al., 2022). Several regeneration methods have been conducted to make the invalid CMS reusable. Chemical regeneration is one the popular methods that necessitate the addition of a chemical (if a regenerating gas or vapor is not already present in the stream), accompanied by eliminating that chemical. Jones and Koros (1994) investigated air feeds saturated with different organic substances on carbon membranes made from aromatic polyimides carbonized at 500°C or 550°C. Additionally, electrothermal regeneration of carbon membranes is most successful in settings devoid of oxidation. The continuous substrate of graphene layers transmits electricity across the membranes of CMS. Including metals into a carbon matrix can boost conductivity (Lie & Hägg, 2006).

12.2.5.3 Application of Carbon Molecular Sieves in Hydrogen Purification

CMS can be used in hydrogen purification because of their ability as a sorbent to selectively adsorb gas molecules based on their size and shape. Araújo et al. (2023) prepared CMS membranes from the carbonization of a wood pulp-derived cellulose-based polymer precursor along with the doping of urea, and they reported that urea induced an increment of structural disorder and pore volumes within the membrane structure. Hydrogen permeability was increased by twofold when only 2.8 wt% urea was doped into the membranes. Also, He (2017) proposed a new method for hydrogen purification formed from biomass fermentation, which utilizes a carbon membrane system working efficiently in dual stages while capturing more than 95% CO_2 (first stage at 20 bar and 120°C and retaining 95% H_2) and obtaining more than 99.5% hydrogen in the second stage (operating at 20 bar and a low temperature of 20°C).

Moreover, Nie et al. (2022) used wood tar for preparing CMS membranes supporting new PC fiber with the wood tar solution dip-coating the outer layers of the PC fiber having different loading with their subsequent pyrolysis at 600°C. The supports and CMS membranes prepared from wood tar solution at 70 wt% showed efficient hydrogen separation even when NA_3-supported membranes had less selectivity and gas permeance. Furthermore, when the pyrolysis temperature varied from 500°C to 700°C, the 70 wt% wood tar solution exhibited the highest selectivity at 600°C. The membrane also exhibited reasonable selectivity of 155 for H_2/N_2 and 340 for H_2/CH_4, with an 86 GPU H_2 permeance. Stable CMS membranes with an extraordinary separation were also prepared from cellulose precursors obtained from ionic liquid. The H_2 and CO_2 selectivities

TABLE 12.2

Progress in CMS Membranes Prepared from Waste for Hydrogen Purification

Carbonaceous Sorbents	Advantages	Disadvantages	References
Activated carbon	Abundant sources; higher potential for commercial applications; higher adsorption performance; low cost; good thermal stability	Complex regeneration and low selectivity	Hassan et al. (2020), Kazeem et al. (2018)
Biochar	High reactivity, high specific surface area; hydrophobic; no regeneration; environmentally friendly, low cost	Inorganic/anionic pollutants with limited adsorption effect; long equilibration time; secondary contamination problems	Weidner et al. (2022), Zhang et al. (2019)

were reported to be extremely high, i.e., more than 2.06×10^5 for H_2/CH_4 and more than 1.46×10^4 for CO_2/CH_4 owing to the pore size precision of the formed membranes (Araújo et al., 2022). In conclusion, very few cases of CMS preparation from waste materials are reported to be applied for hydrogen purification, as given in Table 12.2.

12.3 COMPARISON OF CARBONACEOUS SORBENTS

12.3.1 ADVANTAGES AND DISADVANTAGES OF THE CARBONACEOUS SORBENTS

AC can be created from any cheap carbonaceous material with a significant amount of volatile material, a high carbon content, and low ash content. Although AC is an effective sorbent, its use is limited by its expensive cost and difficult regeneration procedure (Osman et al., 2020a). Increasing energy consumption and manufacturing costs for AC is crucial, even if activation is not required to produce BC (Xu et al., 2016). Moreover, the large pore size variation of all ACs reduces selectivity and adsorption capability for certain gas combinations (Morali et al., 2019).

BC is made from organic materials that have a high specific surface area, are hydrophobic, and are easily regenerable (Osman et al., 2022a). Heat treatment produces highly reactive BC with a porous structure, maintaining surface functional groups and mineral content (Fawzy et al., 2022; Liu et al., 2015). BC has no activation step and uses less energy to generate. Unactivated BC typically exhibits relatively low pore characteristics, which limits its ability to capture CO_2 and store energy. As the utilized BC may be absorbed into the soil to improve its qualities, it can be used as an alternative to CO_2 collection without requiring a regeneration process (Guo et al., 2022). Because the creation of BC from waste biomass is advantageous for carbon abatement and sustainable management, BC is frequently regarded as a practical and sustainable sorbent. However, BC has limited adsorption of anions and inorganic substances (Li et al., 2017). The porous nature and restricted surface functional groups of virgin BC necessitate a lengthy equilibration period (Ling et al., 2017).

CNTs consist of graphene or graphite sheets rolled into tubes with a nanostructure with a large pore structure and are more effective for the adsorption of large molecules (Norizan et al., 2020). CNTs have high strength and stiffness due to their graphitic and tubular structure. However, CNTs have the limitation of high cost (Kumar et al., 2019). CNTs are believed to increase the membrane's resilience to contamination and resolve the common trade-off between permeability and selectivity.

The PC material has an adjustable pore structure, and due to the highly developed porous structure, the PC exhibits good CO_2 adsorption performance at 25°C and ambient pressure (Jang et al., 2018). Porous materials consume less energy and have higher adsorption capacity, which leads to cost-effectiveness (Yang et al., 2017). However, one of the main drawbacks of using PC as a sorbent is the low polarity of the carbon surface, which leads to low selectivity for the adsorption of specific gases (Ren et al., 2017).

Amorphous microporous CMS have a lower pore surface area. Compared to other porous sorbents, CMS have superior selectivity and adsorption ability to adsorb one or more molecules while preventing the adsorption of other compounds in the mixture (Kumar et al., 2022). Moreover, CMS have higher thermal stability and lower regeneration energy at high temperatures (Song et al., 2017). However, CMS can lead to carbon accumulation during gas adsorption, making the pores narrower, and the limitations of high toxicity and high deposition agent cost make CMS not friendly (Lai et al., 2021). Table 12.3 compares the advantages and disadvantages of carbonaceous sorbents.

12.3.2 Factors Affecting Selecting the Suitable Sorbent

The properties of the sorbent, such as surface cation exchange capacity, pore structure, and the quantity and type of functional groups, have a major impact on the effectiveness of adsorption (El-Nahas et al., 2020; Qiu et al., 2021). After carbonization and activation, carbon sorbents develop a variety of pore configurations. Different pore sizes play diverse roles in adsorption, with micropores serving as adsorption sites and mesopores and macropores serving as diffusion channels (Du et al., 2020a). Several types of pollutants can be adsorbed on the surface of carbonaceous sorbents due to variances in the distribution of pore size of carbonaceous sorbents produced from various lignocellulosic substances (Ilomuanya et al., 2017). The larger the pore size, the larger the carbonaceous sorbent can capture larger contaminant molecules.

Moreover, the surface charge of the carbonaceous sorbent affects the electrostatic interactions between the contaminant and the sorbent. The variability of functional groups, the activator of choice, and the length of the sorbent's carbonization all considerably affect the specific surface area of the PC material (Dizbay-Onat et al., 2018; Singh et al., 2017). One of the assessment parameters used to choose the sorbent is the carbonaceous material's total environmental effect. A comparison of the environmental effects of AC, BC, and hydraulic carbon was made, and the use of compounds with higher adsorption capabilities and the regeneration of carbonaceous materials were discovered to be efficient strategies for decreasing the entire environmental impact of various sorbents (Kozyatnyk et al., 2020).

TABLE 12.3

Comparison of the Performance of Carbonaceous Sorbents

Carbonaceous Sorbents	Advantages	Disadvantages	References
Activated carbon	Abundant sources; higher potential for commercial applications; higher adsorption performance; low cost; good thermal stability	Complex regeneration and low selectivity	Hassan et al. (2020), Kazeem et al. (2018)
Biochar	High reactivity, high specific surface area; hydrophobic; no regeneration; environmentally friendly, low cost	Inorganic/anionic pollutants with limited adsorption effect; long equilibration time; secondary contamination problems	Weidner et al. (2022), Zhang et al. (2019)
Carbon nanotubes	Larger pore size; nanostructure; more stable colloidal properties; high chemical and mechanical stability; hydrophobic; high corrosion resistance	The aggregate formation; partial material toxicity; high cost	Kumar et al. (2022), Sajid et al. (2022)
Porous carbon	Adjustable pore structure; easy to regenerate; low cost; easy to prepare; surface functionalization; hydrophobic; easy to modify	High cost; the low polarity of the surface; low selectivity; low heat of adsorption	Ren et al. (2017), Yang et al. (2020c)
Carbon molecular sieve	High selectivity; high thermal stability and low regeneration energy	Cumbersome and costly preparation process; smaller pore surface area; low adsorption efficiency; high toxicity and high deposition agent cost	Kumar et al. (2022), Wang et al. (2021)

12.4 CHALLENGES, OPPORTUNITIES, AND PROSPECTS

Although the long-term cyclic operation and a high adsorption capacity are necessary to ensure the technology's economics and viability, the resilience and stability of biomass-based sorbents have not yet been thoroughly shown. The design of effective green modifiers and modification techniques will determine the future of carbonaceous sorbents. Multifunctional biomass-derived carbon may be produced using several biomass precursors, boosting its applications in industries, including energy, the environment, and medicine (Singh et al., 2019). Improved reuse of agricultural waste can reduce air pollution brought on by plant burning and stop the spread of germs and viruses brought on by agricultural waste's long-term degradation. More cost-effective carbon materials are consistent with a sustainable future.

The toxicity and high cost of some CNTs are the main obstacles that prevent them from being applied. Weakly acting physics is difficult to stabilize the long-term sequestration of contaminants, and different chemical pathways should be developed to modify the carbon sorbent surface. Fine-tuning of pore structure and surface modification can also help to enhance adsorption. Surface modifications such as nitrogen doping, amine modification, oxidation, and metal oxide modification help amplify adsorption. While surface functionalization of CNTs limits their ability to adsorb organic contaminants, it increases their ability to aggregate in water (Chen et al., 2017). The development of next-generation functionalized biohybrid materials that combine reactive PC with inorganic and organic nanostructures improves gas capture under many circumstances.

12.5 CONCLUSION

Carbonaceous sorbents have demonstrated considerable promise for hydrogen purification, providing a possible answer to the rising need for high-purity hydrogen in various sectors. Carbon-based materials, with their large surface area, chemical durability, and variable pore structure, are an interesting option for hydrogen storage and purification applications due to their unique features. Additionally, recent advances in producing functionalized carbonaceous sorbents have resulted in considerable performance enhancements, making their future applications even more intriguing. Despite the numerous benefits of carbonaceous sorbents, there are still obstacles to overcome, such as their relatively limited hydrogen sorption capacity and the need to better optimize their characteristics to satisfy particular application needs. In addition, a deeper understanding of the essential mechanics behind impurities adsorption on carbon-based materials is required. Looking ahead, carbonaceous sorbents will likely play an ever-increasing part in the transition to a sustainable hydrogen economy in the coming years. Hence, further research should emphasize on significantly improving the efficiency of these substances, exploring new functionalization processes, and studying their potential application in large-scale hydrogen generation and storage systems.

REFERENCES

Acevedo, E. R., Franco, C. A., Carrasco-Marín, F., Pérez-Cadenas, A. F., & Cortés, F. B. (2020). Biomass-derived carbon molecular sieves applied to an enhanced carbon capture and storage process (e-CCS) for flue gas streams in shallow reservoirs. *Nanomaterials*, *10*(5), 980. https://doi.org/10.3390/nano10050980

Alengebawy, A., Jin, K., Ran, Y., Peng, J., Zhang, X., & Ai, P. (2021). Advanced pre-treatment of stripped biogas slurry by polyaluminum chloride coagulation and biochar adsorption coupled with ceramic membrane filtration. *Chemosphere*, *267*, 129197. https://doi.org/10.1016/j.chemosphere.2020.129197

Amin, M., Butt, A. S., Ahmad, J., Lee, C., Azam, S. U., Mannan, H. A., Naveed, A. B., Farooqi, Z. U. R., Chung, E., & Iqbal, A. (2023). Issues and challenges in hydrogen separation technologies. *Energy Reports*, *9*, 894–911. Elsevier. https://doi.org/10.1016/j.egyr.2022.12.014

Ao, W., Fu, J., Mao, X., Kang, Q., Ran, C., Liu, Y., Zhang, H., Gao, Z., Li, J., Liu, G., & Dai, J. (2018). Microwave assisted preparation of activated carbon from biomass: A review. *Renewable and Sustainable Energy Reviews*, *92*, 958–979. https://doi.org/10.1016/j.rser.2018.04.051

Araújo, T., Andrade, M., Bernardo, G., & Mendes, A. (2022). Stable cellulose-based carbon molecular sieve membranes with very high selectivities. *Journal of Membrane Science*, *641*, 119852. https://doi.org/10.1016/j.memsci.2021.119852

Araújo, T., Parnell, A. J., Bernardo, G., & Mendes, A. (2023). Cellulose-based carbon membranes for gas separations—Unraveling structural parameters and surface chemistry for superior separation performance. *Carbon*, *204*, 398–410. https://doi.org/10.1016/j.carbon.2022.12.062

Ashtiani, S., Sofer, Z., Průša, F., & Friess, K. (2021). Molecular-level fabrication of highly selective composite ZIF-8-CNT-PDMS membranes for effective CO2/N2, CO2/H2 and olefin/paraffin separations. *Separation and Purification Technology*, *274*, 119003. https://doi.org/10.1016/j.seppur.2021.119003

Ateia, M., Ceccato, M., Budi, A., Ataman, E., Yoshimura, C., & Johnson, M. S. (2018). Ozone-assisted regeneration of magnetic carbon nanotubes for removing organic water pollutants. *Chemical Engineering Journal*, *335*, 384–391. https://doi.org/10.1016/j.cej.2017.10.166

Baamran, K., Al-Naddaf, Q., Lawson, S., Ali Rownaghi, A., & Rezaei, F. (2023). Kinetic process assessment of H2 purification over highly porous carbon sorbents under multicomponent feed conditions. *Separation and Purification Technology*, *306*, 122695. https://doi.org/10.1016/j.seppur.2022.122695

Bak, C. U., Lim, C. J., Lee, J. G., Kim, Y. D., & Kim, W. S. (2019). Removal of sulfur compounds and siloxanes by physical and chemical sorption. *Separation and Purification Technology*, *209*, 542–549. https://doi.org/10.1016/j.seppur.2018.07.080

Banik, C., Lawrinenko, M., Bakshi, S., & Laird, D. A. (2018). Impact of pyrolysis temperature and feedstock on surface charge and functional group chemistry of biochars. *Journal of Environmental Quality*, *47*(3), 452–461. https://doi.org/10.2134/jeq2017.11.0432

Bartocci, P., Russo, G., Yang, H., Hu, S., Skreiberg, Ø., Wang, L., Gallucci, F., Bidini, G., & Fantozzi, F. (2020). Carbon nanotubes for hydrogen purification and storage. In Neha Srivastava, Manish Srivastava, P. K. Mishra, Vijai Kumar Gupta (Eds.), *Green Synthesis of Nanomaterials for Bioenergy Applications* (pp. 211–238). Hoboken, NJ: John Wiley & Sons, Ltd. https://doi.org/10.1002/9781119576785.ch9

Bassyouni, M., Mansi, A. E., Elgabry, A., Ibrahim, B. A., Kassem, O. A., & Alhebeshy, R. (2020). Utilization of carbon nanotubes in removal of heavy metals from wastewater: A review of the CNTs' potential and current challenges. *Applied Physics A: Materials Science and Processing*, *126*(1), 38. https://doi.org/10.1007/s00339-019-3211-7

Bernardo, G., Araújo, T., da Silva Lopes, T., Sousa, J., & Mendes, A. (2020). Recent advances in membrane technologies for hydrogen purification. *International Journal of Hydrogen Energy*, *45*(12), 7313–7338. https://doi.org/10.1016/j.ijhydene.2019.06.162

Bhatt, A., Jain, A., Gurnany, E., Jain, R., Modi, A., & Jain, A. (2016). Carbon nanotubes: A promising carrier for drug delivery and targeting. In A. M. Holban & A. M. Grumezescu (Eds.), *Nanoarchitectonics for Smart Delivery and Drug Targeting* (pp. 465–501). Norwich, NY: William Andrew Publishing. https://doi.org/10.1016/B978-0-323-47347-7.00017-3

Boháčová, M., Zetková, K., Knotek, P., Bouša, D., Friess, K., Číhal, P., Lanč, M., Hrdlička, Z., & Sofer, Z. (2019). Mildly oxidized SWCNT as new potential support membrane material for effective H2/CO2 separation. *Applied Materials Today*, *15*, 335–342. https://doi.org/10.1016/j.apmt.2019.02.014

Bos, H. L., & Broeze, J. (2020). Circular bio-based production systems in the context of current biomass and fossil demand. *Biofuels, Bioproducts and Biorefining*, *14*(2), 187–197. https://doi.org/10.1002/bbb.2080

Brea, P., Delgado, J. A., Águeda, V. I., & Uguina, M. A. (2019). Comparison between MOF UTSA-16 and BPL activated carbon in hydrogen purification by PSA. *Chemical Engineering Journal, 355*, 279–289. https://doi.org/10.1016/j.cej.2018.08.154

Cha, J. S., Park, S. H., Jung, S. C., Ryu, C., Jeon, J. K., Shin, M. C., & Park, Y. K. (2016). Production and utilization of biochar: A review. *Journal of Industrial and Engineering Chemistry, 40*, 1–15. https://doi.org/10.1016/j.jiec.2016.06.002

Chen, C., Geng, X., & Huang, W. (2017). Adsorption of 4-chlorophenol and aniline by nano-sized activated carbons. *Chemical Engineering Journal, 327*, 941–952. https://doi.org/10.1016/j.cej.2017.06.183

Chen, W., He, F., Zhang, S., Xv, H., & Xv, Z. (2018). Development of porosity and surface chemistry of textile waste jute-based activated carbon by physical activation. *Environmental Science and Pollution Research, 25*(10), 9840–9848. https://doi.org/10.1007/s11356-018-1335-5

Chen, W. H., Peng, J., & Bi, X. T. (2015). A state-of-the-art review of biomass torrefaction, densification and applications. *Renewable and Sustainable Energy Reviews, 44*, 847–866. https://doi.org/10.1016/j.rser.2014.12.039

Cheng, G., Dou, K., & Yuan, G. (2022). Hydrogen separation of porous carbon nanotubes: A density functional theory study. *Diamond and Related Materials, 125*, 108986. https://doi.org/10.1016/j.diamond.2022.108986

Cox, M., & Mokaya, R. (2017). Ultra-high surface area mesoporous carbons for colossal pre combustion CO2 capture and storage as materials for hydrogen purification. *Sustainable Energy and Fuels, 1*(6), 1414–1424. https://doi.org/10.1039/c7se00300e

Dai, Y., Zhang, N., Xing, C., Cui, Q., & Sun, Q. (2019). The adsorption, regeneration and engineering applications of biochar for removal organic pollutants: A review. *Chemosphere, 223*, 12–27. https://doi.org/10.1016/j.chemosphere.2019.01.161

Danish, M., & Ahmad, T. (2018). A review on utilization of wood biomass as a sustainable precursor for activated carbon production and application. *Renewable and Sustainable Energy Reviews, 87*, 1–21. https://doi.org/10.1016/j.rser.2018.02.003

Dasappa, S., & Shivapuji, A. M. (2021). Biomass gasification: Thermochemical route to energetic bio-chemicals. In D. Tuli, S. Kasture, & A. Kuila (Eds.), *Advanced Biofuel Technologies: Present Status, Challenges and Future Prospects* (pp. 305–332). Amsterdam, Netherlands: Elsevier. https://doi.org/10.1016/B978-0-323-88427-3.00022-2

Demiral, H., & Demiral, İ. (2018). Preparation and characterization of carbon molecular sieves from chestnut shell by chemical vapor deposition. *Advanced Powder Technology, 29*(12), 3033–3039. https://doi.org/10.1016/j.apt.2018.07.015

Dizbay-Onat, M., Floyd, E., Vaidya, U. K., & Lungu, C. T. (2018). Applicability of industrial sisal fiber waste derived activated carbon for the adsorption of volatile organic compounds (VOCs). *Fibers and Polymers, 19*(4), 805–811. https://doi.org/10.1007/s12221-018-7866-z

Du, Y., Chen, H., Xu, X., Wang, C., Zhou, F., Zeng, Z., Zhang, W., & Li, L. (2020a). Surface modification of biomass derived toluene adsorbent: hierarchically porous characterization and heteroatom doped effect. *Microporous and Mesoporous Materials, 293*, 109831. https://doi.org/10.1016/j.micromeso.2019.109831

Du, Z., Liu, C., Zhai, J., Guo, X., Xiong, Y., Su, W., & He, G. (2021b). A review of hydrogen purification technologies for fuel cell vehicles. *Catalysts, 11*(3) 1–19. https://doi.org/10.3390/catal11030393

El-Nahas, S., Osman, A. I., Arafat, A. S., Al-Muhtaseb, A. H., & Salman, H. M. (2020). Facile and affordable synthetic route of nano powder zeolite and its application in fast softening of water hardness. *Journal of Water Process Engineering, 33*, 101104. https://doi.org/10.1016/j.jwpe.2019.101104

El Gamal, M., Mousa, H. A., El-Naas, M. H., Zacharia, R., & Judd, S. (2018). Bio-regeneration of activated carbon: A comprehensive review. *Separation and Purification Technology*, *197*, 345–359. https://doi.org/10.1016/j.seppur.2018.01.015

Fagnani, H. M. C., da Silva, C. T. P., Pereira, M. M., Rinaldi, A. W., Arroyo, P. A., & de Barros, M. A. S. D. (2019). CO2 adsorption in hydrochar produced from waste biomass. *SN Applied Sciences*, *1*(9), 1–10. https://doi.org/10.1007/s42452-019-1055-6

Far, H. M., Lawson, S., Al-Naddaf, Q., Rezaei, F., Sotiriou-Leventis, C., & Rownaghi, A. A. (2021). Advanced pore characterization and adsorption of light gases over aerogel-derived activated carbon. *Microporous and Mesoporous Materials*, *313*, 110833. https://doi.org/10.1016/j.micromeso.2020.110833

Fawzy, S., Osman, A. I., Mehta, N., Moran, D., Al-Muhtaseb, A. H., & Rooney, D. W. (2022). Atmospheric carbon removal via industrial biochar systems: A techno-economic-environmental study. *Journal of Cleaner Production*, *371*, 133660. https://doi.org/10.1016/j.jclepro.2022.133660

Fuertes, A. B., Ferrero, G. A., Diez, N., & Sevilla, M. (2018). A green route to high-surface area carbons by chemical activation of biomass-based products with sodium thiosulfate. *ACS Sustainable Chemistry & Engineering*, *6*(12), 16323–16331. https://doi.org/10.1021/acssuschemeng.8b03264

Ghani, W. A. W. A. K., Mohd, A., da Silva, G., Bachmann, R. T., Taufiq-Yap, Y. H., Rashid, U., & Al-Muhtaseb, A. H. (2013). Biochar production from waste rubber-wood-sawdust and its potential use in C sequestration: Chemical and physical characterization. *Industrial Crops and Products*, *44*, 18–24. https://doi.org/10.1016/j.indcrop.2012.10.017

Gorbounov, M., Taylor, J., Petrovic, B., & Masoudi Soltani, S. (2022). To DoE or not to DoE? A technical review on & roadmap for optimisation of carbonaceous adsorbents and adsorption processes. *South African Journal of Chemical Engineering*, *41*, 111–128. https://doi.org/10.1016/j.sajce.2022.06.001

Guo, J. H., Li, S. J., Su, Y., & Chen, G. (2020). Theoretical study of hydrogen storage by spillover on porous carbon materials. *International Journal of Hydrogen Energy*, *45*(48), 25900–25911. https://doi.org/10.1016/j.ijhydene.2019.12.146

Guo, S., Li, Y., Wang, Y., Wang, L., Sun, Y., & Liu, L. (2022). Recent advances in biochar-based adsorbents for CO2 capture. *Carbon Capture Science & Technology*, *4*, 100059. https://doi.org/10.1016/j.ccst.2022.100059

Gupta, N., Gupta, S. M., & Sharma, S. K. (2019). Carbon nanotubes: Synthesis, properties and engineering applications. *Carbon Letters*, *29*(5), 419–447. https://doi.org/10.1007/s42823-019-00068-2

Hassan, M. F., Sabri, M. A., Fazal, H., Hafeez, A., Shezad, N., & Hussain, M. (2020). Recent trends in activated carbon fibers production from various precursors and applications: A comparative review. *Journal of Analytical and Applied Pyrolysis*, *145*, 104715. https://doi.org/10.1016/j.jaap.2019.104715

He, X. (2017). Techno-economic feasibility analysis on carbon membranes for hydrogen purification. *Separation and Purification Technology*, *186*, 117–124. https://doi.org/10.1016/j.seppur.2017.05.034

He, B., Liu, J., Zhang, Y., Zhang, S., Wang, P., & Xu, H. (2020). Comparison of structured activated carbon and traditional adsorbents for purification of H2. *Separation and Purification Technology*, *239*, 116529. https://doi.org/10.1016/j.seppur.2020.116529

Hou, J., Xu, L., Han, Y., Tang, Y., Wan, H., Xu, Z., & Zheng, S. (2019). Deactivation and regeneration of carbon nanotubes and nitrogen-doped carbon nanotubes in catalytic per-oxymonosulfate activation for phenol degradation: variation of surface functionalities. *RSC Advances*, *9*(2), 974–983. https://doi.org/10.1039/C8RA07696K

Ihonen, J., Koski, P., Pulkkinen, V., Keränen, T., Karimäki, H., Auvinen, S., Nikiforow, K., Kotisaari, M., Tuiskula, H., & Viitakangas, J. (2017). Operational experiences of PEMFC pilot plant using low grade hydrogen from sodium chlorate production process. *International Journal of Hydrogen Energy*, *42*(44), 27269–27283. https://doi.org/10.1016/j.ijhydene.2017.09.056

Ilomuanya, M., Nashiru, B., Ifudu, N., & Igwilo, C. (2017). Effect of pore size and morphology of activated charcoal prepared from midribs of Elaeis guineensis on adsorption of poisons using metronidazole and Escherichia coli O157:H7 as a case study. *Journal of Microscopy and Ultrastructure*, *5*(1), 32. https://doi.org/10.1016/j.jmau.2016.05.001

Jang, E., Choi, S. W., Hong, S. M., Shin, S., & Lee, K. B. (2018). Development of a cost-effective CO2 adsorbent from petroleum coke via KOH activation. *Applied Surface Science*, *429*, 62–71. https://doi.org/10.1016/j.apsusc.2017.08.075

Jeong, Y.-S., Park, K.-B., & Kim, J.-S. (2020). Hydrogen production from steam gasification of polyethylene using a two-stage gasifier and active carbon. *Applied Energy*, *262*, 114495. https://doi.org/10.1016/j.apenergy.2020.114495

Jones, C. W., & Koros, W. J. (1994). Carbon molecular sieve gas separation membranes-II. Regeneration following organic exposure. *Carbon*, *32*(8), 1427–1432. https://doi.org/10.1016/0008-6223(94)90136-8

Kambo, H. S., & Dutta, A. (2015). Comparative evaluation of torrefaction and hydrothermal carbonization of lignocellulosic biomass for the production of solid biofuel. *Energy Conversion and Management*, *105*, 746–755. https://doi.org/10.1016/j.enconman.2015.08.031

Kaya, M., Atelge, M. R., Bekirogullari, M., Eskicioglu, C., Atabani, A. E., Kumar, G., Yildiz, Y. S., & Unalan, S. (2020). Carbon molecular sieve production from defatted spent coffee ground using ZnCl2 and benzene for gas purification. *Fuel*, *277*, 118183. https://doi.org/10.1016/j.fuel.2020.118183

Kazeem, T. S., Lateef, S. A., Ganiyu, S. A., Qamaruddin, M., Tanimu, A., Sulaiman, K. O., Sajid Jillani, S. M., & Alhooshani, K. (2018). Aluminium-modified activated carbon as efficient adsorbent for cleaning of cationic dye in wastewater. *Journal of Cleaner Production*, *205*, 303–312. https://doi.org/10.1016/j.jclepro.2018.09.114

Kim, T., Song, Y., Kang, J., Kim, S. K., & Kim, S. (2022). A review of recent advances in hydrogen purification for selective removal of oxygen: Deoxo catalysts and reactor systems. *International Journal of Hydrogen Energy*, *47*(59), 24817–24834). https://doi.org/10.1016/j.ijhydene.2022.05.221

Kırbıyık, C. & Kuş, M. (2020). Advanced applications of green materials for gas separation and storage. In *Applications of Advanced Green Materials* (pp. 681–703). Cambridge: Woodhead Publishing. https://doi.org/10.1016/B978-0-12-820484-9.00026-X

Koresh, J., & Soffer, A. (1980). Study of molecular sieve carbons. Part 1—Pore structure, gradual pore opening and mechanism of molecular sieving. *Journal of the Chemical Society, Faraday Transactions 1: Physical Chemistry in Condensed Phases*, *76*, 2457–2471. https://doi.org/10.1039/F19807602457

Kozyatnyk, I., Yacout, D. M. M., Van Caneghem, J., & Jansson, S. (2020). Comparative environmental assessment of end-of-life carbonaceous water treatment adsorbents. *Bioresource Technology*, *302*, 122866. https://doi.org/10.1016/j.biortech.2020.122866

Kumar, D. P., Ramesh, D., Vikraman, V. K., & Subramanian, P. (2022). Synthesis of carbon molecular sieves from agricultural residues: Status, challenges and prospects. *Environmental Research*, *214*, 114022. https://doi.org/10.1016/j.envres.2022.114022

Kumar, G., Shobana, S., Chen, W. H., Bach, Q. V., Kim, S. H., Atabani, A. E., & Chang, J. S. (2017). A review of thermochemical conversion of microalgal biomass for biofuels: Chemistry and processes. *Green Chemistry*, *19*(1), 44–67. https://doi.org/10.1039/c6gc01937d

Kumar Jagadeesan, A., Thangavelu, K., & Dhananjeyan, V. (2020). Carbon nanotubes: Synthesis, properties and applications. In P. Phuong, G. Pratibha, K. Samir, & Y. Kavita (Eds.), *21st Century Surface Science—A Handbook* (pp. 1–21). London: IntechOpen. https://doi.org/10.5772/intechopen.92995

Kumar, P. S., Korving, L., van Loosdrecht, M. C. M., & Witkamp, G.-J. (2019). Adsorption as a technology to achieve ultra-low concentrations of phosphate: Research gaps and economic analysis. *Water Research X, 4,* 100029. https://doi.org/10.1016/j.wroa.2019.100029

Lai, J. Y., Ngu, L. H., & Hashim, S. S. (2021). A review of CO2 adsorbents performance for different carbon capture technology processes conditions. *Greenhouse Gases: Science and Technology, 11*(5), 1076–1117. https://doi.org/10.1002/ghg.2112

Li, Q., Liu, S., Peng, W., Zhu, W., Wang, L., Chen, F., Shao, J., & Hu, X. (2020a). Preparation of biomass-derived porous carbons by a facile method and application to CO2 adsorption. *Journal of the Taiwan Institute of Chemical Engineers, 116,* 128–136. https://doi.org/10.1016/j.jtice.2020.11.001

Li, Q., Lu, T., Wang, L., Pang, R., Shao, J., Liu, L., & Hu, X. (2021). Biomass based N-doped porous carbons as efficient CO2 adsorbents and high-performance supercapacitor electrodes. *Separation and Purification Technology, 275,* 119204. https://doi.org/10.1016/j.seppur.2021.119204

Li, J., Michalkiewicz, B., Min, J., Ma, C., Chen, X., Gong, J., Mijowska, E., & Tang, T. (2019). Selective preparation of biomass-derived porous carbon with controllable pore sizes toward highly efficient CO2 capture. *Chemical Engineering Journal, 360,* 250–259. https://doi.org/10.1016/j.cej.2018.11.204

Li, Q., Qi, Y., & Gao, C. (2015). Chemical regeneration of spent powdered activated carbon used in decolorization of sodium salicylate for the pharmaceutical industry. *Journal of Cleaner Production, 86,* 424–431. https://doi.org/10.1016/j.jclepro.2014.08.008

Li, Y., Wang, Y., Chen, B., Wang, L., Yang, J., & Wang, B. (2022). Nitrogen-doped hierarchically constructed interconnected porous carbon nanofibers derived from polyaniline (PANI) for highly selective CO2 capture and effective methanol adsorption. *Journal of Environmental Chemical Engineering, 10*(6), 108847. https://doi.org/10.1016/j.jece.2022.108847

Li, R., Wang, J. J., Zhou, B., Zhang, Z., Liu, S., Lei, S., & Xiao, R. (2017). Simultaneous capture removal of phosphate, ammonium and organic substances by MgO impregnated biochar and its potential use in swine wastewater treatment. *Journal of Cleaner Production, 147,* 96–107. https://doi.org/10.1016/j.jclepro.2017.01.069

Li, B., Xiong, H., & Xiao, Y. (2020b). Progress on synthesis and applications of porous carbon materials. *International Journal of Electrochemical Science, 15*(2), 1363–1377. https://doi.org/10.20964/2020.02.04

Liao, Y., Jiang, L., Cao, X., Zheng, H., Feng, L., Mao, Y., Zhang, Q., Shen, Q., & Ji, F. (2022). Efficient removal mechanism and microbial characteristics of tidal flow constructed wetland based on in-situ biochar regeneration (BR-TFCW) for rural gray water. *Chemical Engineering Journal, 431,* 134185. https://doi.org/10.1016/j.cej.2021.134185

Lie, J. A., & Hägg, M. B. (2006). Carbon membranes from cellulose: Synthesis, performance and regeneration. *Journal of Membrane Science, 284*(1–2), 79–86. https://doi.org/10.1016/j.memsci.2006.07.002

Ling, L.-L., Liu, W.-J., Zhang, S., & Jiang, H. (2017). Magnesium oxide embedded nitrogen self-doped biochar composites: Fast and high-efficiency adsorption of heavy metals in an aqueous solution. *Environmental Science & Technology, 51*(17), 10081–10089. https://doi.org/10.1021/acs.est.7b02382

Liu, W.-J., Jiang, H., & Yu, H.-Q. (2015). Development of biochar-based functional materials: Toward a sustainable platform carbon material. *Chemical Reviews, 115*(22), 12251–12285. https://doi.org/10.1021/acs.chemrev.5b00195

Liu, J., Goss, J., Calverley, T., Liu, Y., Broomall, C., Kang, J., Golombeski, R., Anaya, D., Moe, B., Mabe, K., Watson, G., & Wetzel, A. (2020a). Carbon molecular sieve fiber with 3.4-4.9 Angstrom effective micropores for propylene/propane and other gas separations. *Microporous and Mesoporous Materials, 305*, 110341. https://doi.org/10.1016/j.micromeso.2020.110341

Liu, B., Yu, X., Shi, W., Shen, Y., Zhang, D., & Tang, Z. (2020b). Two-stage VSA/PSA for capturing carbon dioxide (CO2) and producing hydrogen (H2) from steam-methane reforming gas. *International Journal of Hydrogen Energy, 45*(46), 24870–24882. https://doi.org/10.1016/j.ijhydene.2020.06.264

Llosa Tanco, M. A., Medrano, J. A., Cechetto, V., Gallucci, F., & Pacheco Tanaka, D. A. (2021). Hydrogen permeation studies of composite supported alumina-carbon molecular sieves membranes: Separation of diluted hydrogen from mixtures with methane. *International Journal of Hydrogen Energy, 46*(37), 19758–19767. https://doi.org/10.1016/j.ijhydene.2020.05.088

Manyà, J. J., & Gascó, G. (2020). Biochar as a sustainable resource to drive innovative green technologies. In *Biochar as a Renewable-Based Material* (pp. 1–33). World Scientific (Europe). Singapore: Toh Tuck Link. https://doi.org/10.1142/9781786348975_0001

Marques, S. C. R., Mestre, A. S., Machuqueiro, M., Gotvajn, A. Ž., Marinšek, M., & Carvalho, A. P. (2018). Apple tree branches derived activated carbons for the removal of β-blocker atenolol. *Chemical Engineering Journal, 345*, 669–678. https://doi.org/10.1016/j.cej.2018.01.076

Márquez, P., Benítez, A., Chica, A. F., Martín, M. A., & Caballero, A. (2022). Evaluating the thermal regeneration process of massively generated granular activated carbons for their reuse in wastewater treatments plants. *Journal of Cleaner Production, 366*, 132685. https://doi.org/10.1016/j.jclepro.2022.132685

Marsh, H., & Rodríguez-Reinoso, F. (2006). Activated carbon (origins). In H. Marsh & F. Rodríguez-Reinoso (Eds.), *Activated Carbon* (Vol. 32, Issue 9, pp. 13–86). Amsterdam: Elsevier. https://doi.org/10.1016/b978-008044463-5/50016-9

Maulana Kusdhany, M. I., & Lyth, S. M. (2021). New insights into hydrogen uptake on porous carbon materials via explainable machine learning. *Carbon, 179*, 190–201. https://doi.org/10.1016/j.carbon.2021.04.036

McQuillan, R. V., Stevens, G. W., & Mumford, K. A. (2018). The electrochemical regeneration of granular activated carbons: A review. *Journal of Hazardous Materials, 355*, 34–49. https://doi.org/10.1016/j.jhazmat.2018.04.079

Moon, D. K., Park, Y., Oh, H. T., Kim, S. H., Oh, M., & Lee, C. H. (2018). Performance analysis of an eight-layered bed PSA process for H2 recovery from IGCC with pre-combustion carbon capture. *Energy Conversion and Management, 156*, 202–214. https://doi.org/10.1016/j.enconman.2017.11.013

Morali, U., Demiral, H., & Sensoz, S. (2019). Synthesis of carbon molecular sieve for carbon dioxide adsorption: Chemical vapor deposition combined with Taguchi design of experiment method. *Powder Technology, 355*, 716–726. https://doi.org/10.1016/j.powtec.2019.07.101

Mukherjee, A., Zimmerman, A. R., & Harris, W. (2011). Surface chemistry variations among a series of laboratory-produced biochars. *Geoderma, 163*(3–4), 247–255. https://doi.org/10.1016/j.geoderma.2011.04.021

Nacu, G., Bulgariu, D., Cristina Popescu, M., Harja, M., Toader Juravle, D., & Bulgariu, L. (2016). Removal of Zn(II) ions from aqueous media on thermal activated sawdust. *Desalination and Water Treatment, 57*(46), 21904–21915. https://doi.org/10.1080/19443994.2015.1128366

Nasruddin, M. N., Fahmi, M. R., Abidin, C. Z. A., & Yen, T. S. (2018). Regeneration of spent activated carbon from wastewater treatment plant application. *Journal of Physics: Conference Series, 1116*(3), 032022. https://doi.org/10.1088/1742-6596/1116/3/032022

Niazi, L., Lashanizadegan, A., & Sharififard, H. (2018). Chestnut oak shells activated carbon: Preparation, characterization and application for Cr (VI) removal from dilute aqueous solutions. *Journal of Cleaner Production, 185*, 554–561. https://doi.org/10.1016/j.jclepro.2018.03.026

Nie, J., Okada, F., Kita, H., Tanaka, K., Mihara, T., Kondo, D., Yamashita, Y., & Yahagi, N. (2022). Fabrication of carbon molecular sieve membranes supported on a novel porous carbon fiber. *Energy and Fuels, 36*(13), 7147–7157. https://doi.org/10.1021/acs.energyfuels.2c00907

Norizan, M. N., Moklis, M. H., Ngah Demon, S. Z., Halim, N. A., Samsuri, A., Mohamad, I. S., Knight, V. F., & Abdullah, N. (2020). Carbon nanotubes: Functionalisation and their application in chemical sensors. *RSC Advances, 10*(71), 43704–43732. https://doi.org/10.1039/D0RA09438B

Nour, M., Berean, K., Balendhran, S., Ou, J. Z., Plessis, J. Du, McSweeney, C., Bhaskaran, M., Sriram, S., & Kalantar-zadeh, K. (2013). CNT/PDMS composite membranes for H2 and CH4 gas separation. *International Journal of Hydrogen Energy, 38*(25), 10494–10501. https://doi.org/10.1016/j.ijhydene.2013.05.162

Oginni, O., Singh, K., Oporto, G., Dawson-Andoh, B., McDonald, L., & Sabolsky, E. (2019). Influence of one-step and two-step KOH activation on activated carbon characteristics. *Bioresource Technology Reports, 7*, 100266. https://doi.org/10.1016/j.biteb.2019.100266

Olabi, A. G., Alami, A. H., Ayoub, M., Aljaghoub, H., Alasad, S., Inayat, A., Abdelkareem, M. A., Chae, K. J., & Sayed, E. T. (2023). Membrane-based carbon capture: Recent progress, challenges, and their role in achieving the sustainable development goals. *Chemosphere, 320*, 137996. https://doi.org/10.1016/j.chemosphere.2023.137996

Omorogie, M. O., Babalola, J. O., & Unuabonah, E. I. (2016). Regeneration strategies for spent solid matrices used in adsorption of organic pollutants from surface water: A critical review. *Desalination and Water Treatment, 57*(2), 518–544. https://doi.org/10.1080/19443994.2014.967726

Osman, A. I., Abdelkader, A., Farrell, C., Rooney, D., & Morgan, K. (2019a). Reusing, recycling and up-cycling of biomass: A review of practical and kinetic modelling approaches. *Fuel Processing Technology, 192*, 179–202. https://doi.org/10.1016/j.fuproc.2019.04.026

Osman, A. I., Blewitt, J., Abu-Dahrieh, J. K., Farrell, C., Al-Muhtaseb, A. H., Harrison, J., & Rooney, D. W. (2019b). Production and characterisation of activated carbon and carbon nanotubes from potato peel waste and their application in heavy metal removal. *Environmental Science and Pollution Research, 26*(36), 37228–37241. https://doi.org/10.1007/s11356-019-06594-w

Osman, A. I., Farrell, C., Al-Muhtaseb, A. H., Harrison, J., & Rooney, D. W. (2020a). The production and application of carbon nanomaterials from high alkali silicate herbaceous biomass. *Scientific Reports, 10*(1), 2563. https://doi.org/10.1038/s41598-020-59481-7

Osman, A. I., Fawzy, S., Farghali, M., El-Azazy, M., Elgarahy, A. M., Fahim, R. A., Maksoud, M. I. A. A., Ajlan, A. A., Yousry, M., Saleem, Y., & Rooney, D. W. (2022a). Biochar for agronomy, animal farming, anaerobic digestion, composting, water treatment, soil remediation, construction, energy storage, and carbon sequestration: A review. *Environmental Chemistry Letters, 20*(4), 2385–2485. https://doi.org/10.1007/s10311-022-01424-x

Osman, A. I., Mehta, N., Elgarahy, A. M., Hefny, M., Al-Hinai, A., Al-Muhtaseb, A. H., & Rooney, D. W. (2022b). Hydrogen production, storage, utilisation and environmental impacts: A review. *Environmental Chemistry Letters, 20*(1), 153–188. https://doi.org/10.1007/s10311-021-01322-8

Osman, A. I., O'Connor, E., McSpadden, G., Abu-Dahrieh, J. K., Farrell, C., Al-Muhtaseb, A. H., Harrison, J., & Rooney, D. W. (2020b). Upcycling brewer's spent grain waste into activated carbon and carbon nanotubes for energy and other applications via two-stage activation. *Journal of Chemical Technology & Biotechnology, 95*(1), 183–195. https://doi.org/10.1002/jctb.6220

Pacheco, M. J., Vences, L. J., Moreno, H., Pacheco, J. O., Valdivia, R., & Hernández, C. (2021). Review: Mixed-matrix membranes with CNT for CO2 separation processes. *Membranes*, *11*(6), 457. https://doi.org/10.3390/membranes11060457

Pallarés, J., González-Cencerrado, A., & Arauzo, I. (2018). Production and characterization of activated carbon from barley straw by physical activation with carbon dioxide and steam. *Biomass and Bioenergy*, *115*, 64–73. https://doi.org/10.1016/j.biombioe.2018.04.015

Pei, P., Wang, M., Chen, D., Ren, P., & Zhang, L. (2020). Key technologies for polymer electrolyte membrane fuel cell systems fueled impure hydrogen. *Progress in Natural Science: Materials International*, *30*(6), 751–763. https://doi.org/10.1016/j.pnsc.2020.08.015

Pishnamazi, M., Nakhjiri, A. T., Taleghani, A. S., Marjani, A., Rezakazemi, M., & Shirazian, S. (2020). Molecular investigation into the effect of carbon nanotubes interaction with CO2 in molecular separation using microporous polymeric membranes. *Scientific Reports*, *10*(1), 13285. https://doi.org/10.1038/s41598-020-70279-5

Qiu, B., Tao, X., Wang, H., Li, W., Ding, X., & Chu, H. (2021). Biochar as a low-cost adsorbent for aqueous heavy metal removal: A review. *Journal of Analytical and Applied Pyrolysis*, *155*, 105081. https://doi.org/10.1016/j.jaap.2021.105081

Qiu, H., & Yang, J. (2017). Structure and properties of carbon nanotubes. In H. Peng, Q. Li, & T. Chen (Eds.), *Industrial Applications of Carbon Nanotubes* (pp. 47–69). Amsterdam: Elsevier. https://doi.org/10.1016/B978-0-323-41481-4.00002-2

Rashidi, N. A., & Yusup, S. (2017). A review on recent technological advancement in the activated carbon production from oil palm wastes. *Chemical Engineering Journal*, *314*, 277–290. https://doi.org/10.1016/j.cej.2016.11.059

Regmi, C., Ashtiani, S., Hrdlička, Z., & Friess, K. (2021). CO2 /CH4 and H2 /CH4 gas separation performance of CTA-TNT@CNT hybrid mixed matrix membranes. *Membranes*, *11*(11), 862. https://doi.org/10.3390/membranes11110862

Ren, X., Li, H., Chen, J., Wei, L., Modak, A., Yang, H., & Yang, Q. (2017). N-doped porous carbons with exceptionally high CO2 selectivity for CO2 capture. *Carbon*, *114*, 473–481. https://doi.org/10.1016/j.carbon.2016.12.056

Şahin, Ö., Kaya, M., & Saka, C. (2018). Preparation and characterization of small pore carbon molecular sieves by chemical vapor deposition of pistachio shells. *Analytical Letters*, *51*(15), 2429–2440. https://doi.org/10.1080/00032719.2018.1432630

Sajid, M., Asif, M., Baig, N., Kabeer, M., Ihsanullah, I., & Mohammad, A. W. (2022). Carbon nanotubes-based adsorbents: Properties, functionalization, interaction mechanisms, and applications in water purification. *Journal of Water Process Engineering*, *47*, 102815. https://doi.org/10.1016/j.jwpe.2022.102815

Sevilla, M., Díez, N., & Fuertes, A. B. (2021). More sustainable chemical activation strategies for the production of porous carbons. *ChemSusChem*, *14*(1), 94–117. https://doi.org/10.1002/cssc.202001838

Shamsudin, I. K., Abdullah, A., Idris, I., Gobi, S., & Othman, M. R. (2019). Hydrogen purification from binary syngas by PSA with pressure equalization using microporous palm kernel shell activated carbon. *Fuel*, *253*, 722–730. https://doi.org/10.1016/j.fuel.2019.05.029

Shao, X., Feng, Z., Xue, R., Ma, C., Wang, W., Peng, X., & Cao, D. (2011). Adsorption of CO2, CH4, CO2/N2 and CO2/CH4 in novel activated carbon beads: Preparation, measurements and simulation. *AIChE Journal*, *57*(11), 3042–3051. https://doi.org/10.1002/aic.12515

Sharif, F., Gagnon, L. R., Mulmi, S., & Roberts, E. P. L. (2017). Electrochemical regeneration of a reduced graphene oxide/magnetite composite adsorbent loaded with methylene blue. *Water Research*, *114*, 237–245. https://doi.org/10.1016/j.watres.2017.02.042

Shi, W., Yang, H., Shen, Y., Fu, Q., Zhang, D., & Fu, B. (2018). Two-stage PSA/VSA to produce H2 with CO2 capture via steam methane reforming (SMR). *International Journal of Hydrogen Energy*, *43*(41), 19057–19074. https://doi.org/10.1016/j.ijhydene.2018.08.077

Singh, R., & Gupta, S. M. (2016). *Introduction to Nanotechnology*. New Delhi: Oxford University Press.

Singh, G., Kim, I. Y., Lakhi, K. S., Joseph, S., Srivastava, P., Naidu, R., & Vinu, A. (2017). Heteroatom functionalized activated porous biocarbons and their excellent performance for CO2 capture at high pressure. *Journal of Materials Chemistry A, 5*(40), 21196–21204. https://doi.org/10.1039/C7TA07186H

Singh, G., Lakhi, K. S., Sil, S., Bhosale, S. V., Kim, I. Y., Albahily, K., & Vinu, A. (2019). Biomass derived porous carbon for CO2 capture. *Carbon, 148*, 164–186. https://doi.org/10.1016/j.carbon.2019.03.050

Song, X., Wang, L., Ma, X., & Zeng, Y. (2017). Adsorption equilibrium and thermodynamics of CO2 and CH4 on carbon molecular sieves. *Applied Surface Science, 396*, 870–878. https://doi.org/10.1016/j.apsusc.2016.11.050

Song, M., Zhou, Y., Ren, X., Wan, J., Du, Y., Wu, G., & Ma, F. (2019). Biowaste-based porous carbon for supercapacitor: The influence of preparation processes on structure and performance. *Journal of Colloid and Interface Science, 535*, 276–286. https://doi.org/10.1016/j.jcis.2018.09.055

Srinivas, G., Krungleviciute, V., Guo, Z. X., & Yildirim, T. (2014). Exceptional CO2 capture in a hierarchically porous carbon with simultaneous high surface area and pore volume. *Energy and Environmental Science, 7*(1), 335–342. https://doi.org/10.1039/c3ee42918k

Steel, K. M., & Koros, W. J. (2003). Investigation of porosity of carbon materials and related effects on gas separation properties. *Carbon, 41*(2), 253–266. https://doi.org/10.1016/S0008-6223(02)00309-3

Sun, Z., Wang, M., Fan, J., Zhou, Y., & Zhang, L. (2020). Regeneration Performance of activated carbon for desulfurization. *Applied Sciences, 10*(17), 6107. https://doi.org/10.3390/app10176107

Tian, X., Wang, Y., Zeng, Z., Dai, L., Peng, Y., Jiang, L., Yang, X., Yue, L., Liu, Y., & Ruan, R. (2021). Study on the mechanism of co-catalyzed pyrolysis of biomass by potassium and calcium. *Bioresource Technology, 320*, 124415. https://doi.org/10.1016/j.biortech.2020.124415

Tomczyk, A., Sokołowska, Z., & Boguta, P. (2020). Biochar physicochemical properties: Pyrolysis temperature and feedstock kind effects. *Reviews in Environmental Science and Biotechnology, 19*(1), 191–215. https://doi.org/10.1007/s11157-020-09523-3

Toński, M., Paszkiewicz, M., Dołżonek, J., Flejszar, M., Bielicka-Giełdoń, A., Stepnowski, P., & Białk-Bielińska, A. (2021). Regeneration and reuse of the carbon nanotubes for the adsorption of selected anticancer drugs from water matrices. *Colloids and Surfaces A: Physicochemical and Engineering Aspects, 618*, 126355. https://doi.org/10.1016/j.colsurfa.2021.126355

Tripathi, M., Sahu, J. N., & Ganesan, P. (2016). Effect of process parameters on production of biochar from biomass waste through pyrolysis: A review. *Renewable and Sustainable Energy Reviews, 55*, 467–481. https://doi.org/10.1016/j.rser.2015.10.122

Tsai, W.-T., & Jiang, T.-J. (2018). Mesoporous activated carbon produced from coconut shell using a single-step physical activation process. *Biomass Conversion and Biorefinery, 8*(3), 711–718. https://doi.org/10.1007/s13399-018-0322-x

Wahby, A., Silvestre-Albero, J., Sepúlveda-Escribano, A., & Rodríguez-Reinoso, F. (2012). CO2 adsorption on carbon molecular sieves. *Microporous and Mesoporous Materials, 164*, 280–287. https://doi.org/10.1016/j.micromeso.2012.06.034

Wan, S., Wu, J., Zhou, S., Wang, R., Gao, B., & He, F. (2018). Enhanced lead and cadmium removal using biochar-supported hydrated manganese oxide (HMO) nanoparticles: Behavior and mechanism. *Science of the Total Environment, 616–617*, 1298–1306. https://doi.org/10.1016/j.scitotenv.2017.10.188

Wang, Y., Chen, J., Ihara, H., Guan, M., & Qiu, H. (2021). Preparation of porous carbon nanomaterials and their application in sample preparation: A review. *TrAC: Trends in Analytical Chemistry, 143*, 116421. https://doi.org/10.1016/j.trac.2021.116421

Wang, L., & Hu, X. (2018). Recent advances in porous carbon materials for electrochemical energy storage. *Chemistry: An Asian Journal, 13*(12), 1518–1529). https://doi.org/10.1002/asia.201800553

Wasajja, H., Lindeboom, R. E. F., van Lier, J. B., & Aravind, P. V. (2020). Techno-economic review of biogas cleaning technologies for small scale off-grid solid oxide fuel cell applications. *Fuel Processing Technology, 197*, 106215. https://doi.org/10.1016/j.fuproc.2019.106215

Wei, J., Tu, C., Yuan, G., Liu, Y., Bi, D., Xiao, L., Lu, J., Theng, B. K. G., Wang, H., Zhang, L., & Zhang, X. (2019). Assessing the effect of pyrolysis temperature on the molecular properties and copper sorption capacity of a halophyte biochar. *Environmental Pollution, 251*, 56–65. https://doi.org/10.1016/j.envpol.2019.04.128

Weidner, E., Karbassiyazdi, E., Altaee, A., Jesionowski, T., & Ciesielczyk, F. (2022). Hybrid metal oxide/biochar materials for wastewater treatment technology: A review. *ACS Omega, 7*(31), 27062–27078. https://doi.org/10.1021/acsomega.2c02909

Xu, Y., Chen, X., Wu, D., Luo, Y., Liu, X., Qian, Q., Xiao, L., & Chen, Q. (2018). Carbon molecular sieves from soybean straw-based activated carbon for CO2/CH4 separation. *Carbon Letters, 25*(1), 68–77. https://doi.org/10.5714/CL.2018.25.068

Xu, L., Rungta, M., & Koros, W. J. (2011). Matrimid(r) derived carbon molecular sieve hollow fiber membranes for ethylene/ethane separation. *Journal of Membrane Science, 380*(1–2), 138–147. https://doi.org/10.1016/j.memsci.2011.06.037

Xu, X., Schierz, A., Xu, N., & Cao, X. (2016). Comparison of the characteristics and mechanisms of Hg(II) sorption by biochars and activated carbon. *Journal of Colloid and Interface Science, 463*, 55–60. https://doi.org/10.1016/j.jcis.2015.10.003

Xu, F., Tang, Z., Huang, S., Chen, L., Liang, Y., Mai, W., Zhong, H., Fu, R., & Wu, D. (2015). Facile synthesis of ultrahigh-surface-area hollow carbon nanospheres for enhanced adsorption and energy storage. *Nature Communications, 6*(1), 1–12. https://doi.org/10.1038/ncomms8221

Yaashikaa, P. R., Kumar, P. S., Varjani, S., & Saravanan, A. (2020). A critical review on the biochar production techniques, characterization, stability and applications for circular bioeconomy. *Biotechnology Reports, 28*, e00570. https://doi.org/10.1016/j.btre.2020.e00570

Yahya, M. A., Al-Qodah, Z., & Ngah, C. W. Z. (2015). Agricultural bio-waste materials as potential sustainable precursors used for activated carbon production: A review. *Renewable and Sustainable Energy Reviews, 46*, 218–235. https://doi.org/10.1016/j.rser.2015.02.051

Yang, Y., Han, Y., Pang, R., & Ho, W. S. W. (2020a). Amine-containing membranes with functionalized multi-walled carbon nanotubes for CO2/H2 separation. *Membranes, 10*(11), 1–15. https://doi.org/10.3390/membranes10110333

Yang, Y., Narayanan Nair, A. K., & Sun, S. (2020b). Adsorption and diffusion of carbon dioxide, methane, and their mixture in carbon nanotubes in the presence of water. *Journal of Physical Chemistry C, 124*(30), 16478–16487. https://doi.org/10.1021/acs.jpcc.0c04325

Yang, P., Rao, L., Zhu, W., Wang, L., Ma, R., Chen, F., Lin, G., & Hu, X. (2020c). Porous carbons derived from sustainable biomass via a facile one-step synthesis strategy as efficient CO2 adsorbents. *Industrial and Engineering Chemistry Research, 59*(13), 6194–6201. https://doi.org/10.1021/acs.iecr.0c00073

Yang, J., Yue, L., Hu, X., Wang, L., Zhao, Y., Lin, Y., Sun, Y., DaCosta, H., & Guo, L. (2017). Efficient CO2 capture by porous carbons derived from coconut shell. *Energy & Fuels, 31*(4), 4287–4293. https://doi.org/10.1021/acs.energyfuels.7b00633

You, Y. W., Lee, D. G., Yoon, K. Y., Moon, D. K., Kim, S. M., & Lee, C. H. (2012). H2 PSA purifier for CO removal from hydrogen mixtures. *International Journal of Hydrogen Energy, 37*(23), 18175–18186. https://doi.org/10.1016/j.ijhydene.2012.09.044

Yu, K. L., Lau, B. F., Show, P. L., Ong, H. C., Ling, T. C., Chen, W. H., Ng, E. P., & Chang, J. S. (2017). Recent developments on algal biochar production and characterization. *Bioresource Technology, 246*, 2–11. https://doi.org/10.1016/j.biortech.2017.08.009

Yuan, P., Wang, J., Pan, Y., Shen, B., & Wu, C. (2019). Review of biochar for the management of contaminated soil: Preparation, application and prospect. *Science of the Total Environment, 659*, 473–490. https://doi.org/10.1016/j.scitotenv.2018.12.400

Zhai, M., Xu, Y., Guo, L., Zhang, Y., Dong, P., & Huang, Y. (2016). Characteristics of pore structure of rice husk char during high-temperature steam gasification. *Fuel, 185*, 622–629. https://doi.org/10.1016/j.fuel.2016.08.028

Zhang, X., Bao, C., Zhou, F., & Lai, N.-C. (2023). Modeling study on a two-stage hydrogen purification process of pressure swing adsorption and carbon monoxide selective methanation for proton exchange membrane fuel cells. *International Journal of Hydrogen Energy*. https://doi.org/10.1016/j.ijhydene.2023.01.138

Zhang, Y., Hao, F., Xiao, H., Liu, C., Shi, X., & Chen, X. (2016). Hydrogen separation by porous phosphorene: A periodical DFT study. *International Journal of Hydrogen Energy, 41*(48), 23067–23074. https://doi.org/10.1016/j.ijhydene.2016.10.108

Zhang, Z., Zhu, Z., Shen, B., & Liu, L. (2019). Insights into biochar and hydrochar production and applications: A review. *Energy, 171*, 581–598. https://doi.org/10.1016/j.energy.2019.01.035

Zhao, S. X., Ta, N., & Wang, X. D. (2017). Effect of temperature on the structural and physicochemical properties of biochar with apple tree branches as feedstock material. *Energies, 10*(9), 1293. https://doi.org/10.3390/en10091293

Zhou, X. L., Zhang, H., Shao, L. M., Lü, F., & He, P. J. (2021). Preparation and application of hierarchical porous carbon materials from waste and biomass: A review. *Waste and Biomass Valorization, 12*(4), 1699–1724. https://doi.org/10.1007/s12649-020-01109-y

Zulkefli, N. N., Masdar, M. S., Wan Isahak, W. N. R., Jahim, J. M., Md Rejab, S. A., & Lye, C. C. (2019). Removal of hydrogen sulfide from a biogas mimic by using impregnated activated carbon adsorbent. *PLoS One, 14*(2), e0211713. https://doi.org/10.1371/journal.pone.0211713

13 Metal–Organic Frameworks for Hydrogen Purification

*Jyoti Kaushal, Madhu Khatri, and
Shailendra Kumar Arya*

13.1 INTRODUCTION

MOFs, i.e., metal–organic frameworks, are crystalline and porous inorganic–organic hybrid chemical materials. The two main components required for forming MOFs, namely metal ions and organic ligands, are assembled in numerous combinations with the help of coordinating bonds to design multipurpose 3D crystalline structures that exhibit exceptional intrinsic and topological properties. Unlike all the other existing nanomaterials, the ease of tailoring MOFs with desired characteristics and functional properties has led to their extensive studies in the past few decades (Furukawa et al., 2013; Kumar et al., 2015). MOFs have now been established as quickly tunable functional materials with advanced and exceptional properties, including diversity in structure and function, large surface area, regular porosity, optical properties, catalytic properties, and electrochemical and magnetic characteristics (Lee et al., 2009; Cui et al., 2012). The application of MOFs has been explored in varied fields like gas storage, gas separation, molecular sensing, bioimaging, and drug delivery agents (Suh et al., 2012). Owing to their large surface area, porous structure, and presence of free functional groups, MOFs have been recently explored in the field of bio-immobilization of various bio-active compounds, including enzymes, DNA, various proteins, and drugs, thus possessing the ability to extend their application in various fields (Wang et al., 2014; Cui et al., 2015) and in the purification of hydrogen for use as a renewable energy source.

With a growing population, the demand for existing energy sources (fossil fuels) has increased dramatically, which has led to the potential idea of replacing this nonrenewable energy source with renewable ones, one of which is hydrogen-based systems (Smolinski & Howaniec, 2020). This system employs hydrogen as an energy carrier to transfer energy from diverse sources. Hydrogen is a promising energy source that allows the delivery of a zero-carbon economy and is found in abundance on Earth. It is a clean energy source as its emissions are harmless and, therefore, it can convert chemical energy to electrical with efficiency when utilized in a fuel cell. Moreover, the property of high energy density/unit mass for hydrogen allows it to be efficiently used without fluctuations in supply (Mao et al., 2021).

DOI: 10.1201/9781003382522-16

Unfortunately, the limited presence of hydrogen on earth forces us to extract hydrogen from compounds containing hydrogen in elemental form, including water sources and hydrocarbons, which require purification after extraction (Holladay et al., 2009). MOFs have received attention over the past few decades as materials for hydrogen purification following their extraction. The most common format in which the MOF is used for hydrogen purification is in the form of a membrane. A MOF membrane allows the hydrogen extracted from varied sources to pass through its intricate structure, thereby restricting any impurity from passing. MOF, therefore, acts as a selective obstruction that allows only the required hydrogen molecule to pass through (Lu et al., 2007).

Generally speaking, the separation of a gas from a mixture of gases follows four main mechanisms: (i) the size and shape-related sieving effect of the molecules of gas; (ii) different surface communications leading to the equilibrium effect of thermodynamics; (iii) different diffusion rates of the gases causing the kinetic effect in separation; and (iv) light molecules causing the quantum effect. It is, therefore, necessary to look into the pore size, the kinetic parameters of the gas molecules, and the environment of the surface of the MOF pore before the separation of hydrogen gas from any mixture. Hydrogen gas particularly has a meager absorptive strength compared to the other gases. Therefore, for hydrogen gas to penetrate a membrane of MOF, the material should have very small micropores in its structure. Primarily, the selectivity in gas adsorption of various MOFs was assessed using gas adsorption isotherms with a single component with the help of commercial instruments. This was done because of the complexity of the mixture of gases present for separation.

This chapter provides insight into the synthesis strategies, MOFs' preparation mode, and the mechanism behind using them as membranes for hydrogen purification.

13.2 SYNTHESIS OF METAL–ORGANIC FRAMEWORK

The synthesis procedures for preparing different MOFs have been explored worldwide ever since the applicability of this material in various processes has been investigated. Many of these routes are fast, cost-effective, follow green synthesis methods, and are environmentally friendly. The main routes for MOF synthesis are classified as follows: (i) solvothermal, (ii) slow evaporation, (iii) electrochemical, (iv) mechanochemical, (v) microwave-assisted, and (vi) sonochemical (Figure 13.1) (Stock et al., 2012).

Among these techniques, the solvothermal method is the most used methodology for MOF synthesis (Dey et al., 2014). However, the method of slow evaporation is the most suitable route of MOF preparation since it requires no external energy for the MOF preparation but takes several hours or even days for the preparation, making it slightly time-consuming as compared to other methods (Tranchemontagne et al., 2008). The mechanochemical and electrochemical methods allow adding more freedom while synthesizing MOFs. In contrast, the microwave-assisted procedure is applied to chemically inert metal ions to enhance the crystalline nature of the MOFs since the temperature is homogeneously maintained over a particular zone (Martinez et al., 2012). Ultrasonic waves are used in MOF synthesis in the case of the

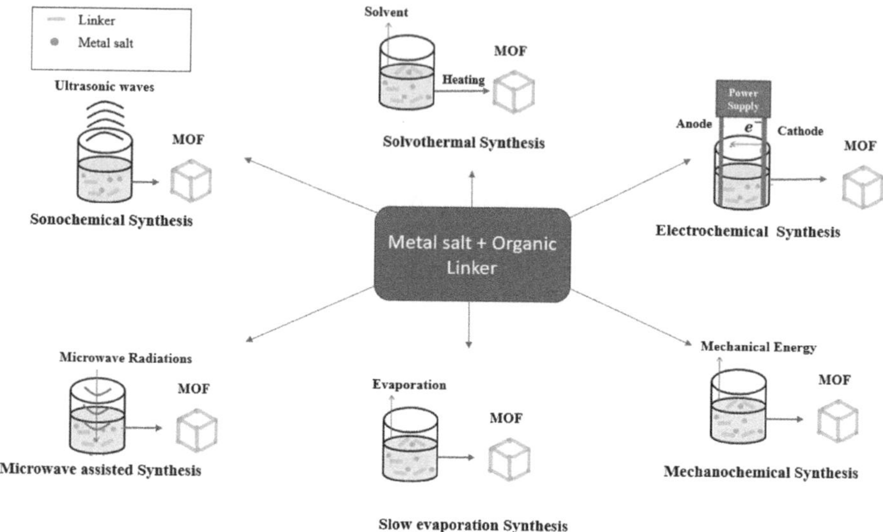

FIGURE 13.1   Synthesis methods for MOFs.

sonochemical synthesis method, where the cavitation created by these waves causes a rapid increase or decrease of pressure or temperature, leading to the creation of rapid MOF crystal formation.

13.3   MECHANISM OF HYDROGEN PURIFICATION THROUGH MOF MEMBRANE

The most common conformation in which MOF is used for the separation and purification of hydrogen is a membrane, which allows only selective molecules (in this case, hydrogen) to pass through while leaving behind any impurities.

Several membrane separation mechanisms have been explored, the main ones being molecular sieving, surface diffusion, condensation through the capillary, and surface diffusion. The two main mechanisms to separate hydrogen include solution diffusion and sieving.

The solution–diffusion method relies mainly on the mobility and solubility of the gas molecules. Molecules easily condensed tend to be more soluble, whereas smaller molecules diffuse quickly and efficiently. In the molecular sieve method of separation, the main influential factor is the molecule size, since the smaller the molecule, the faster the diffusion rate (Koros & Fleming, 1993). It is, therefore, understood that the most common factor that plays an essential role in hydrogen purification through MOF membrane is the size of the molecules (impurities) compared to the membrane pore size and the variable solubility of the molecules in the membrane. Furthermore, certain other factors, including target gas concentration, temperature, electric potential, and partial pressure of gas molecules, strongly influence the membrane separation mechanism (Tao et al., 2015).

13.3.1 Parameters for the Adequate Performance of MOF Membranes

Selectivity and permeability are the two main known parameters that affect the performance of any membrane system, in this case, MOF membrane. To put it in simple words, permeability is the ability of a membrane to let certain-sized molecules pass through it. The following formula defines it:

$$P = D \times S \tag{13.1}$$

where P is the permeability (cm³ (STP)cm cm⁻²·s cmHg), D is the diffusion coefficient (cm² s⁻¹), and S is the solubility coefficient. D represents the kinetic variable and depends on the size of the gas molecule to be separated. S represents the thermodynamic variable, which considers the gas's condensation parameter.

13.3.2 Designing of MOFs for Hydrogen Separation

MOF membranes, to date, are built from many organic/inorganic units that have been used to separate molecules as small as ions into gaseous molecules, organic dyes, or even specific biomolecules. Using membranes to separate gaseous components has many advantages over conventional purification techniques in relation to friendliness with the environment, low consumption of energy, and ease of operation. Overall, traditional polymeric membranes demonstrate low duplicability and have stability concerns since they are sensitive to changes in temperature and severe chemical environments.

Furthermore, they have a trade-off association between permeability and selectivity. A low permeability/selectivity requires a large membrane area and numerous separation steps. In that way, significant efforts have been dedicated to synthesizing porous membranes as a substitute due to their robustness and higher permeability/selectivity. Notably, the issue of plasticization with polymers can be overcome by using MOFs. Since the discovery of Cu-based MOFs in 1989, MOFs have grown in importance as membrane materials owing to their sturdiness, higher porosity, more extensive area of surface, tunable functionality, and more significant ability to exchange ions, making them a fit candidate for separation of molecular gases (38). MOF membranes, in general, can be classified into two types. The first one comprises pure MOFs grown directly as a membrane with the polycrystalline structure on support (referred to as the polycrystalline MOF membranes), whereas the second one is made of MOF crystals spread in numerous polymers as a matrix for the formation of a hybrid mixture, known as MOF mixed matrix membranes (MMMs).

However, the large surface area of MOF-based films faced many challenges linked to the strength of bonding with its substrate. To evade poor connection between MOF membrane and substrate, numerous seeding methodologies have been established, including the contra-diffusion technique, reactive seeding, electrospinning, and gel–vapor deposition (GVD). Nevertheless, structural flaws and arbitrarily oriented crystals make it challenging to sustain selective permeation and flux, which adds to the weakness of this approach. Also, scaling up the production of MOF membranes to meet separation at the industrial scale remains a challenge. MMMs, on the other hand, overcome these confines, as in this approach, pre-formed MOFs are spread into several matrices, ensuring a composite structure of the membrane. A significant property of MMMs is the well-organized contacts between the two materials without accumulating the MOF units. Besides, the morphology, the MOF filler size, and the fraction of MOFs in the matrix are all design contemplations for different MMMs.

13.3.2.1 Flexible MOFs

Recently, soft porous crystals (SPCs), first proposed in 1998, have been developed as progressive flexible MOFs (f MOFs) with crystalline order and supportive structural change. These MOFs have shown numerous exciting properties. Their structural change is a retort to multiple exterior chemical and physical stimuli like guest adsorption/desorption, photochemical stimulus, thermal changes, pressure changes, temperature changes, and contact with gases or solvents. During the phase transformation in f MOFs, there is a drastic change in their porosity, which benefits the careful guest adsorption due to the enhanced molecular sieving outcome. The noteworthy changes in cell volume and rotations of organic linkers have been confirmed by powdered X-ray diffraction and single-crystal X-ray examination (Lim et al., 2021).

13.4 MOFS FOR HYDROGEN PURIFICATION

MOFs, including pellets, a membrane, and crystalline structures, have separated multiple gases and their purification. Hydrogen separation and purification have been done using common MOFs like ZIF-7, MIL-53, COF, Cu-BTC, MOF-5, ZIF-8, etc. The various types of MOFs and their contribution to hydrogen gas purification have been further discussed.

13.4.1 MOFs WITH MULTIPLE CARBOXYLATE GROUPS

Ligands with multiple carboxylate groups have been used as organic linkers to prepare MOFs for hydrogen gas separation. Most commonly, BDC (Benzene-1,4-dicarboxylate), benzene-1,3- dicarboxylate, and BTC (benzene-1,3,5-tricarboxylate) have been used as organic ligands for the preparation of different MOFs falling under this category. Certain materials in this category and those utilized for hydrogen purification include MOF-5, MIL-53, and Cu-BTC (Janiak and Vieth, 2010).

MOF-5 ($Zn_4O(BDC_3)$) was first used as an agent for hydrogen gas separation and purification, and it was identified that the procedure followed the Knudsen diffusion pattern. The permeance value was evaluated as 3×10^{-6} mol s^{-1} m^{-2} Pa^{-1} (Liu et al., 2009). Further research on the same MOF for separating and purifying hydrogen gas led to the synthesis of MOF-5 membranes prepared by a different method. This method led to the formation of MOF-5 with lower permeance than before, which was considered desirable for hydrogen gas separation since the hydrogen particles are tiny. Zhao et al. (2011) reported another paper on MOF-5 being used as a separation tool, with its permeance being lower than that reported by Liu et al. (2009).

Recently, MOF-5 used for the separation of hydrogen gas from impurities was reported by Kloutse et al. (2018). They used a powder consistency material for hydrogen gas separation from ternary materials like H_2:CO_2:N_2 = 45%:10%:45% and H_2:CO_2:CH_4 = 42.5%:42.5%:15% under different pressures ranging from 100 to 1,000 kPa at a temperature of 297 K. The order in which gases were adsorbed onto the MOF-5 material revealed that H_2 was the least adsorbed, leading to its separation from the other gases. The adsorption also rose with a rise in the pressure of the system.

MIL-53, an aluminum-based MOF, is another common MOF in this category. Its framework consists of $AlO_4(OH)_2$ chains, forming an octahedral flexible structure. This flexibility helps adjust the reversible cell volume, leading to better interaction between gas molecules and the MOF framework. This MOF is also chemically and thermally stable (Loiseau et al., 2004). NH_2-MIL-53 has shown the most remarkable ability among the other MOFs in the MIL-53 family to separate hydrogen gas (Li et al., 2015). This material had a large pore size, leading to its high permeance.

Cu-BTC, a copper-based MOF, is the second most common MOF used for hydrogen purification, following ZIF. This MOF typically has a large pore size that efficiently provides a large permeance. A Cu-BTC MOF with the largest permeance was reported by Li et al. (2014b). The permeance, as recorded, was 7×10^{-5} mol m^{-2} s^{-1} Pa^{-1}, and the MOF was made to grow on polyacrylonitrile, forming a fibrous hollow membrane through certain chemical modifications. The H_2/CO_2 separation factor was evaluated as 7.1 at a temperature of 20°C. Another report of a Cu-BTC membrane with a higher separation factor of 13.6 at a temperature of 40°C was given by Zhou et al. (2012).

The separation factor by Cu-BTC MOF for purification of H_2 from N_2 has been reported as 13.7 at ambient room temperature (Shah et al., 2013). In this regard, a swift deposition technique was employed to prepare the Cu-BTC MOF membrane, which was thermally stable and required less effort than other such membranes. Besides separating from N_2, this material could also separate H_2 from CH_4 with a high separation factor of 8.9. However, the prepared membrane's permeance was somewhat poor compared to other such membranes. Another Cu-BTC MOF, which used stainless steel as a support material for the preparation of the MOF, showed a high separation factor (11.2) at a temperature of 25°C for the separation of H_2 from CH_4. This stainless-steel Cu-BTC membrane was elevated in its permeance value and was found to be just 13 mm in breadth. The separation factor was highest for the purification of H_2 from CH_4 compared to N_2 and CO_2 (Ben et al., 2012).

13.4.2 MOF-74

MOF-74 is a mono- or bimetallic MOF prepared using different metal ions (Mg, Mn, Zn, Co, Ni, MnNi, MnCo, and MnZn) and is known for its excellent features of remarkable properties, including the presence of open metal positions, easy tunability, and highly porous structure (Díaz-García et al., 2014). However, the two main MOF-74 preparations that are used for hydrogen gas separation reported in the literature belong to the category of monometallic MOF-74, including MOF-74-Mg and MOF-74-Ni, because of the limitation of the pore size that does not allow the methodology of sieving effect for gas separation to take place. Among the two types, the MOF-74-Mg was reported to be less stable than the Ni-based counterpart for gas separation application as it lost substantial adsorption capacity in the presence of water (Liu et al., 2011). Furthermore, reports by other authors suggest a significant selectivity of the MOF-74-Ni for the separation of hydrogen gas (Al-Naddaf et al., 2018).

13.4.3   UiO-66

UiO-66 with the molecular formula $Zr_6O_4(OH)_4$ $(BDC)_6$ is a MOF material with advantages such as thermal stability, mechanical firmness, and excellent chemical structure, which enables it to be used as a support material for the purification of the hydrogen gas. Various MOFs belonging to this category were testified for H_2 gas purification. The MOF's (Zr)UiO-66, UiO-66 (Zr)-Br, and UiO-67(Zr) were studied, and the results revealed that (Zr)UiO-66 had the highest stability and ability to purify the hydrogen from various mixtures effectively. The MOF's large pore size and volume were the reasons for its high purifying activity (Banu et al., 2013).

13.4.4   ZEOLITIC IMIDAZOLATE FRAMEWORKS

The Zeolitic Imidazolate Frameworks (ZIFs) are a type of MOF that is isomorphic in zeolites' topology. Ligands based on Imidazole coordinate the metal ions (typically transition metals like Zn^{2+}/ Co^{2+}) to the organic linker through a tetrahedral bond formation. MOFs belonging to this class are the most widely used materials for separating and purifying H_2 gas. The main ZIF MOFs used for H_2 gas purification have been discussed further.

13.4.4.1   ZIF-7

ZIF-7 is a type of ZIF formed by the combination of zinc cations with anions of benzimidazole. ZIF-7 typically has a sodalite topology and is widely used for applying H_2 gas purification. The highest value of permeance reported to date is 2.3 × 10^{-6} mol/sm²Pa, which was achieved with the help of a membrane-based ZIF MOF supported on a PVDF fiber (Li et al., 2014a).

The separation ratio of hydrogen gas from nitrogen (1:1 ratio) was found to be 35 using this ZIF framework (Cacho-Bailo et al., 2015). The researchers developed a MOF membrane of the ZIF framework using a microfluidic mechanism to cause direct crystallization. The membrane was supported on a hollow polysulfone fiber and was approximately 2.4 mm thick. Furthermore, the author reports a separation ratio for H_2/CH_4 (1:1) from the same material to be ~35. In the case of the purification of H_2 from CO_2, the ZIF-7 framework showed a separation ratio of 23 (Banu et al., 2013). This ZIF material was supported on the graphene oxide, forming a membrane structure with the MOF embedded within the layers of the membrane. The material was further supported on a PVDF fiber. However, the permeance of this material was relatively low even though its thickness was shallow. Therefore, this material needs to improve its permeance to support the purification process efficiently.

13.4.4.2 ZIF-8

ZIF-8 is a very robust and widely used MOF for the purification of hydrogen gas. It is prepared with Zn ions and linked with Imidazole ligands. The essential characteristics of this MOF, large pore size and a hydrophobic character, are responsible for its applicability as a purifying material (Bux et al., 2009). The hydrogen permeance reported for ZIF-8 had the highest value of 5.7×10^{-5} mol/sm^{2}Pa. The ZIF-8 material used for this purpose was a membrane structure prepared at room temperature using Al_2O_3 substrate tubing in a contra-diffusional methodology. The separation ratios for hydrogen purification from CO_2 and N_2 were reported as 15.5 and 17 (Xie et al., 2012). A team of researchers reported a separation ratio of as high as 34 for the separation of H_2 from CO_2 (1:1) ratio, and that for separation and purification from CH_4 was 38. Again, the contra-diffusional methodology was used to prepare the MOF membrane. However, this membrane with 0.5 mm thickness had low permeance for H_2, which could be attributed to the smaller pore size achieved (Zhang et al., 2017). This study was further supported by Hou et al. (2018), who also demonstrated that the preparation of a ZIF membrane at room temperature could be used for hydrogen purification but was limited by the thickness of the membrane prepared. The permeance appeared to be lower than that achieved by Zang et al. (2017) corresponding to an even lower membrane thickness.

13.4.4.3 ZIF-9

Compared to the other two MOFs discussed above, the number of papers suggesting the use of ZIF-9 MOF in the purification of hydrogen gas is quite limited. Zhang et al. (2013) reported the highest permeance for hydrogen separation as 1.4×10^{-5} mol sm^{2}Pa^{-1}. This MOF was a hybrid and was represented as ZIF-9–67. The ZIF was supported on the Al_2O_3 support membrane. However, the separation ratio was still meager, with a value of 8.9 at room temperature, limiting its use as a H_2 gas purification material.

A team of researchers stated the separation ratio of hydrogen from CO_2 was 40 in 2015 (Huang et al., 2015). A layer-on-layer deposition methodology was used to synthesize the hybrid membrane of ZIF-9, which was deposited on an Al_2O_3 support.

13.5 CONCLUSION

MOFs have been used in various applications and recently in the purification of hydrogen gas from multiple mixtures. MOFs comprise a coordination polymer formed by the interaction of a metal ion and an organic linker with properties like high porosity, water stability, large surface area, and the ability to be embedded in a membrane structure. Different MOFs used in hydrogen purification include MOFs with multiple carboxylate groups and the ZIF, which have mainly been reported. Cu-BTC MOFs have been reported to show the maximum positive results for hydrogen gas purification among all other MOFs. However, the MOF preparation cost still needs to be improved and lowered down for this application to show any significance in real-world applications. Furthermore, the strength and stability of the MOFs need significant improvement for the commercial use of these materials in hydrogen gas purification.

REFERENCES

Al-Naddaf, Q., Thakkar, H., & Rezaei, F. (2018). Novel zeolite-5A@ MOF-74 composite adsorbents with core-shell structure for H2 purification. *ACS Applied Materials & Interfaces*, *10*(35), 29656–29666.

Banu, A. M., Friedrich, D., Brandani, S., & Düren, T. (2013). A multiscale study of MOFs as adsorbents in H2 PSA purification. *Industrial & Engineering Chemistry Research*, *52*(29), 9946–9957.

Ben, T., Lu, C., Pei, C., Xu, S., Qui, S. (2012). Polymer-supported and free-standing metal organic framework membrane. *Chemistry: A European Journal*, *18*, 10250–10253.

Bux, H., Liang, F., Li, Y., Cravillon, J., Wiebcke, M., & Caro, J. (2009). Zeolitic imidazolate framework membrane with molecular sieving properties by microwave-assisted solvothermal synthesis. *Journal of the American Chemical Society*, *131*(44), 16000–16001.

Cacho-Bailo, F., Catalan-Aguirre, S., Etxeberria-Benavides, M., Karvan, O., Sebastian, V., Tellez, C., & Coronas, J. (2015). Metal-organic framework membranes on the inner-side of a polymeric hollow fiber by microfluidic synthesis. *Journal of Membrane Science*, *476*, 277–285.

Cui, L., Wu, J., Li, J., & Ju, H. (2015). Electrochemical sensor for lead cation sensitized with a DNA functionalized porphyrinic metal-organic framework. *Analytical Chemistry*, *87*(20), 10635–10641.

Cui, Y., Xu, H., Yue, Y., Guo, Z., Yu, J., Chen, Z., Gao, J, Yang, Y., Qian, G. & Chen, B. (2012). A luminescent mixed-lanthanide metal-organic framework thermometer. *Journal of the American Chemical Society*, *134*(9), 3979–3982.

Dey, C., Kundu, T., Biswal, B. P., Mallick, A., & Banerjee, R. (2014). Crystalline metal-organic frameworks (MOFs): Synthesis, structure and function. *Acta Crystallographica Section B: Structural Science, Crystal Engineering and Materials*, *70*(1), 3–10.

Díaz-García, M., Mayoral, A., Diaz, I., & Sanchez-Sanchez, M. (2014). Nanoscaled M-MOF-74 materials prepared at room temperature. *Crystal Growth & Design*, *14*(5), 2479–2487.

Furukawa, H., Cordova, K. E., O'Keeffe, M., & Yaghi, O. M. (2013). The chemistry and applications of metal-organic frameworks. *Science*, *341*(6149), 1230444.

Holladay, J. D., Hu, J., King, D. L., & Wang, Y. (2009). An overview of hydrogen production technologies. *Catalysis Today*, *139*(4), 244–260.

Hou, J., Wei, Y., Zhou, S., Wang, Y., & Wang, H. (2018). Highly efficient H2/CO2 separation via an ultrathin metal-organic framework membrane. *Chemical Engineering Science*, *182*, 180–188.

Huang, Y., Xiao, Y., Huang, H., Liu, Z., Liu, D., Yang, Q., & Zhong, C. (2015). Ionic liquid functionalized multi-walled carbon nanotubes/zeolitic imidazolate framework hybrid membranes for efficient H2/CO2 separation. *Chemical Communications*, *51*(97), 17281–17284.

Janiak, C., & Vieth, J. K. (2010). MOFs, MILs and more: Concepts, properties and applications for porous coordination networks (PCNs). *New Journal of Chemistry*, *34*(11), 2366–2388.

Kloutse, F. A., Hourri, A., Natarajan, S., Benard, P., & Chahine, R. (2018). Hydrogen separation by adsorption: Experiments and modelling of H2-N2-CO2 and H2-CH4-CO2 mixtures adsorption on CuBTC and MOF-5. *Microporous and Mesoporous Materials*, *271*, 175–185.

Koros, W. J., & Fleming, G. K. (1993). Membrane-based gas separation. *Journal of Membrane Science*, *83*(1), 1–80.

Kumar, P., Bansal, V., Deep, A., & Kim, K. H. (2015). Synthesis and energy applications of metal organic frameworks. *Journal of Porous Materials*, *22*(2), 413–424.

Lee, J., Farha, O. K., Roberts, J., Scheidt, K. A., Nguyen, S. T., & Hupp, J. T. (2009). Metal-organic framework materials as catalysts. *Chemical Society Reviews, 38*(5), 1450–1459.

Li, W., Meng, Q., Li, X., Zhang, C., Fan, Z., & Zhang, G. (2014a). Non-activation ZnO array as a buffering layer to fabricate strongly adhesive metal-organic framework/PVDF hollow fiber membranes. *Chemical Communications, 50*(68), 9711–9713.

Li, W., Su, P., Zhang, G., Shen, C., & Meng, Q. (2015). Preparation of continuous NH2-MIL-53 membrane on ammoniated polyvinylidene fluoride hollow fiber for efficient H2 purification. *Journal of Membrane Science, 495*, 384–391.

Li, W., Yang, Z., Zhang, G., Fan, Z., Meng, Q., Shen, C., & Gao, C. (2014b). Stiff metal-organic framework-polyacrylonitrile hollow fiber composite membranes with high gas permeability. *Journal of Materials Chemistry A, 2*(7), 2110–2118.

Lim, D. W., Ha, J., Oruganti, Y., & Moon, H. R. (2021). Hydrogen separation and purification with MOF-based materials. *Materials Chemistry Frontiers, 5*(11), 4022–4041.

Liu, J., Benin, A. I., Furtado, A. M., Jakubczak, P., Willis, R. R., & LeVan, M. D. (2011). Stability effects on CO2 adsorption for the DOBDC series of metal-organic frameworks. *Langmuir, 27*(18), 11451–11456.

Liu, Y., Ng, Z., Khan, E. A., Jeong, H. K., Ching, C. B., & Lai, Z. (2009). Synthesis of continuous MOF-5 membranes on porous α-alumina substrates. *Microporous and Mesoporous Materials, 118*(1–3), 296–301.

Loiseau, T., Serre, C., Huguenard, C., Fink, G., Taulelle, F., Henry, M., Bataille, T. & Férey, G. (2004). A rationale for the large breathing of the porous aluminum terephthalate (MIL-53) upon hydration. *Chemistry: A European Journal, 10*(6), 1373–1382.

Lu, G. Q., Da Costa, J. D., Duke, M., Giessler, S., Socolow, R., Williams, R. H., & Kreutz, T. (2007). Inorganic membranes for hydrogen production and purification: A critical review and perspective. *Journal of Colloid and Interface Science, 314*(2), 589–603.

Mao, D., Griffin, J. M., Dawson, R., Fairhurst, A., & Bimbo, N. (2021). Metal organic frameworks for hydrogen purification. *International Journal of Hydrogen Energy, 46*(45), 23380–23405.

Martinez Joaristi, A., Juan-Alcañiz, J., Serra-Crespo, P., Kapteijn, F., & Gascon, J. (2012). Electrochemical synthesis of some archetypical Zn^{2+}, Cu^{2+}, and Al^{3+} metal organic frameworks. *Crystal Growth & Design, 12*(7), 3489–3498.

Shah, M. N., Gonzalez, M. A., McCarthy, M. C., & Jeong, H. K. (2013). An unconventional rapid synthesis of high performance metal-organic framework membranes. *Langmuir, 29*(25), 7896–7902.

Smolinski, A., & Howaniec, N. (2020). Hydrogen energy, electrolyzers and fuel cells: The future of modern energy sector. *International Journal of Hydrogen Energy, 45*(9), 5607–5607.

Stock, N., & Biswas, S. (2012). Synthesis of metal-organic frameworks (MOFs): Routes to various MOF topologies, morphologies, and composites. *Chemical Reviews, 112*(2), 933–969.

Suh, M. P., Park, H. J., Prasad, T. K., & Lim, D. W. (2012). Hydrogen storage in metal-organic frameworks. *Chemical Reviews, 112*(2), 782–835.

Tao, Z., Yan, L., Qiao, J., Wang, B., Zhang, L., & Zhang, J. (2015). A review of advanced proton-conducting materials for hydrogen separation. *Progress in Materials Science, 74*, 1–50.

Tranchemontagne, D. J., Hunt, J. R., & Yaghi, O. M. (2008). Room temperature synthesis of metal-organic frameworks: MOF-5, MOF-74, MOF-177, MOF-199, and IRMOF-0. *Tetrahedron, 64*(36), 8553–8557.

Wang, X., Makal, T. A., & Zhou, H. C. (2014). Protein immobilization in metal-organic frameworks by covalent binding. *Australian Journal of Chemistry, 67*(11), 1629–1631.

Xie, Z., Yang, J., Wang, J., Bai, J., Yin, H., Yuan, B., Lu, J., Zhang, Y., Zhou, L. & Duan, C. (2012). Deposition of chemically modified α-Al2O3 particles for high performance ZIF-8 membrane on a macroporous tube. *Chemical Communications*, *48*(48), 5977–5979.

Zhang, S., Wang, Z., Ren, H., Zhang, F., & Jin, J. (2017). Nanoporous film-mediated growth of ultrathin and continuous metal-organic framework membranes for high-performance hydrogen separation. *Journal of Materials Chemistry A*, *5*(5), 1962–1966.

Zhang, C., Xiao, Y., Liu, D., Yang, Q., & Zhong, C. (2013). A hybrid zeolitic imidazolate framework membrane by mixed-linker synthesis for efficient CO2 capture. *Chemical Communications*, *49*(6), 600–602.

Zhao, Z., Ma, X., Li, Z., & Lin, Y. S. (2011). Synthesis, characterization and gas transport properties of MOF-5 membranes. *Journal of Membrane Science*, *382*(1–2), 82–90.

Zhou, S., Zou, X., Sun, F., Zhang, F., Fan, S., Zhao, H., Schiestel, T. & Zhu, G. (2012). Challenging fabrication of hollow ceramic fiber supported Cu3 (BTC)2 membrane for hydrogen separation. *Journal of Materials Chemistry*, *22*(20), 10322–10328.

14 Swing Technologies for Hydrogen Purification

A.O. Ogunyinka, A.P.I. Popoola, S.L. Pityana,
E.R. Sadiku, and O.M. Popoola

14.1 INTRODUCTION

Purification and recovery of gases are viable methods for removing impurities and contaminants. Purification and recovery techniques will ensure the high quality, safety, and economic viability of gas products. There are several methods in the literature employed in the purification and recovery of gases, including distillation, membrane separation, and swing adsorption. However, the choice of a method depends on the specific gas mixture and the high level of purity required, respectively. Gas distillation is a common method of purification and recovery. It operates by separating a mixture of chemicals at varying boiling points into their constituents. Separating these components can be achieved by heating the mixture to a temperature at which one or more components vaporize while the others remain liquid. As the vapour condenses, it becomes liquid and is consequently distilled. Due to the low boiling points of many gases, it is easy to separate them from other components. Natural gas is typically purified through distillation. The distillation process is a versatile and effective technique that has become increasingly important in modern industries such as oil and gas, chemical manufacturing, and pharmaceuticals (Lei et al., 2007).

Membrane separation is one of the techniques typically used to separate and purify distinct components in a mixture by using thin barriers such as microfiltration, ultrafiltration, and nano-filtration. This process is widely applicable in various industries such as pharmaceuticals, food and beverages, and water treatment. Membrane separation is considered a cost-effective and environmentally friendly alternative to traditional separation methods (Gan et al., 2023).

Swing adsorption is a separation process that exploits the selective adsorption properties of certain materials to isolate different gases or liquids from a mixture. It involves cyclically exposing the mixture to a carefully selected adsorbent material, such as zeolite or activated carbon, which selectively adsorbs one or more of the components of the mixture while allowing the other components to pass through. During adsorption, the selective adsorbent material adsorbs one or more of the components from the mixture onto the material surface, thereby separating it from the other components (Prabhansu et al., 2015; de Oliveira et al., 2023).

The adsorbed components can then be desorbed from the material by changing the temperature or pressure, which releases the adsorbed component back into the gas or liquid phase.

DOI: 10.1201/9781003382522-17

The principle of selective adsorption is what allows swing adsorption to be an effective separation process. Moreover, adsorbent materials have respective selectivity for different components of a mixture, which can selectively remove one or more components and retain others. Therefore, swing adsorption is a more versatile process utilized for a wide range of separation applications (Zhou et al., 2022; Tingelinhas et al., 2023).

14.2 SWING ADSORPTION SEPARATION AND PURIFICATION

Gases are produced from several energy sources, including natural gas, oil, coal, biomass, and hydrocarbons (Saraceno, Ruocco and Palma, 2023). The gas product from these sources comes along with hydrogen and various levels of impurities, depending on the feedstock and production method. Aside from the gas product, the hydrogen production unit contains a variety of minute compositions, such as steam methane, carbon monoxide, nitrogen, sulphur, and carbon dioxide. To obtain a pure hydrogen content, separation and purification techniques are required to remove contaminants and impurities. The retentate impurities remain in the voids of a microporous or mesoporous sorbent until a separation process is conducted under varying conditions (pressure). As the sorbent becomes saturated, the gas pressure decreases while the regenerated impurity is released. Lin et al. evaluated the rapid pressure swing adsorption (RPSA) process for ammonia. The author used various parameters and conditions to determine the most efficient and cost-effective approach for ammonia separation. A high recovery of pure ammonia was produced under optimized conditions (Ahmed et al., 2021; Shi et al., 2021).

The swing technologies for hydrogen purification are based on the physical binding of gas molecules to adsorbent material (Panda, Sahoo, and Ramakrishnan, 2023). As gas molecules interact with adsorbent materials, the force between them depends on the gas component, types of adsorbent material, partial pressure, and operating temperature. It is the differential forces of the adsorbent material that cause the separation effect. In pressure swing adsorption (PSA), the sorption process is performed using the alternating effect of partial pressure and pressure swing, where the temperature remains constant. Moreover, a short range of cycles can be achieved. The process is cost-effective when removing a large portion of impurities (Panda, Sahoo, and Ramakrishnan, 2023).

Furthermore, many factors need to be considered in the choice of an efficient purification technology. These include the feedstock of synthesis gas (syngas) and the hydrogen gas specifications, as well as economic viability and environmental impact. Additionally, process integration will also play a vital role. Because using multiple purification techniques will support achieving a high level of purity and consequently increase the efficiency of hydrogen production at a lower cost. Adequate research needs to be conducted in order to select the best technologies for specific needs (Carminati, Araújo and De Medeiros, 2023).

There are many effective adsorption techniques involved in separation, including solid-phase, liquid-phase, and gas-phase adsorption (Neubert et al., 2017). The solid phase occurs when the molecules or ions are adhering to the surface of the solid material (Saheed, Azeez and Suah, 2022). This surface provides the active site for attraction, which may be in the form of an irregular site, pores, or a chemical functional group. Solid adsorption is applicable in various fields, such as catalysts, which can facilitate chemical reactions, gas separation, and purification processes. It is also applicable in the pharmaceutical and petrochemical industries, as well as water treatment plants (Neubert et al., 2017). Liquid-phase adsorption is widely used in chromatography processing plants (Ocaña-González et al., 2023). Gas-phase adsorption is utilized in adsorption technologies such as activated filters, where pollutants and impurities are removed from air or gases.

Swing pressure adsorption is highly variable in performance and effectiveness due to several key factors. These include adsorption selection, PSA kinetics, temperature adsorption equilibrium, bed design, mass transfer, regeneration efficiency, system pollution, and scale-up considerations (Durán, Rubiera and Pevida, 2022).

a. The adsorbent affinity and selectivity towards the target component in the mixture are crucial in determining the efficiency of separation.
b. An efficient pressure swing depends on its magnitude, frequency, and duration. Therefore, a high optimum pressure swing will be required to maximize the efficiency of separation.
c. A rapid kinetic in the adsorption and desorption phases will increase the efficiency of the swing pressure. While the extent of separation between the adsorbed and desorbed can be determined at the equilibrium level.
d. The separation efficiency would increase when mass transfer between gas phases and adsorbents is effective.
e. Bed design and geometry configuration would ensure flowability and optimum contact between the mixture and adsorbent.

Hydrogen purification and separation systems need to be structurally designed in a geometrical form to ensure efficient and effective operation and safety (Ghorbani, Zendehboudi and Afrouzi, 2023). Factors like flow dynamics, material selection, pressure, temperature, and maintenance requirements need to be considered (Angikath et al., 2023; Tafete and Habtu, 2023). This type of design will facilitate a smooth flow of hydrogen and other gases through each layer of purification and separation (Tafete and Habtu, 2023). Each segment of the components (adsorption bed, membrane, and heat exchanger) needs to be optimized for efficient gas flow while minimizing the pressure drop. There must be safety measures incorporated into the design, such as pressure relief valves, rupture discs, and flame arrestors, to mitigate the potential hazards and detect leakages and abnormalities, which would require adequate ventilation and monitoring systems. Modular design and accessibility for maintenance must also be incorporated to ensure ease of inspection, maintenance, replacement, and scalability (Durán, Rubiera and Pevida, 2022; Ghorbani, Zendehboudi and Afrouzi, 2023).

Moreover, if there is any residual hydrogen concentration in the chamber processor, it will undergo a membrane process where water vapour, oxygen, nitrogen, noble gases, and carbon dioxide diffuse through the hollow fibre membrane at various speeds depending on their molecular structure. Oxygen and nitrogen will be separated at the membrane based on their differential diffusion grades. The final stage of

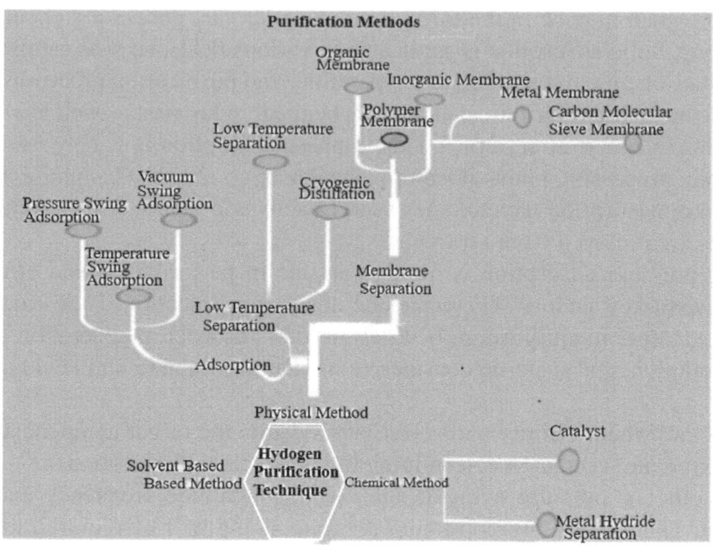

FIGURE 14.1 A gas purification and separation process (Du et al., 2021).

purification (as shown in Figure 14.1) will involve cryogenic separation, which will determine the purity and quality of the product. There are two methods by which cryogenic separation can be categorized: physical and chemical methods. These categories can also be subdivided into pressure swing, temperature swing, and vacuum adsorption. The low-temperature separation method can be an adsorption-based system or it can be cryogenic distillation among the membrane-based methods, thus depending on the types of membranes, either inorganic or organic membranes (Lin, Hsieh and Malmali, 2021).

Figure 14.1 shows various techniques involved in gas purification. A considerable selection of these techniques will depend on many factors, such as the nature of impurities, their concentration, and the intended application of the purified gas. Some of the commonly employed swing adsorption processes and the choice of the appropriate method will depend on the specific application and the gases involved, and this includes temperature swing adsorption (TSA), PSA, and vacuum swing adsorption (VSA).

14.2.1 Temperature Swing Adsorption

The TSA technique requires a long heating and cooling cycle and multiple layers to process the feeding gases while regenerating the feed product. In TSA, a sorbent material is exposed to a gas stream containing the target gas at a lower temperature, in which the sorbent selectively adsorbs the target gas. The gas stream is removed and then the sorbent is heated to a higher temperature, which causes desorption of the target gas from the sorbent. Finally, the deserted gas is recovered later (Berg et al., 2019; Gholami, Van Assche and Denayer, 2023; Zhao et al., 2023).

TSA is a promising technique for gas separation because it can operate at ambient pressure and temperature, requires no additional chemicals or solvents, and is scalable for industrial applications. It is widely used for separating CO_2 from flue gas emissions and natural gas processing as depicted in Figure 14.2 (Liu et al., 2023; Masuda et al., 2023). One of the challenges in TSA is selecting the appropriate sorbent material with high selectivity and capacity for the target gas. In addition, optimizing the operating conditions, such as temperature and gas flow rate, is critical for achieving high separation efficiency and reducing energy consumption (Park et al., 2021; van Paasen et al., 2021).

Temperature–vacuum swing adsorption (TVSA) is a relatively new concept in gas separation, and it has shown significant promise and potential in laboratory experiments and simulations. The growing interest in TVSA is due to its potential advantages over existing technologies, such as its ability to separate gases with similar boiling points, high selectivity, and its potential promise for energy savings. However, neither industrial commercialization nor extensive research has been done on TVSA in the literature, owing to a lack of systematic investigation of TVSA by researchers. Additionally, TVSA's technology is still in its infancy, and consequently, several technical challenges are yet to be addressed that have hampered its scalability for industrial use. One such critical issue is the development of efficient and cost-effective adsorbents that can withstand the high temperatures and vacuum conditions required for TVSA (Zhao et al., 2019; Zhu et al., 2021). As a result, a systematic investigation and development of TVSA will be necessary in the future as researchers and industry seek to improve gas separation processes and reduce energy consumption (Bos, Pietersen and Brilman, 2019; Dawood, Anda and Shafiullah, 2020).

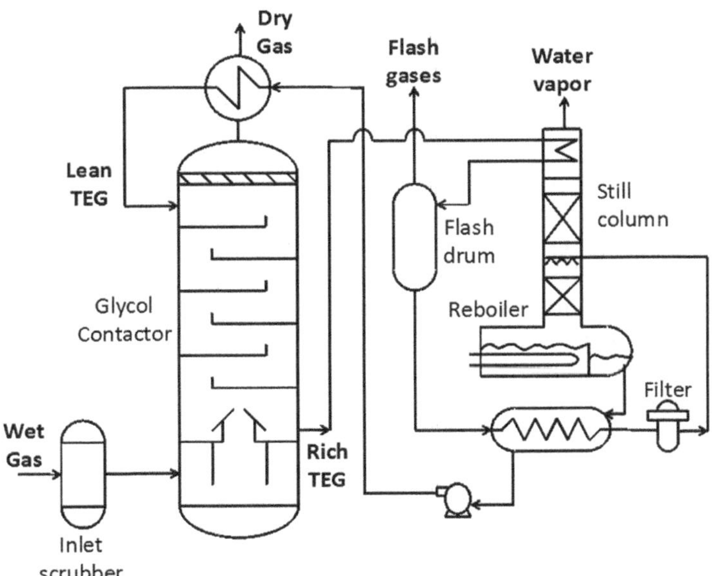

FIGURE 14.2 Temperature swing adsorption (Al Wahedi, Saleh and Dadach, 2020).

A combined TVSA along with TSA processes is commonly used for the purification and recovery of volatile organic compounds (VOCs) using activated carbon as an adsorbent. These processes involve the adsorption and desorption behaviours of volatile compounds such as benzene and ethene. To develop an optimally efficient adsorption process, appropriate adsorbents and optimized designs would be needed. This is necessary because of the growing demand for efficiency by chemical and allied industries to meet the demand of consumers and also reduce the impact of environmental pollution in society (Tingelinhas et al., 2023).

Nonetheless, some critical parameters such as temperature, duration, gas flow rate, and the concentration of volatile gases affect the performance of combined TVSA and TSA. Optimizing these parameters would increase the efficiency of the purification and recovery processes. Besides, TVSA was developed to get the highest grade of purity from a desorbing product, which is the case for carbon capture, where carbon dioxide is the most strongly adsorbed compound (Hedin et al., 2013; Sui et al., 2016; Li et al., 2023).

14.2.2 Pressure Swing Adsorption

PSA occurs when pressure increases due to loading at a constant temperature. This method usually operates on a cycle basis and removes all the impurities in a single step. Pressure can be changed much faster than temperature to achieve the same separation. Consequently, PSA is considered to be more advantageous and cost-effective to maintain compared to TSA (Gomez-Rueda et al., 2022). The pressure swing can separate various types of gases from a mixture of gases under higher pressure and ambient temperature, depending on the molecular behaviour and affinity of the gas to an adsorbent material. As such, PSA exhibits good isothermal behaviour and is widely used for hydrogen purification. Moreover, PSA has a high capability for producing ultra-pure hydrogen with a high percentage of hydrogen recovery (Adhikari and Fernando, 2006; Liemberger et al., 2017).

In the operational system, the gas molecules are attracted towards the surface area of the sorbent. Since each gas has a different attraction, affinity, and bonding with surface area, this results in a high exothermic reaction, causing a large amount of heat to escape. The heat released will accelerate the equilibrium pressure and also regulate the thermal management, thereby keeping the reaction rate at its optimum. The thermodynamic limit of loading on the adsorbent will be determined at a certain composition of an equilibrium pressure and temperature. At equilibrium, the adsorbent material is in a state of dynamic equilibrium, where the rate of adsorption of gas molecules on the surface of the adsorbent is equal to the rate of desorption of gas molecules from the surface (Zhao et al., 2019). When a change in the loading of the adsorbent material occurs, the isothermal forces arise. The difference in adsorption affinity between the gas components causes the more strongly adsorbed component to be retained on the adsorbent material, while the less strongly adsorbed component is desorbed and passes through the adsorbent bed (Liu, Ritter and Kaul, 2000; Ivanova and Lewis, 2012; Lin, Hsieh and Malmali, 2021). The parameters obtained at this equilibrium, such as temperature and pressure, will be required in the process design and system analysis. As the gas is adsorbed at high pressure, the process will swing to low pressure to desorb the adsorbed gas for reuse (Shigaki et al., 2022). At the final stage, the hydrogen will be secured at higher pressure while the impurities are obtained at ambient pressure, as depicted in Figure 14.3.

The feedstock of gas supplied determines the quality of the final gas, so integrating the system with steam methane reforming for hydrogen purification is necessary to establish a continuous process. The energy for this will come from the PSA units tail gas. While more layers are supplied to create a continuous refinery, the entire system will be subject to batch processing.

Liemberger et al. (2017) described this process as follows: adsorption, highest pressure, depressurization, light recycling, and pressurization. The performance indicator will also demonstrate the process plants ability to be recycled, a high percentage of hydrogen purity, adequate energy consumption, and high productivity. Since one sorbent will be insufficient to remove all contaminants, multiple absorbents will be used, greatly increasing energy efficiency. As a result, the choice of absorbent is necessitated by a specific component that requires removal from the input gas. The PSA unit design determines the order in which the impurities are removed. The most widely used materials in PSA are alumina, silica gel, molecular sieves, zeolite, and activated carbon. Lastly, these different materials have different affinity layers that sorbents employ to improve purity (Sant Anna et al., 2017).

Water, CO_2, CH_4, N_2, and Ar are the first impurities and contaminants eliminated sequentially. Because C3$^+$ is heavily adsorbed, it is removed using silica gel before it can reach the next layer of activated carbon. The adsorption order for activated carbon is water, CO_2, CH_4, CO, N_2, and H_2, and the kinetics of the adsorption process is micropore diffusion, which demonstrates the sequential order of H_2, N_2, CO, and CH_4 (Ma et al., 2019). Hydrogen remains unabsorbed and is collected. Water is the first impurity that requires removal from the first layer. Because alumina has a higher loading capacity than silica gel, it is followed by aromatic hydrocarbons (Moon, Lee and Lee, 2016).

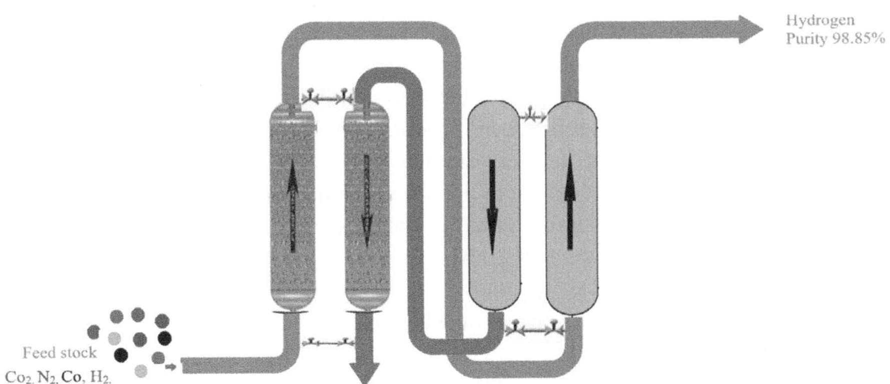

FIGURE 14.3 Pressure swing adsorption process for hydrogen purification. (Adapted from Martínez García et al., 2022.)

While activated carbon lies between the alumina and silica gel, it has a low volumetric capacity. Once the water content is removed, the next layer would be a higher hydrocarbon because the third layer is activated carbon, which is poisonous. Hence, it is very complex to dissolve from the third layer. To this end, the hydrocarbons must be eliminated in the second layer, which employs silica gel, before moving on to the third layer. The third layer involves activated carbon, widely regarded for its affinity for various impurities. Several impurities can be extracted using activated carbon in adsorption order: water, carbon dioxide, methane, carbon monoxide, nitrogen, and hydrogen (Van Acht et al., 2020).

The benefits of using activated carbon as a molecular sieve for gas separation were highlighted by Golmakani, Fatemi and Tamnanloo (2017) and Ma et al. (2019). The authors listed some of the advantages it has over the conventional approach, particularly for the adsorption of smaller gas molecules (nitrogen and oxygen) while allowing larger molecules to permeate, such as methane and carbon dioxide. Because of its substantial surface area and pore volume, activated carbon is a highly porous material used in the separation process. As a result, the size of the molecules being separated and the size of the pores in the activated carbon both play a significant role in the mechanism. Macropores, mesopores, and micropores are the three types of pores that can be found in this material. Among these, micropores show excellent performance because the diffusion process is driven by a composition gradient. This diffusion occurs in materials with pore sizes of less than two nanometres, such as zeolites, activated carbon, and other types of membranes. One advantage of activated carbon is that it is less expensive, can be regenerated, versatile, and isothermal, which is less non-linear than zeolites, and the cost is comparatively less. This property makes it a reliable and predictable absorbent for industrial use. However, it has low adsorption for carbon monoxide (CO) due to the effect of partial pressure. The fourth layer required the removal of CO content using zeolite. As a result, the layer will absorb all the components of gases—CO, CO_2, and moisture. While the retentate will be collected at the outlet, the process is efficient and economical.

Kumar et al. stated various requirements for absorbent materials used in a PSA unit. The author mentioned some characteristics of high stability in the presence of numerous impurities that might be present in the feed stream. These include high capability, low cost, high loading capacity, desorption without vacuum or high temperature, high mechanical stability, and productivity (Kumar et al., 2022).

Kerry et al. reported that the performance of PSA is dependent on the type of sorbent used, such as the percentage of adsorbent loaded with impurities. Using multiple absorbent layers will certainly enhance the efficiency and increase the effectiveness of the separation process, as it allows for greater flexibility in the design of the adsorption bed. As the process becomes more expensive and complex, more sorbent and additional equipment will be needed to control the flow of gas. This method is applicable where a tremendously high percentage of pure hydrogen is recovered from steam methane reforming, partial oxidation, methane decomposition, gasification in a petrochemical or refinery, or coke oven gas (Kerry, 2007; Imteyaz, Tahir and Habib, 2021). The PSA method has a high energy efficiency compared with other carbon dioxide removal methods and is regarded as best-in-class for some reasons, the first of which is:

   i. Its fast-cycle design allows for a highly compact unit with lower life cycle costs.

   ii. The vacuum recycling process increases its recovery to over 95%.

   iii. The PSA system is reliable and offers high performance with a long life cycle.

   iv. A significant amount of hydrogen will be recovered.

14.2.3 Vacuum Swing Adsorption

The VSA process utilizes a cycle of pressure changes to selectively adsorb one gas while releasing the others. This process usually separates both oxygen and nitrogen from the air. In this case, an activated carbon or zeolite bed will be exposed to an air mixture, and the nitrogen is desorbed before the collection of a stream of oxygen. The process is energy-efficient and can be easily scaled up or down, depending on the application (Golmakani, Fatemi and Tamnanloo, 2017). Due to the complexity of the processes, it makes the assessment difficult to evaluate. It is applicable in various industries, such as medical oxygen production, wastewater treatment, and petrochemical refining. As part of the design phase of VSA processes (as shown in Figure 14.4), various variables are optimized (such as adsorbent selection, operating pressure, flow rate, and cycle time) to achieve maximum efficiency and cost-effectiveness (Krishnamurthy, 2022; Gholami, Van Assche and Denayer, 2023). A linear isotherm can be modelled as a swing pressure adsorption system. Based on this linearity, one can determine the heat of adsorption and optimal operating conditions. There are many factors that can influence the VSA processes, including evacuation pressure, purity, recovery, and energy consumption; for example, gas recovery usually decreases with increases in evacuation pressure (Li et al., 2019). Having a high evacuation pressure can cause more gas molecules to become stuck on the surface due to adsorption or condensation, making it harder to recover them. To increase the performance of the vacuum pressure swing, the condition of operation needs to be optimized (Zhou et al., 2022). Sometimes optimizing these conditions often requires combining theoretical analysis, experimentation, and practical experience. The goal is to strike a balance that maximizes gas recovery while minimizing energy consumption and process complexity.

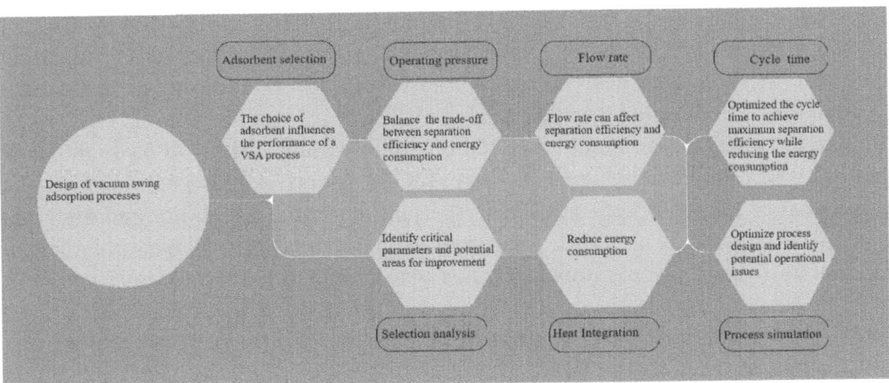

FIGURE 14.4 Vacuum swing adsorption (Zhang et al., 2022).

TABLE 14.1

Advantages and Disadvantages of Swing Adsorption (Amin et al., 2023)

Swing Adsorption Method	Advantages	Disadvantages
TSA	High selectivity	High capital costs
	Low energy consumption	Energy-intensive
	scalability	Limited selectivity
	Low emissions	Adsorbent degradation
	Flexibility	Complex operation
PSA	Effective for weakly adsorbent gas species that will require high purity	High initial capital investment
	Cost-effective	Energy-intensive process
	Easy to operate	Limited application
	Environmentally friendly	Adsorbent bed saturation
VSA	Featuring a high purity level	Low capacity
	Ease of operation	High capital cost
	Environmentally friendly	Adsorbent degradation
		Energy consumption
		A finite range of applications

Table 14.1 provides a summary of the merits and demerits of the respective swing adsorption methods discussed in Subsection 3.2. Additionally, Table 14.2 highlights the challenges of the swing adsorption separation methods in the context of adsorption selection, temperature control, pressure drop, and regeneration efficiency.

14.3   MEMBRANE-BASED SEPARATION

Membrane separation is a simple, economical, and efficient method for selectively attaching molecules to surfaces. It uses thin films without pinholes and can be classified into microfiltration, ultrafiltration, ion exchange, and reverse osmosis. Membrane materials include metallic, carbon-based, polymeric, or inorganic membranes as depicted in Figure 14.5. The process separates hydrogen gas from contaminants in the feed stream. It is frequently used for gas separation because of its exceptional chemical, thermal, and mechanical stability while yet displaying great performance (Adhikari and Fernando, 2006; Ahmed et al., 2021).

Metallic membranes, made of palladium and platinum metals, offer high density and catalytic properties but can cause irreversible defects. Alignments with copper or gold enhance their properties, offering thermal-mechanical stability and high catalytic activity (Wang et al., 2022; Jokar et al., 2023). Glassy and rubbery polymers are utilized for hydrogen separation, with glassy and rubbery having contrary properties of high selectivity and low flux, respectively (Nagel et al., 2000; Grönniger et al., 2023). Polysulfide-swing and polyamide are used for membrane synthesis for hydrogen separation (Han et al., 2023).

Ceramic separation for hydrogen adsorption uses zeolite and activated carbon as specialized ceramic materials with large surface areas and specific pore patterns

TABLE 14.2

Various Challenges Exist in Swing Adsorption Separation (Amin et al., 2023)

Challenges	TSA	PSA	VSA
Adsorbent selection	Adsorbent materials have the following properties: high selectivity, high capacity, and good thermal stability. These materials can degrade when exposed to high temperatures, corrosive gases, and other environmental factors, resulting in increased operating costs	May not always be sufficient to achieve the desired separation, especially when the gas mixture contains components that have similar physical properties	The choice of adsorbent is critical for the success of VSA. It must have high selectivity and capacity for the gas being separated. Strong enough to withstand cyclic loading and unloading without significant degradation in performance
Temperature control	The temperature swing rate can affect the separation efficiency and must be optimized for each specific application	Heat-transfer limitations can affect the efficiency of the process	
Pressure drop	It can affect process efficiency and increase operating costs. Careful design and optimization of the bed geometry and flow rate can help minimize pressure drop	It can reduce the flow rate and the overall efficiency of the process	The pressure drop across the bed of adsorbent can increase as the bed becomes saturated with the adsorbate. This can reduce the efficiency of the process and increase the operating costs
Regeneration efficiency-	Incomplete desorption can lead to a reduced adsorption capacity over time, except through optimization	The adsorption capacity of the adsorbent is limited, and it may decrease over time due to contamination and aging, and the use of purge gas can result in significant energy consumption and operating costs	The adsorbent must be periodically regenerated to remove the adsorbed gas and restore its capacity. Though it reduces the pressure or increases the temperature, which requires energy and can increase the operating costs
System design	Scaling up may require modifications to the process design and increase the cost of production	Scaling up may require modifications to the process design and increase the capital costs	It requires a complex system of valves, compressors, and heat exchangers to control the flow of gas and the temperature and pressure within the adsorbent bed and can be capital-intensive

to selectively absorb hydrogen gas from mixed gas streams (Gondolini et al., 2023). Synthetic zeolites like Faujasite, ZSM-5, and Mordenite are suitable for gas purification due to their high chemical composition and pore size control (Amin et al., 2023). But the main challenges in permeability are geometric configuration, size, and

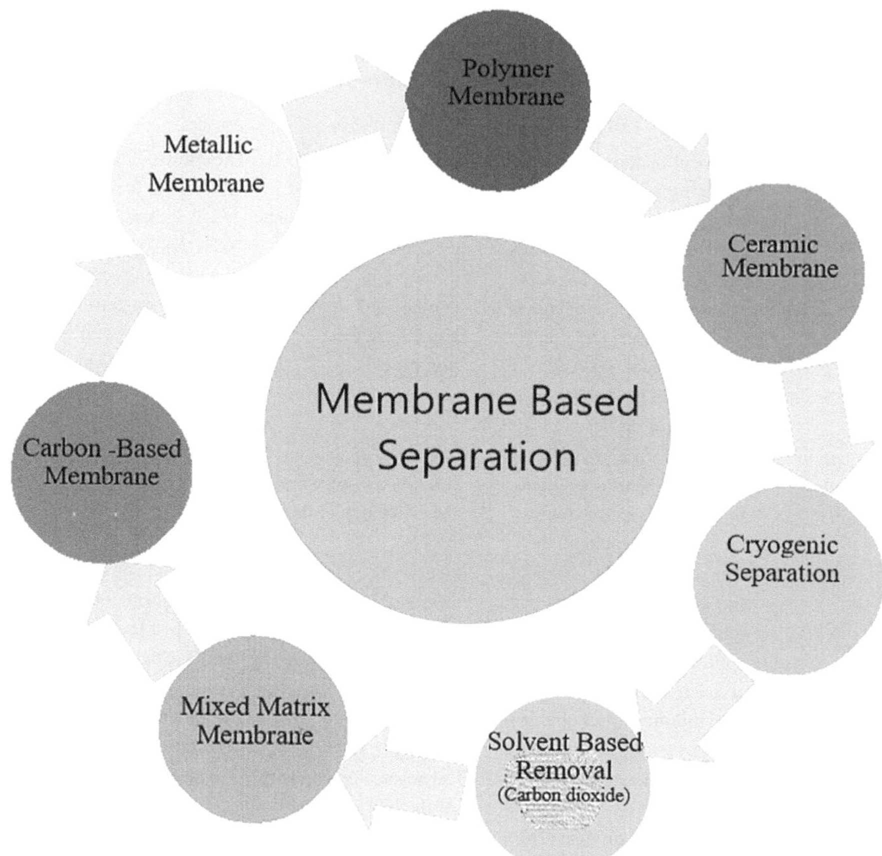

FIGURE 14.5   Membrane-based separation (Singla et al., 2022).

polarity (Huang et al., 2023). *Solution diffusion* is the primary separation mechanism in carbon-based membranes. Examples include carbon nanotubes and molecular sieves (Baamran et al., 2023; Wawrzyńczak et al., 2023). *Mixed matrix membranes* are composite materials employed in separation, purification, and filtration processes (gas and water purification). It is composed of a polymer matrix and dispersed filler particles, which provide unique properties and enhance its performance (Al-Rowaili et al., 2023; Feng et al., 2023).

Hydrogen purification uses a *solvent-based removal technique*, combining hybrid carbon dioxide solvent-based removal with PSA. Common sorbents include monoethanolamine, diethanolamine, triethanolamine, diisopropanolamine, methyldiethanolamine, and diglycolamine (He et al., 2023; Naquash et al., 2023). *Cryogenic separation* is a low-temperature technique used to separate components of gaseous mixtures using their boiling points (Zafanelli et al., 2023).

Finally, membrane-based hydrogen purification systems are being improved for industrial applications to increase efficiency and selectivity.

14.4 CONCLUSION

This section focuses on hydrogen purification and separation. One of the methods used is the chemical solvent-based removal method, which is similar to PSA. It can achieve a high level of purity of 95%–97% and is often used in large-scale hydrogen purification. Furthermore, it necessitates a large amount of steam for solvent regeneration. The improvement of this method will involve the use of hybrid equipment that will be integrated with a smaller PSA unit. The VSA process can be used for a variety of gas separation applications, including the purification of natural gas and biogas, the separation of oxygen and nitrogen from the air, and the production of high-purity hydrogen. TSA is an efficient separation technique for gas mixtures with low concentrations of the target gas, whereas other separation techniques may not be economical. It has been widely used in industrial applications, including carbon capture and storage, natural gas processing, and air separation. Cryogenic separation is another method for hydrogen purification, which requires very low temperatures and low pressure. To this end, it is capital-intensive due to the large amount of energy required. This method will need a compressor for hydrogen purification, and in addition, it can produce a high percentage of purity of about 90%–99%. Membrane-based separation methods are usually ideal for both small- and medium-scale purification, and their recovery will depend on the type of membrane used, although a considerable percentage of purity will also be recorded. Other techniques involve the use of hybrid systems, which combine two systems, such as PSA with membrane-based separation or solvent-based removal methods (carbon dioxide removal techniques). The pair of these will completely increase the cost of production while improving hydrogen recovery, but they are capital-intensive due to high capital expenditure (CAPEX), operational expenditure (OPEX), and the compressor. Conclusively, a well-designed system would contribute to efficient hydrogen production, distribution, and utilization, leading to a more sustainable energy future.

REFERENCES

Adhikari, S. and Fernando, S. (2006) 'Hydrogen membrane separation techniques', *Industrial and Engineering Chemistry Research*, 45(3), pp. 875–881. doi:10.1021/IE050644L.

Ahmed, Z. et al. (2021) 'Recent advances in MXene-based separation membranes', *ChemBioEng Reviews*. doi:10.1002/cben.202000026.

Al-Rowaili, F.N. et al. (2023) 'Mixed matrix membranes for H2/CO2 gas separation: A critical review', *Fuel*, 333, p. 126285. doi:10.1016/J.FUEL.2022.126285.

Al Wahedi, F.S., Saleh, M.H. and Dadach, Z.E. (2020) 'Black powder in sales gas pipelines: Sources and technical recommendations', *World Journal of Engineering and Technology*, 8(01), pp. 60–73. doi:10.4236/wjet.2020.81007.

Amin, M. et al. (2023) 'Issues and challenges in hydrogen separation technologies', *Energy Reports*, 9, pp. 894–911. doi:10.1016/J.EGYR.2022.12.014.

Angikath, F. et al. (2023) 'Technoeconomic assessment of hydrogen production from natural gas pyrolysis in molten bubble column reactors', *International Journal of Hydrogen Energy* [Preprint]. doi:10.1016/J.IJHYDENE.2023.07.308.

Baamran, K. et al. (2023) 'Kinetic process assessment of H2 purification over highly porous carbon sorbents under multicomponent feed conditions', *Separation and Purification Technology*, 306, p. 122695. doi:10.1016/J.SEPPUR.2022.122695.

Berg, F. et al. (2019) 'Temperature swing adsorption in natural gas processing: A concise overview', *ChemBioEng Reviews*. doi:10.1002/cben.201900005.

Bos, M.J., Pietersen, S. and Brilman, D.W.F. (2019) 'Production of high purity CO2 from air using solid amine sorbents', *Chemical Engineering Science: X*, 2, p. 100020. doi:10.1016/J.CESX.2019.100020.

Carminati, H.B., Araújo, O. de Q.F. and De Medeiros, J.L. (2023) 'Swing technologies for syngas purification'. In: *Advances in Synthesis Gas: Methods, Technologies and Applications: Volume 2: Syngas Purification and Separation*, pp. 133–178. Elsevier ISBN: 0323918778,9780323918770. doi:10.1016/B978-0-323-91877-0.00011-8.

Dawood, F., Anda, M. and Shafiullah, G.M. (2020) 'Hydrogen production for energy: An overview', *International Journal of Hydrogen Energy*, 45(7), pp. 3847–3869. doi:10.1016/J.IJHYDENE.2019.12.059.

de Oliveira, D.C. et al. (2023) 'Gas cleaning systems for integrating biomass gasification with Fischer-Tropsch synthesis—A review of impurity removal processes and their sequences', *Renewable and Sustainable Energy Reviews*, 172, p. 113047. doi:10.1016/J.RSER.2022.113047.

Du, Z. et al. (2021) 'A review of hydrogen purification technologies for fuel cell vehicles', *Catalysts*. doi:10.3390/catal11030393.

Durán, I., Rubiera, F. and Pevida, C. (2022) 'Modeling a biogas upgrading PSA unit with a sustainable activated carbon derived from pine sawdust. Sensitivity analysis on the adsorption of CO2 and CH4 mixtures', *Chemical Engineering Journal*, 428, p. 132564. doi:10.1016/J.CEJ.2021.132564.

Huang, F., Largier, T., Luo, L., Zheng, W., Cornelius, C.J. (2023) 'Molecular insight into 6FD polyimide-branched poly(phenylene) copolymers: Synthesis, block compatibility, and gas transport study', *ACS Applied Polymer Materials*, 5(1), pp. 290–301.

Feng, X. et al. (2023) 'Mixed-matrix membranes based on novel hydroxamate metal-organic frameworks with two-dimensional layers for CO2/N2 separation', *Separation and Purification Technology*, 305, p. 122476. doi:10.1016/J.SEPPUR.2022.122476.

Gan, G. et al. (2023) 'Adsorption and membrane separation for removal and recovery of volatile organic compounds', *Journal of Environmental Sciences*, 123, pp. 96–115. doi:10.1016/J.JES.2022.02.006.

Gholami, M., Van Assche, T.R. and Denayer, J.F. (2023) 'Temperature vacuum swing, a combined adsorption cycle for carbon capture', *Current Opinion in Chemical Engineering*, 39, p. 100891. doi:10.1016/J.COCHE.2022.100891.

Ghorbani, B., Zendehboudi, S. and Afrouzi, Z.A. (2023) 'Multi-objective optimization of an innovative integrated system for production and storage of hydrogen with net-zero carbon emissions', *Energy Conversion and Management*, 276, p. 116506. doi:10.1016/J.ENCONMAN.2022.116506.

Golmakani, A., Fatemi, S. and Tamnanloo, J. (2017) 'Investigating PSA, VSA, and TSA methods in SMR unit of refineries for hydrogen production with fuel cell specification', *Separation and Purification Technology*, 176, pp. 73–91. doi:10.1016/J.SEPPUR.2016.11.030.

Gomez-Rueda, Y. et al. (2022) 'Rapid temperature swing adsorption using microwave regeneration for carbon capture', *Chemical Engineering Journal*, 446, p. 137345. doi:10.1016/J.CEJ.2022.137345.

Gondolini, A. et al. (2023) 'Development and hydrogen permeation of freeze-cast ceramic membrane', *Journal of Membrane Science*, 684, p. 121865. doi:10.1016/J.MEMSCI.2023.121865.

Grönniger, B. et al. (2023) 'Water sorption in rubbery and glassy polymers, nifedipine, and their ASDs', *Molecular Pharmaceutics*, 20(4), pp. 2194–2206. doi:10.1021/acs.molpharmaceut.3c00006.

Han, S. et al. (2023) 'Ionic liquid modified COF nanosheet interlayered polyamide membranes for elevated nanofiltration performance', *Desalination*, 548, p. 116300. doi:10.1016/J.DESAL.2022.116300.

He, C. et al. (2023) 'Thermodynamic and kinetic synergetic separation of CO2/C2H2 in an ultramicroporous metal-organic framework', *Separation and Purification Technology*, 304, p. 122318. doi:10.1016/J.SEPPUR.2022.122318.

Hedin, N. et al. (2013) 'Adsorbents for the post-combustion capture of CO2 using rapid temperature swing or vacuum swing adsorption', *Applied Energy*. doi:10.1016/j.apenergy.2012.11.034.

Imteyaz, B., Tahir, F. and Habib, M.A. (2021) 'Thermodynamic assessment of membrane-assisted premixed and non-premixed oxy-fuel combustion power cycles', *Journal of Energy Resources Technology, Transactions of the ASME*, 143(5). doi:10.1115/1.4049463.

Ivanova, S. and Lewis, R. (2012) 'Producing nitrogen via pressure swing adsorption', *Chemical Engineering Progress*, 108(6), pp. 38–42.

Jokar, S.M. et al. (2023) 'The recent areas of applicability of palladium based membrane technologies for hydrogen production from methane and natural gas: A review', *International Journal of Hydrogen Energy*, 48(16), pp. 6451–6476. doi:10.1016/J.IJHYDENE.2022.05.296.

Kerry, F.G. (2007) *Industrial Gas Handbook: Gas Separation and Purification*. Boca Raton: CRC Press. doi:10.1201/9781420008265.

Krishnamurthy, S. (2022) 'Vacuum swing adsorption process for post-combustion carbon capture with 3D printed sorbents: Quantifying the improvement in productivity and specific energy over a packed bed system through process simulation and optimization', *Chemical Engineering Science*, 253, p. 117585. doi:10.1016/J.CES.2022.117585.

Kumar, A. et al. (2022) 'Absorption based solid state hydrogen storage system: A review', *Sustainable Energy Technologies and Assessments*, 52, p. 102204. doi:10.1016/J.SETA.2022.102204.

Li, H. et al. (2019) 'Modelling and simulation of two-bed PSA process for separating H2 from methane steam reforming', *Chinese Journal of Chemical Engineering*, 27(8), pp. 1870–1878. doi:10.1016/J.CJCHE.2018.11.022.

Li, Y. et al. (2023) 'Process analysis of temperature swing adsorption and temperature vacuum swing adsorption in VOCs recovery from activated carbon', *Chinese Journal of Chemical Engineering*, 53, pp. 346–360. doi: 10.1016/j.cjche.2022.01.029.

Liemberger, W. et al. (2017) 'Experimental analysis of membrane and pressure swing adsorption (PSA) for the hydrogen separation from natural gas', *Journal of Cleaner Production*, 167, pp. 896–907. doi:10.1016/J.JCLEPRO.2017.08.012.

Lin, B., Hsieh, I. and Malmali, M. (2021) 'Rapid pressure swing adsorption for small scale ammonia separation: A proof-of-concept', *Journal of Advanced Manufacturing and Processing*, 3(2), pp. 1–15. doi:10.1002/amp2.10077.

Liu, W. et al. (2023) 'Analysis on temperature vacuum swing adsorption integrated with heat pump for efficient carbon capture', *Applied Energy*, 335, p. 120757. doi:10.1016/J.APENERGY.2023.120757.

Liu, Y., Ritter, J.A. and Kaul, B.K. (2000) 'Simulation of gasoline vapor recovery by pressure swing adsorption', *Separation and Purification Technology*, 20(1), pp. 111–127. doi:10.1016/S1383-5866(00)00066-6.

Lei, Z., Li, C. and Chen, B. (2007) 'Extractive distillation: A review', *Separation & Purification Review*, 32(2), pp. 121–213. doi:doi.org/10.1081/SPM-120026627.

Ma, S. et al. (2019) 'Hydrogen purification layered bed optimization based on artificial neural network prediction of breakthrough curves', *International Journal of Hydrogen Energy*, 44(11), pp. 5324–5333. doi:10.1016/J.IJHYDENE.2018.12.142.

Martínez García, M. et al. (2022) 'Simulation and state feedback control of a pressure swing adsorption process to produce hydrogen', *Mathematics*, 10(10), pp. 1–22. doi:10.3390/math10101762.

Masuda, S. et al. (2023) 'High-purity CO2 recovery following two-stage temperature swing adsorption using an internally heated and cooled adsorber', *Separation and Purification Technology*, 309, p. 123062. doi:10.1016/J.SEPPUR.2022.123062.

Moon, D.K., Lee, D.G. and Lee, C.H. (2016) 'H2 pressure swing adsorption for high pressure syngas from an integrated gasification combined cycle with a carbon capture process', *Applied Energy*, 183, pp. 760–774. doi:10.1016/J.APENERGY.2016.09.038.

Nagel, C. et al. (2000) 'Free volume distributions in glassy polymer membranes: Comparison between molecular modeling and experiments', *Macromolecules*, 33(6), pp. 2242–2248. doi:10.1021/ma990760y.

Naquash, A. et al. (2023) 'Separation and purification of syngas-derived hydrogen: A comparative evaluation of membrane- and cryogenic-assisted approaches', *Chemosphere*, 313, p. 137420. doi:10.1016/J.CHEMOSPHERE.2022.137420.

Neubert, M. et al. (2017) 'Experimental comparison of solid phase adsorption (SPA), activated carbon test tubes and tar protocol (DIN CEN/TS 15439) for tar analysis of biomass derived syngas', *Biomass and Bioenergy*, 105, pp. 443–452. doi:10.1016/J.BIOMBIOE.2017.08.006.

Ocaña-González, J.A. et al. (2023) 'Solid supports and supported liquid membranes for different liquid phase microextraction and electromembrane extraction configurations. A review', *Journal of Chromatography A*, 1691, p. 463825. doi:10.1016/J.CHROMA.2023.463825.

Park, J. et al. (2021) 'Improving energy efficiency for a low-temperature CO2 separation process in natural gas processing', *Energy*, 214, p. 118844. doi:10.1016/J.ENERGY.2020.118844.

Prabhansu et al. (2015) 'A review on the fuel gas cleaning technologies in gasification process', *Journal of Environmental Chemical Engineering*, 3(2), pp. 689–702. doi:10.1016/J.JECE.2015.02.011.

Panda, P.K., Sahoo, B. and Ramakrishnan, S. (2023) 'Hydrogen production, purification, storage, transportation, and their applications: A review', *Energy Technology*, 11(7), pp. 1–11. doi:10.1002/ente.202201434.

Saheed, I.O., Azeez, S.O. and Suah, F.B.M. (2022) 'Imidazolium based ionic liquids modified polysaccharides for adsorption and solid-phase extraction applications: A review', *Carbohydrate Polymers*, 298, p. 120138. doi:10.1016/J.CARBPOL.2022.120138.

Sant Anna, H.R. et al. (2017) 'Machine learning model and optimization of a PSA unit for methane-nitrogen separation', *Computers and Chemical Engineering*, 104, pp. 377–391. doi:10.1016/j.compchemeng.2017.05.006.

Saraceno, E., Ruocco, C. and Palma, V. (2023) 'A review of coal and biomass hydrogasification: Process layouts, hydrogasifiers, and catalysts', *Catalysts*, 13(2), pp. 1–38. doi:10.3390/catal13020417.

Shi, F. et al. (2021) 'MXene versus graphene oxide: Investigation on the effects of 2D nanosheets in mixed matrix membranes for CO2 separation', *Journal of Membrane Science*, 620, pp. 1–10. doi:10.1016/j.memsci.2020.118850.

Shigaki, N. et al. (2022) 'Performance evaluation of gas fraction vacuum pressure swing adsorption for CO2 capture and utilization process', *International Journal of Greenhouse Gas Control*, 120, p. 103763. doi:10.1016/J.IJGGC.2022.103763.

Singla, S. et al. (2022) 'Hydrogen production technologies—Membrane based separation, storage and challenges', *Journal of Environmental Management*. doi:10.1016/j.jenvman.2021.113963.

Sui, H. et al. (2016) 'Novel off-gas treatment technology to remove volatile organic compounds with high concentration', *Industrial and Engineering Chemistry Research*, 55(9), pp. 2594–2603. doi:10.1021/acs.iecr.5b02662.

Tafete, G.A. and Habtu, N.G. (2023) 'Reactor configuration, operations and structural catalyst design in process intensification of catalytic reactors: A review', *Chemical Engineering and Processing: Process Intensification*, 184, p. 109290. doi:10.1016/J. CEP.2023.109290.

Tingelinhas, J. et al. (2023) 'Pillared clays as cost-effective adsorbents for carbon capture by pressure swing adsorption processes in the cement industry', *Industrial & Engineering Chemistry Research*, 62(13), pp. 5613–5623. doi:10.1021/acs.iecr.2c04209.

Van Acht, S.C.J. et al. (2020) 'Simulation of integrated novel PSA/EHP/C process for high-pressure hydrogen recovery from Coke Oven Gas', *International Journal of Hydrogen Energy*, 45(30), pp. 15196–15212. doi:10.1016/J.IJHYDENE.2020.03.211.

van Paasen, S. et al. (2021) 'Development of the solid sorbent technology for post combustion CO2 capture towards commercial prototype', *International Journal of Greenhouse Gas Control*, 109, pp. 1–12. doi:10.1016/j.ijggc.2021.103368.

Wang, M. et al. (2022) 'Externally self-supported metallic nickel hollow fiber membranes for hydrogen separation', *Journal of Membrane Science*, 653, p. 120513. doi:10.1016/J. MEMSCI.2022.120513.

Wawrzyńczak, D., Majchrzak-Kucęba, I., Pevida, C., Bonura, G., Nogueira, R., De Falco, M. (ed.) (2023) *'The Carbon Chain in Carbon Dioxide Industrial Utilization Technologies'*. New York: Taylor & Francis Group, pp. 1–173. Available at: https://library.oapen.org/ handle/20.500.12657/60523 (Accessed 27 March, 2023).

Zafanelli, L.F.A.S. et al. (2023) 'A novel cryogenic fixed-bed adsorption apparatus for studying green hydrogen recovery from natural gas grids', *Separation and Purification Technology*, 307, p. 122824. doi:10.1016/J.SEPPUR.2022.122824.

Zhang, C. et al. (2022) 'Vacuum pressure swing adsorption for producing fuel cell grade hydrogen from IGCC', *Energy*, 257. doi:10.1016/j.energy.2022.124715.

Zhao, C. et al. (2023) 'Sulfur-resistant MoO2/TS-1 zeolite armored PdCu alloy composite membrane for hydrogen separation under H2S containing steam', *Journal of Membrane Science Letters*, 3(1), p. 100037. doi:10.1016/J.MEMLET.2023.100037.

Zhao, R. et al. (2019) 'Thermodynamic exploration of temperature vacuum swing adsorption for direct air capture of carbon dioxide in buildings', *Energy Conversion and Management*, 183, pp. 418–426. doi:10.1016/J.ENCONMAN.2019.01.009.

Zhou, L. et al. (2022) 'Design of refinery hydrogen networks with pressure swing adsorption unit configuration under uncertainty: Economy and flexibility aspects', *Industrial & Engineering Chemistry Research*, 61(21), pp. 7322–7334. doi:10.1021/acs. iecr.1c04487.

Zhu, X. et al. (2021) 'Design of steam-assisted temperature vacuum-swing adsorption processes for efficient CO2 capture from ambient air', *Renewable and Sustainable Energy Reviews*, 137, p. 110651. doi:10.1016/J.RSER.2020.110651.

Section IV

Other Technologies for Hydrogen Purification

15 Plasma-Assisted Technology for Hydrogen Purification

Kexin Yin, Yinglong Wang, Jifu Zhang,
Jingxue Wang, and Mengjin Zhou

15.1 INTRODUCTION

Energy development has always been an indispensable key movement for people's normal life and social production activities power resources. However, the excessive consumption of fossil fuels leads to the shortage of non-renewable energy sources, which brings about the negative impact of excessive greenhouse gas emissions, aggravated global climate change and rising sea level. Therefore, it is inevitable to find green, clean and renewable energy sources to replace traditional fuels. Hydrogen energy is one of the most promising energy sources to resolve the energy dilemma, the primary path of global energy transformation to sustainable development and the major direction of clean and green energy as a significant secondary energy source in the future (Filippov and Yaroslavtsev, 2021). The European Union has formulated the Energy Technology Roadmap 2050, which recognizes hydrogen energy as a vital part of the energy system, combined with fuel cells as an essential element for the structural transformation of the future energy system (Bednarczyk et al., 2022). In recent years, countries all over the world have set the development of hydrogen energy as their national strategic plans, defined their action plans and mapped out their development routes to explore the future direction of energy (Zhao et al., 2022). Since the 1980s, the size of the global hydrogen energy market has ulteriorly broaden, and countries have launched important hydrogen energy program one after another. In 2013, the world's first hybrid power plant was built to store hydrogen energy in Brandenburg, Germany. The hydrogen obtained from electrolysis is burned to drive the electricity generated by the generator, which is then continued to be used for hydrogen production by electrolysis. In 2015, the Mainz Energy Project, the world's largest hydrogen plant, was officially launched (Wang, 2018b). The main objective of the project is to convert clean and renewable energy sources into hydrogen for use and storage in order to effectively reduce fluctuations caused by renewable energy systems. In 2018, Germany's hydrogen-powered train went online, running on a 100-km route between Cuxhaven and Buxtehude, becoming one of the first demonstration projects to combine hydrogen energy with fuel cells (Li et al., 2021a). As a kind of renewable energy, the hydrogen industry is being developed all over the world. In 2021, Japan's New Energy Technology Development Agency set up

DOI: 10.1201/9781003382522-19

a "Green innovation fund" totaling 2 trillion yen. Eight of the 17 announced projects so far involve hydrogen energy, such as the construction of a large hydrogen supply chain, the use of renewable energy electrolytic water to produce hydrogen, the construction of a fuel ammonia supply chain and the use of hydrogen in the steelmaking process (Riera et al., 2023).

Hydrogen is the most abundant element in nature and the most abundant material in the universe, accounting for about 75%. Hydrogen sources on Earth are also very wide (Figure 15.1). Water, fossil fuels and all living organisms contain hydrogen, which can be used to produce hydrogen for energy purposes. Unlike fossil energy sources such as oil, natural gas and coal, which are only distributed in a few parts of the world, hydrogen production is not limited by time and region (Blanco et al., 2022).

This indicates that a new type of green and easy to produce high energy fuel has been discovered. This is hydrogen, an element that is going to have a big impact on humanity today and in the future, and it could be the next "main energy source". The main advantages of hydrogen energy are as follows:

a. Hydrogen energy is no carbon energy, is the most environmentally friendly energy and leaves no CO_2 in the environment during its life cycle. It has the advantages of zero pollution and zero emission during use. As an alternative to fossil fuels, hydrogen can reduce greenhouse gas emissions.
b. Hydrogen energy is a bridge of renewable energy, which can transform unstable renewable energy into stable energy. The utilization of renewable resources such as water, wind, light and nuclear energy is the current research hotspot (Orhan et al., 2017).
c. Except nuclear fuel, hydrogen has the highest heat value among all fossil fuels, chemical fuels and bio-fuels, up to 142,351 kJ/kg, the heat per kilogram of hydrogen combustion, which is about 3 times of gasoline, 3.9 times of alcohol and 4.5 times of coke.

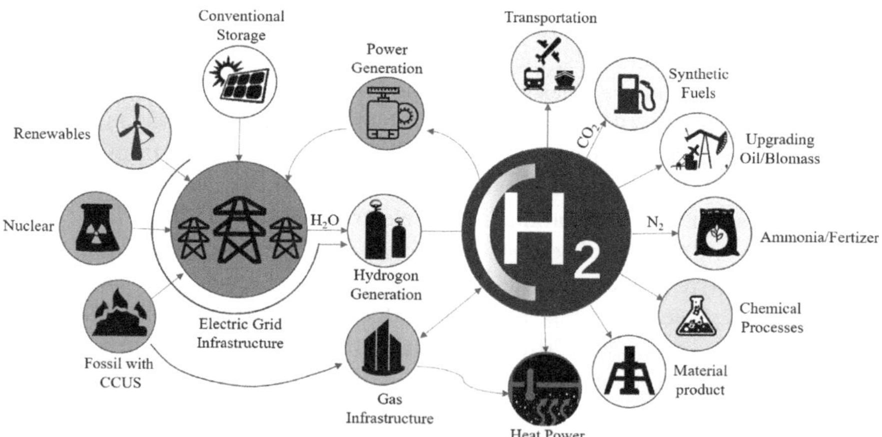

FIGURE 15.1   Sources and utilization of hydrogen energy.

d. Hydrogen is non-toxic compared with other fuels. When hydrogen is burned, it produces water and a small amount of hydrogen nitride, but does not produce harmful substances such as carbon monoxide, carbon dioxide, hydrocarbons, lead and dust particles. After proper treatment of small amounts of nitrogen and hydrogen, the environment will not be polluted and the water produced by combustion can be recycled to produce hydrogen.

e. Hydrogen energy is also a secondary energy source, which is produced by using other energy sources through certain methods. Unlike coal, oil and natural gas, which can be directly exploited, it is almost entirely produced by fossil fuels. If the waste hydrogen can be recovered from the certain project, it can be recovered to about 100 million cubic meters every year, which is quite considerable.

f. Hydrogen has become an important industrial raw material, which can be used in clinical and medical supplies. About 2/3 of the hydrogen produced in the world is used in the ammonia synthesis industry (Chen et al., 2022). Hydrogen can be used for melting and cutting metal in oxygen–hydrogen flame up to 3,000°C high temperatures. Hydrogen bombs can be produced using energy from the fusion of atoms of the hydrogen isotopes deuterium and tritium. Liquid hydrogen, the ideal rocket fuel, is mixed with liquid oxygen to produce a specific impulse of about 350. The above applications of hydrogen only play a part in the role of hydrogen, but we have not realized the truly powerful applications, such as controlled nuclear fusion technology, which is an epoch-making application of hydrogen. Once we realize controlled nuclear fusion technology, we can not only get rid of the polluting chemical energy of human beings, but also gradually improve the ecological environment of the earth (Ongena and Ogawa, 2016). In addition, the powerful energy generated by nuclear clustering can also enable mankind to achieve epoch-making progress in space technology, such as the realization of the initial sub-light speed flight.

To sum up, hydrogen is a kind of clean energy in the future, which occupies a certain position in reducing greenhouse gases. It is versatile and can be used as an "ideal energy" for power generation, heating and transportation fuel. Intensive development of hydrogen energy is essential to sustainable development strategies.

15.2 HYDROGEN PURIFICATION METHODS

At present, several main ways of purifying hydrogen include by-product hydrogen in the chlor-alkali industry, hydrogen production by electrolytic water, hydrogen production from chemical raw materials (methanol cracking, ethanol cracking and liquid ammonia cracking), hydrogen production from petrochemical resources (petroleum cracking and water gas process) and new hydrogen production methods (biomass and photochemistry) (Das et al., 2022). In the traditional hydrocarbon hydrogen purification process, large-scale hydrogen purification is mainly based on methane hydrogen production, and small- and medium-sized hydrogen production is mainly based on methanol hydrogen production (Labanca et al., 2020). Currently, the on-board hydrogen production methods are mainly based on methanol hydrogen production. The hydrocarbon is reformed by steam or oxygen under high temperatures (300~1,000°C) and heavy metal catalyst is used to improve the reaction speed (Li et al., 2021b). Impurities of hydrocarbons and carbon deposition at high temperatures will deactivate the catalyst. And it takes a certain amount of time for the whole hydrogen production system to be heated to the reaction temperature, resulting in low flexibility of catalytic hydrogen production process. In addition, traditional hydrogen production technology requires a large amount of primary energy or production of raw materials, which leads to a series of negative impacts such as excessive greenhouse gas emissions, global climate change and sea level rise (Cho et al., 2023). Plasma-assisted hydrogen purification technology refers to the ionization of air, N_2, Ar or other gasification agents under high pressure, resulting in high temperature above 1,650°C, organic solid waste decomposition into combustible gases such as H_2, CO and CH_4, and non-combustible solid waste into harmless residue such as vitreous solid residue. At the same time, the incorporation of the arc into plasma significantly reduces the emission of harmful gases into the ambient atmosphere, such as furan and dioxins. This technique can solve or avoid problems encountered by traditional methods (Chung et al., 2018). It can be seen in Figure 15.2 that the principle of plasma-assisted hydrogen purification is roughly the same as the traditional principle, but the difference is that the active substance that stimulates the chemical reaction is different. The active substance of the traditional method is the catalyst, while the active substance of the plasma method is high-energy electrons and free radicals. The plasma continuously absorbs energy from the outside world through the gas and dissociates into ions to form hydrogen. The basic components of plasma are electrons and heavy particles, including positive and negative ions and neutral particles. According to the particle temperature of plasma, plasma is usually divided into heat equilibrium plasma and non-equilibrium plasma (Krasheninnikov et al., 2022). When the temperature of heavy particles is close to the temperature of electrons, it is called the heat equilibrium plasma. At this time, the electron density of plasma is high, mainly in the form of arc and plasma torch discharge. When the temperature of heavy particles is much lower than that of electrons, it is called non-equilibrium plasma. At this time, the electron density is about $10^{10}\,m^{-3}$. But the electron temperature is the same as that of the heat equilibrium plasma, which is about $10^4\,K$ (Jian and Qing, 2016), while the heavy particles are generally not much warmer than room temperature. Non-equilibrium plasma mainly has glow, microwave, corona and other discharge forms (Valentino et al., 2020). Because of the use of electricity, plasma reaction has a high degree of control and can adjust the gas rate and composition over a wide range to achieve optimization of the reaction. With the help of highly active particles such as electrons, ions and excited substances, plasma can greatly improve the chemical reaction speed, or provide energy for endothermic reaction and avoid the use of heterogeneous catalysts.

These advantages and the short reaction time caused by the high energy density of the plasma provide the possibility to reduce the size and weight of the hydrogen production reactor. In addition, plasma conversion hydrogen production does not have high requirements for raw materials. As long as hydrogen-containing substances, such as natural gas, gasoline, diesel, heavy oil, alcohol, biofuels and even water, can be used as raw materials for plasma hydrogen production. Plasma hydrogen reactor equipment input is small; its main component-electrode is only metal or graphite material (Neeraj et al., 2022). Plasma hydrogen also has the advantages of fast reaction rate, low reaction temperature and flexible parameter control in the process of converting hydrocarbons into hydrogen. Compared with the thermochemical method, the device has small size, fast start-up, low energy consumption and large range of operating parameters, which is especially suitable for the on-board hydrogen production system and small distributed hydrogen production system using natural gas as raw materials. In view of the above characteristics, the plasma method is suitable for hydrogen purification in various scales, including dispersed distribution and changing production conditions (Sun et al., 2017).

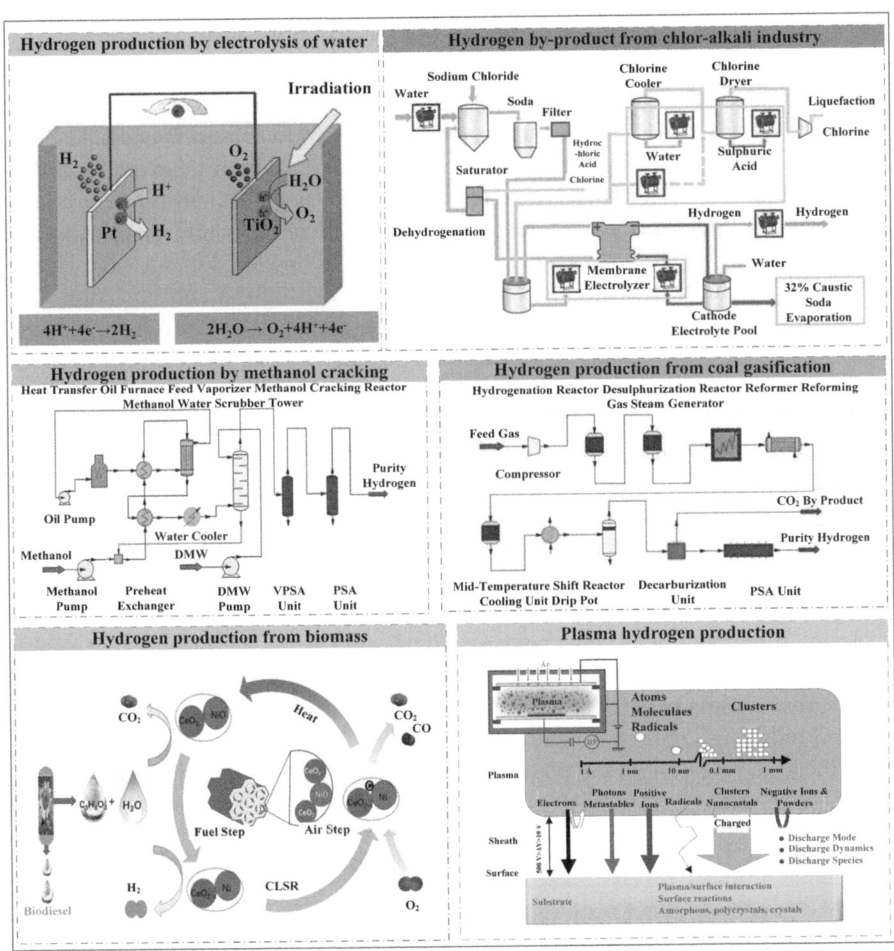

FIGURE 15.2 Principles of different hydrogen purification processes (Dou et al., 2021; Zhang et al., 2020)

Because of these unique advantages of plasma reaction, it is superior to other technologies in hydrogen purification. In recent years, many researchers have carried out experiments and designs on hydrogen purification by plasma chemistry. At present, plasma has been widely used. It has opened up potential application fields include metal refining, hazardous waste treatment and plasma chemical industry. In the field of plasma chemical industry, such as hydrogen plasma cracking coal to acetylene, plasma coal gasification and plasma calcium carbide, these technologies are closely related to hydrogen purification. Many researchers have applied plasma hydrogen technology to various carriers. At present, there are more studies on alcohol plasma hydrogen production, such as plasma ethanol reforming hydrogen purification. However, there are also many researchers using methane plasma to produce hydrogen, plasma decomposition of ammonia to produce hydrogen and other compounds, such as dimethyl ether (DME) to heat organic waste plasma hydrogen. And few researchers are using new microwave liquid plasma equipment for hydrogen purification. Lytle et al. (2019) studied, analyzed and simulated the technology of hydrogen purification by microwave plasma and applied it to practical production in a novel way. Steam near the tungsten electrode was simulated to produce hydrogen. The simulation of steam mass fragmentation was performed in a vacuum concealed reaction vessel with tungsten electrode using plasma unit and electromagnetic COMSOL module. Models are used to derive kinetic results for electron density, electron temperature, plasma velocity and material interactions. The concentration dynamics of other free components in the reaction geometry of the whole domain were studied, and it was concluded that the electrode tip needed the highest concentration. Chen et al. (2014) reviewed the conversion of methanol, ethanol and DME using different types of cold plasma. Hydrogen is the main product and its by-products vary with the reaction conditions. The cold plasma system is considered to have good potential for hydrogen production. It is important to improve the reaction conditions and reactor design for future applications. Wang et al. (2012b) described a heterogeneous non-thermal plasma reactor that used electrical discharge to co-convert liquid fuel and natural gas. Using a 500 kJ/kg cetane energy input, the plasma process collectively converted 9.36% cetane and 20% methane (by mass). And through molecular dissociation mechanism modeling, hydrogen production was 34.8 kWh/kg-H_2 and greenhouse gas emissions were minimal. The chemical conversion efficiency of plasma was about 30%. This conversion process was more efficient and reduced greenhouse gas emissions compared to conventional technologies. Therefore, it can be seen that plasma hydrogen has great advantages in improving hydrogen production and reducing greenhouse gas emissions, so it is necessary and far-reaching to strengthen the research on the plasma hydrogen method.

15.3 PRINCIPLE OF PLASMA HYDROGEN PURIFICATION

15.3.1 Liquid-Phase Discharge Plasma Hydrogen

Liquid-phase discharge plasma, also known as "electro-hydraulic effect", usually refers to the plasma formed by discharge in a liquid, rather than the plasma itself being liquid. The properties of plasma formed in liquid are similar to those in gas. The biggest difference is that gas is insulated, whereas liquid conducts electricity, making it more difficult to generate plasma in liquid than in gas (Sun, 2013).

TABLE 15.1
Characteristics of Different Liquid-Phase Discharge Plasmas (Xin, 2018)

	Numerical Value		
Parameter	**Pulse Corona**	**Pulse Spark**	**Pulsed Arc**
Frequency (HZ)	$10^2\sim103$	$10^{-2}\sim102$	$10^{-3}\sim10^{-2}$
Peak current (A)	$10\sim102$	$10^2\sim103$	$10^3\sim105$
Peak voltage (V)	$10^4\sim106$	$10^4\sim105$	$10^3\sim104$
Voltage rise time (s)	$10^{-9}\sim10^{-7}$	$10^{-9}\sim10^{-5}$	$10^{-6}\sim10^{-5}$
Pressure wave	Weak; Medium	Medium	Strong
Ultraviolet light	Weak; Medium	Medium	Strong

Liquid-phase discharge plasma will simultaneously produce four effects: strong electric field, strong oxidizing free radical, ultraviolet light and shock wave. Table 15.1 shows the characteristics of liquid-phase discharge plasma (Sun, 2013; Romat et al., 2007). Regardless of the type of discharge form, liquid-phase discharge is accompanied by intense electrical regions, impact shock waves and ultraviolet radiation. The voltage rise time is at the level of microsecond or even nanosecond, and the peak current usually exceeds 100 A. Such a rapidly generated strong current will inevitably generate a mass of high-energy electrons that collide with surrounding molecules, causing abundant free radicals and other neutral particles to be produced (Xin, 2018).

The free radicals generated by discharge in water mainly include $\cdot$OH, $\cdot$O and $\cdot$H, among which $\cdot$OH and $\cdot$O are highly oxidizing and play a strong role in decomposing organic matter (Locke et al., 2006), and the formation of $\cdot$H is the key to the synthesis of hydrogen. It can be seen that the free radicals generated in liquid-phase discharge are very suitable for the application of alcohol decomposition to produce hydrogen. In addition to free radicals, liquid-phase discharge will also produce O_3, H_2O_2 and other neutral particles. Both O_3 and H_2O_2 have strong oxidation properties and can survive in liquid for a long time (O_3 can be stable in liquid for 1~10 min, while H_2O_2 can be stable in liquid for tens of minutes; Sun, 2013), which will certainly promote the decomposition of alcohols.

15.3.2 Hydrogen Purification from Rotary Arc Discharge Plasma Weighted Ethanol

Rotary arc has the advantages of uniform plasma discharge, an electrode that is not easy to corrode, etc., making it suitable for fuel reforming; compared with the general sliding arc discharge, the plasma area formed by rotary arc discharge is larger, which is conducive to the reaction.

Air as a working gas is mainly composed of O_2 and N_2. Bombard O_2 with high-energy electrons to separate O_2:

$$O_2 + e \rightarrow O + O + e \tag{15.1}$$

O radicals produced by ionization dissociation have a strong oxidizing ability, and they can react with ethanol or other free radicals to form oxygenated compounds. O radicals can also trap H atoms in certain particles, forming –OH groups and breaking the C=C bond of some C_2 (Yan and Du, 2017).

The main turnkey reactions involved in the experiment are as follows:

Partial oxidation of ethanol:

$$C_2H_5OH + 0.5O_2 \rightarrow 2CO + 3H_2 \tag{15.2}$$

$$C_2H_5OH + 1.5O_2 \rightarrow 2CO_2 + 3H_2 \tag{15.3}$$

Thermal decomposition of ethanol:

$$C_2H_5OH \rightarrow CH_4 + H_2 + CO \tag{15.4}$$

In addition, the direct entry of ethanol from the discharge end will cause charcoal deposition in the plasma generation area, resulting in unstable plasma generation and electrode contamination (Czylkowski et al., 2015). Therefore, ethanol vapor is injected directly into the plasma flame. This air intake method can effectively inhibit carbon deposition in the plasma generation area (Guo et al., 2022).

15.3.3 ROTATING SLIDING ARC PLASMA WEIGHT SHAPING HYDROGEN

Common rotary sliding arc reactors are mostly a conical inner electrode and a cylindrical outer electrode structure. When thousands of volts of voltage are applied between high- and low-voltage electrodes, a strong electric field will be generated between the two electrodes, and the electric field strength gradually increases as the voltage increases. When the electric field strength is greater than or equal to the breakdown field strength of the reactant, the plasma medium is broken down to form an arc at the narrowest gap (usually through 2~8 mm; Sreethawong et al., 2007). The arc rotates around the inner electrode driven by tangential airflow and stretches continuously in the process of rotation until it disappears.

For liquid fuels (such as ethanol and methanol), reforming hydrogen production technology mainly includes four types: CO_2 reforming (dry reforming), partial oxidation reforming, steam reforming and self-heating reforming.

CO_2 reforming is usually used to make syngas (the main components are H_2 and CO); because the reaction is endothermic, it needs to absorb heat from the outside to maintain the efficient progress of the reaction and its chemical reaction formula is as follows (Oliveira-Vigier et al., 2008):

$$C_xH_yO_z + (x-z)CO_2 \rightarrow (2x-z)CO + y/2H_2 \tag{15.5}$$

The raw materials for partial oxidation reforming are liquid fuel and oxygen or air. When oxygen is sufficient (complete oxidation), the main products are H_2O and CO_2.

When oxygen is insufficient (partial oxidation), the main products are CO and H_2. The chemical reaction formula of the partial oxidation reaction is as follows (Mattos and Noronha, 2005):

$$C_xH_yO_z + \left(x/2 - z/2\right)O_2 \rightarrow xCO + y/2H_2 \tag{15.6}$$

$$C_xH_yO_z + \left(x + y/4 - z/2\right)O_2 \rightarrow xCO_2 + y/2H_2O \tag{15.7}$$

The steam reforming reaction is a strong endothermic reaction, which needs to absorb a large amount of heat from the outside world, and the chemical reaction formula is as follows (Seelam et al., 2012):

$$C_xH_yO_z + (2x - z)H_2O \rightarrow xCO_2 + \left(2x - z + y/2\right)H_2 \tag{15.8}$$

The self-heating reforming reaction is a special form of reforming in equilibrium that does not require additional heat absorption from the outside world, and hydrogen selectivity is high during the reaction. Usually, an appropriate amount of steam and air is added to increase the production of hydrogen and reduce the heat required in the reaction process, and the chemical reaction formula is as follows (Kugai et al., 2006):

$$C_xH_yO_z + wO_2 + (2x - 2w - z)H_2O \tag{15.9}$$

$$xCO_2 + \left(2x - 2w - z + y/2\right)H_2 \tag{15.10}$$

15.4 PLASMA MEDIUM

Usually, argon, hydrogen, nitrogen and other gases are used as plasma medium; however, steam mixed with nitrogen or argon can also be used as plasma medium. The plasma medium should be selected with low cost and high heat value (HHV). The type and nature of the plasma medium affect the heating and momentum transfer of the plasma. Oxygen is considered a good plasma medium to reduce the total flow rate and impurity content of the gas in the reaction vessel, which is more favorable in some common applications. Carbon dioxide and nitrogen can also be considered plasma mediators because higher arc voltages increase the actual power of the ejection. Steam plasmas can achieve a similar effect, but materials such as hydrogen, oxygen and hydroxyl radicals can cause extreme electrode corrosion to the metal.

In some cases, plasma reactor electrodes should have a long service life without corroding the electrodes with the gases used, such as in direct-current/alternating-current (DC/AC) plasma systems where argon and nitrogen are commonly used. Contrary to expectations, the resulting gas contains a mass of nitrogen and argon, which reduces the calorific value of the resultant syngas. If pure oxygen or water vapor is used as a plasma gas, the heat value of the syngas is increased. Electrode corrosion does not occur in radio frequency (RF) and microwave plasma systems, so oxidizing gases such as steam and oxygen can be used as plasma medium. The characteristics of common plasma medium are shown in Figure 15.3.

Researchers have demonstrated that when plasma treats appropriate carbonaceous materials in the presence of steam, hydrogen and carbon monoxide are the main products (Nishikawa et al., 2004; Ivan and Boris, 1992). Essentially, steam as a plasma medium produces more carbon monoxide and significantly increases hydrogen concentration and total gas production.

Du et al. (2016) studied the effect of actual discharge power, plasma medium type and high-water content on vaporization effect, and finally showed that when plasma medium type is flowing air, gas production is large, CO selectivity and H_2/CO are the highest, H_2/CO ratio in total gas-liquid end product is also the highest, but carbon conversion is the lowest. When the plasma medium type is nitrogen, the reaction rate is the fastest, the carbon conversion rate is the highest and the CO selectivity is similar to that when the plasma medium is air. The lower the water content, the more unfavorable it is to further improve the synthesis gas yield; the slower the chemical

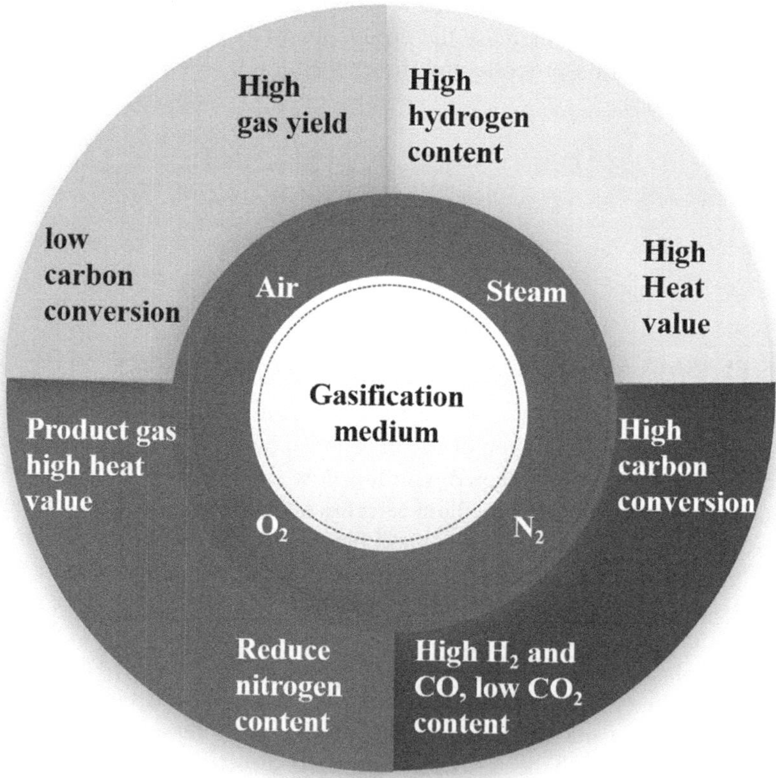

FIGURE 15.3 Characteristics of common gasification medium (Nishikawa et al., 2004; Tendler et al., 2005; Pu et al., 2013; Huang and Tang, 2009).

reaction speed, carbon conversion and H_2/CO will also decrease with the decrease of water content over time. This can increase the high moisture content to help reduce CH_4 and further improve CO selectivity.

Oost et al. (2006) studied the use of gas mixture of steam and a small amount of argon as carrier gas for pyrolysis waste. The main advantages of using a water-stabilized arc plasma torch are the plasma element and slow rate of flow, as well as low levels of other gases more commonly found in gas stabilization systems. Synthetic gas with high calorific value, high hydrogen content and low carbon dioxide content can be synthesized. Huang et al. (2006) used steam plasma to pyrolyze medical waste. The results showed that when the equilibrium composition of C-H-O in the system is C/O=1 and the temperature range is 1,400 ~2,000 K, carbon monoxide and hydrogen are the main gas components. The volume ratio of other components (CO_2, C_2H_4, C_2H_2 and CH_4) is less than 1% and the conversion degree of raw materials is nearly 100%. In contrast to air plasma, steam plasma treatment does not produce nitrogen oxides if the raw material does not contain nitrogen. Huang and Tang (2009) studied the pyrolysis treatment of waste tire particle using a capacitively coupled RF plasma reactor. The use of nitrogen is considered a carrier gas. The response time is 10 min. The results show that the final products are H_2, CO, CH_4 and CO_2. Finally, the actual power of RF is the main driving factor. When the actual RF power is further increased from 1,600 to 2,000 W, the solid material conversion rate is further increased from 40% to 76.8%. Hydrogen production rate further increased from 36.58 to 84.53 mL/min. Mohsenian et al. (2015) researched the treatment of polymeric garbage using dual DC thermal plasma torches. Argon gas is a carrier gas. The reaction time is 3 min. Raw materials processed include polypropylene (PP), polyethylene (PE), Polyvinyl chloride (PVC) and Acrylonitrile Butadiene Styrene (ABS) plastic. The final results indicated that the hydrogen density and yield obtained by treating ABS samples were higher than those of other samples. Carbon black was the most important solid product of hot plasma pyrolysis polymer.

15.5 PLASMA GENERATOR AND WORKING PRINCIPLE

There are many ways to classify plasma, but the commonly used classification method is generally divided by temperature or discharge form. According to temperature division, plasma can be divided into high-temperature plasma and low-temperature plasma (Bogaerts et al., 2002; Schumacher, 2007). High-temperature plasma, also known as complete equilibrium plasma or highly ionized plasma, has gas temperature, ion temperature and electron temperature that are completely equal, with the general temperature is about 1.5×10^7 K (Charles, 2014). Low-temperature plasma is a mildly ionized plasma in which the ion temperature is generally lower than the electron temperature, and it can be further divided into hot plasma and cold plasma. Hot plasma is also known as local heat balance plasma and its electron temperature is almost the same as the ion temperature, which is about 2,000~20,000 K (Venkatramani, 1995). The cold plasma is called non-heat equilibrium plasma and its electron temperature is much higher than the ion temperature, which is about 10~100 times (Thirumdas et al., 2017). The ion temperature of this plasma is about 100~1,000 K. In particular, this chapter focuses on the field of plasma hydrogen, and the current research mainly focuses on the hydrogen production by different discharge forms of low-temperature plasma, which will be highlighted in this section.

At present, the discharge forms of gas-phase discharge plasma hydrogen mainly include glow discharge, corona discharge, spark discharge, arc discharge, dielectric barrier discharge (DBD) and microwave discharge. The basic principle of gas-phase discharge is shown in Figure 15.4.

Since the discharge medium is gas, the raw materials for hydrogen production are mainly gaseous hydrogen-containing raw materials, including methane and DME. Ghorbanzadeh et al. (2009) used pulsed glow discharge plasma for hydrogen purification by CH_4 and CO_2; the products produced were H_2, CO, C_2H_2, C_2H_4 and C_2H_6. When the peak voltage is 19 kV, discharge frequency 2 kHz, CH_4: $CO_2=1$, H_2 selectivity is 71%, $H_2/CO=1.2$. Aleknaviciute et al. (2013) used corona discharge to decompose methane to produce CO_x-free H_2; the by-products are mainly acetylene and ethylene. With the increase of discharge power, hydrogen production increases linearly; hydrogen flow can reach 62.7 mL/min, and hydrogen percentage concentration can reach 34.13%. Wang et al. (2014) used gas phase sliding arc discharge to reform DME to produce hydrogen. By accelerating the feeding speed of DME, reduce the production and energy consumption of hydrogen production. The maximum hydrogen flow is about 13.2 mL/min and the energy consumption of hydrogen production is 1.7×10 kWh/m³H_2. Khadir et al. (2017) used DBD to reform methane to produce hydrogen under normal pressure; the discharge frequency was 50 kHz, and the CH_4 conversion rate and H_2/CH_4 could reach 58.03% and 28.75%, respectively.

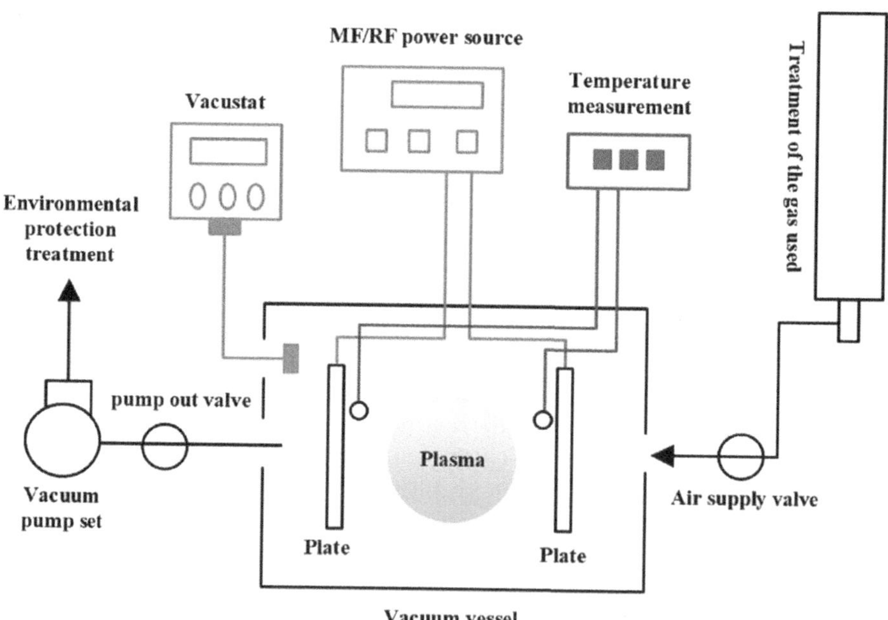

FIGURE 15.4 Basic principles of gas-phase discharge (Wang et al., 2018a).

However, when the raw materials for hydrogen production are mainly hydrogen-containing liquid raw materials, the use of vapor-phase discharge requires gasification of the liquid raw materials or through the carrier gas into the discharge area. The liquid raw materials used include ethanol, methanol and toluene. Arkhipenko et al. (2016) used DC corona discharge to reform ethanol to produce hydrogen under normal pressure. The carrier gas was air and the flow rate was set to 0.4 L/min. The carrier gas brought the ethanol aqueous solution into the discharge zone and the percentage concentration of hydrogen can reach 58.66%. Rincin et al. (2019) used argon gas as the carrier gas to decompose ethanol to hydrogen under normal pressure. The experiments showed that when using microwave vapor-phase discharge to produce hydrogen, the ethanol conversion rate can reach 99.6%, the hydrogen selectivity can reach 89%, the hydrogen flow rate is 20.5 mL/min and the energy consumption of hydrogen production is about 241 kWh/m^{3}H$_2$. Czylkowski et al. (2015) realized the hydrogen production of microwave vapor-phase discharge reforming ethanol under atmospheric pressure; the ethanol conversion rate can reach 99%, the hydrogen production flow rate is 12,500 mL/min, and the energy consumption is about 4.1 kWh/m^{3}H$_2$.

Table 15.2 summarizes some applications of gas-phase discharge plasma in the field of hydrogen purification. In general, the use of gas-phase discharge plasma system to hydrogen has slightly higher energy consumption, slightly lower hydrogen production and low hydrogen percentage concentration, which has not yet met the needs of hydrogen production devices used in vehicles and ships. At the same time, from the perspective of output, the use of alcohol (methanol and ethanol) to produce

TABLE 15.2

Applications of Hydrogen Produced by Discharge Plasma in Gas Phase

Plasma Type	Raw Material	Hydrogen Flow (mL/min)	Percentage Concentration of Hydrogen (%)	Energy Consumption for Hydrogen Purification (kWh mH$_2$$^{-1}$)	References
Corona	Methane	62.7	34.1	–	Wang et al. (2018a)
Corona	Dimethyl ether	42.6	–	5.9	Zou et al. (2007)
Corona	Ethanol	400	58.7	–	Arkhipenko et al. (2016)
Sparkle	Methane	75	45	–	Moshrefi and Rashidi (2014)
Electric arc	Methanol	280.5	–	2.99	Zhang et al. (2016)
Media blocking	Methane	12	–	27.2	Sarmiento et al. (2007)
Microwave	Ethanol	20.5	–	241	Rincón et al. (2014)
Microwave	Ethanol	12500	–	4.1	Czylkowski et al. (2015)

hydrogen is slightly higher than the use of other gas fuels to produce hydrogen, but when using alcohol gas-phase discharge to produce hydrogen, it is necessary to bring alcohol substances into the discharge zone through carrier gas (Ar, He and N_2) or external heating source (Ricard et al., 2016; Yang et al., 2023), which undoubtedly increases the cost of using the method and is not conducive to the miniaturization and convenience of the device. In addition, gas-phase discharge in alcohols is also prone to carbon deposition phenomenon; carbon attached to the electrode will affect the discharge and reduce the hydrogen production. When the carbon deposition is too much, the plasma quenching will block hydrogen production. It may be the way to solve these problems that trigger plasma system to produce hydrogen in the alcohol directly.

Liquid-phase discharge has matured after 30 years of development and its reactor types are also diverse. The types of liquid-phase discharge plasma reactors mainly include needle–plate type, line–plate type, DBD reactor, glow discharge reactor and high-frequency discharge reactor. Needle–plate reactors are widely used in liquid-phase discharge (Lee et al., 2023). Its basic structure is shown in Figure 15.5. The needle electrode is usually connected to high voltage, while the plate electrode is grounded and the plasma is initiated at the tip of the needle electrode. Multi-needle–plate structures, needle–needle type, rod–rod and other structures derived from the needle–plate structure are also commonly used in liquid-phase discharge applications. The advantage of needle–plate reactors is that plasma is relatively easy to initiate and can directly produce liquid substances in liquids, without the problem of gas–liquid mass transfer. Different from expectations, the relatively low transfer reaction rate in the liquid phase will cause rapid changes in the ambient temperature and concentration between the plasma region and the bulk solution, resulting in the recombination and quenching of bioactive functional groups in the discharge region, which is also a common problem of

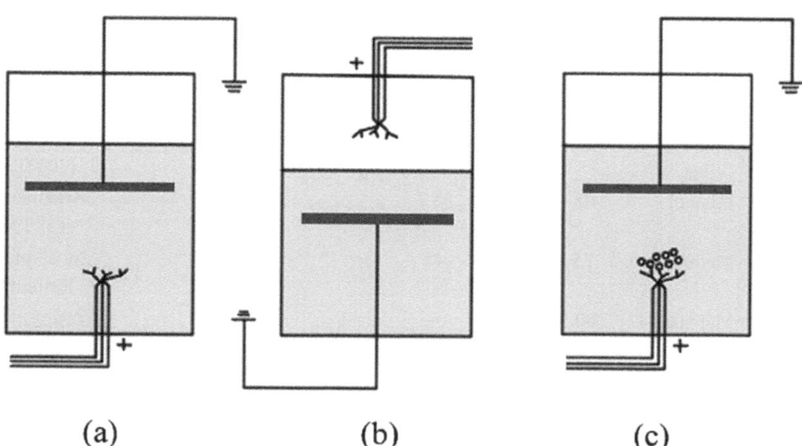

(a)   (b)   (c)

FIGURE 15.5 Structures of needle–plate reactor for discharge plasma in liquid (Vasilev et al., 2022): (a) direct liquid-phase discharge; (b) gas–liquid two-phase discharge; and (c) liquid bubbling discharge.

direct liquid-phase discharge (Xiang et al., 2019). In general, the use of needle–plate reactor is simpler and more convenient, and it is easy to initiate and study liquid plasma.

The wire–plate reactor consists of one or more wire electrodes and one plate electrode, and its basic structure is shown in Figure 15.6. Research by Guo et al. (2006) shows that the morphology of plasma in water is similar to that in air when the wire–plate structure is used. The advantage of the wire–plate reactor is that the discharge range is large; more active substances will be produced to act on the bulk liquid, but the energy of initiating plasma is relatively higher (Li et al., 2007). The plate–plate reactor associated with the multi-wire–plate reactor and the plate–pore-plate reactor generated by the further optimization have also been widely used in the study of liquid-phase discharge plasma.

DBDs are discharges of dielectric insulation inserted into the gap of discharge electrodes to generate non-thermal equilibrium ionized gas (cold plasma), also known as silent discharge. There is more than one electrode structure for DBD (Jiang et al., 2020; Liu et al., 2021a,b). Figure 15.7 shows the basic electrode structure of DBDs. DBDs are filled with a certain working gas between two discharge electrodes, and one or two electrodes are covered with an insulating medium, or the medium can be directly suspended in the discharge space or filled with a granular medium. When a high enough AC voltage is applied between the two electrodes, the gas between the electrodes will be broken down resulting in a discharge known as DBD.

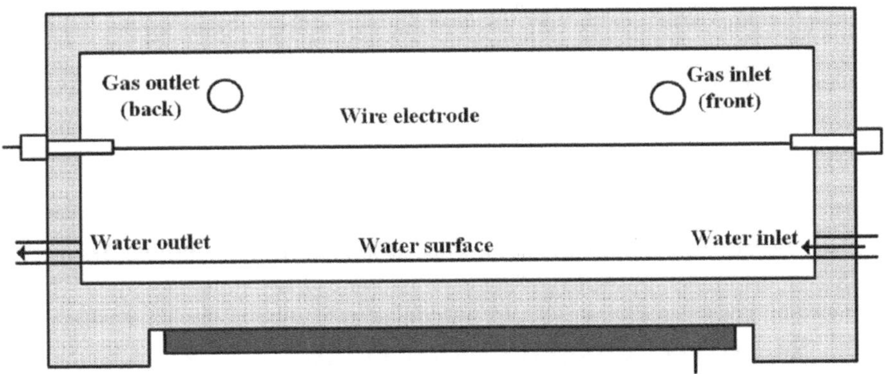

FIGURE 15.6 Structure of wire–plate reactor for discharge plasma in liquid (Guo et al., 2006).

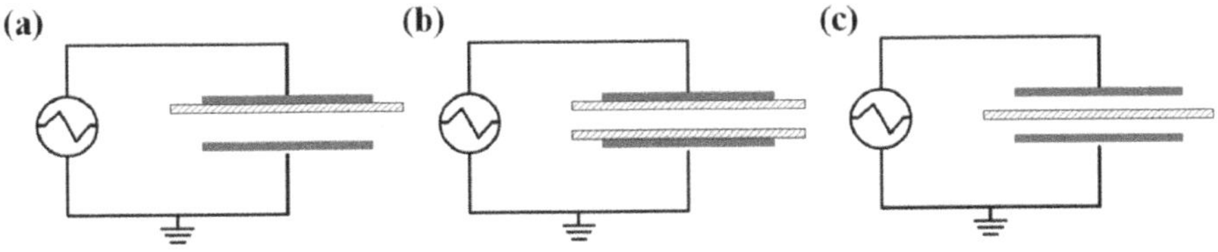

FIGURE 15.7 Structures of dielectric barrier discharge electrode (Wang et al., 2012a).

DBD is usually driven by sine-wave AC high-voltage power supply (Neretti et al., 2022). With the increase of input electric field voltage, the state of reaction gas in the system will undergo three stages of change: from insulation state to breakdown and finally discharge. When the voltage of the input electric field is relatively low, although some gases will have some ionization and free diffusion, the current is zero because the current is too insufficient to cause the gas in the reaction area to undergo plasma reaction. With the gradual increase of the input electric field voltage, the charged particles in the reaction zone increase. However, when the breakdown voltage of the reaction gas is not reached, the electric field between the two electrodes is low, which cannot provide enough energy to cause an inelastic collision of the gas molecules. Due to the lack of inelastic collision, the number of charged particles cannot increase substantially, so the reaction gas is still in the state of insulation and cannot generate discharge. If the input voltage continues to be increased, when the electric field between the two electrodes is large enough for the gas molecules to conduct inelastic collision, the discharge gas will increase a large number of charged particles. When the electron density in space is higher than a certain critical value and reaches the breakdown voltage, many microdischarge filamentous channels between the two poles will be generated, and the luminescence phenomenon can be obviously observed in the system. At this point, the current will increase rapidly as the applied voltage increases. Due to the dielectric barrier in the discharge space, the current growth is limited, so as not to form sparks or arcs. Generally, DBD can be divided into filament discharge, uniform discharge and spot discharge. The specific form depends on the type of discharge gas, the nature of the medium, the voltage value of the input electric field and the frequency of the high-voltage power supply (Ran et al., 2018; Richard et al., 2019). In general, dielectric barrier glow discharge is unstable and will be transformed into filament discharge after a period of time. Although DBD has been widely used, its theoretical research is limited to microdischarge or a specific part of the whole discharge process. So far, none of the DBD theories can be applied to various situations. The reason is that the working conditions of various DBDs are very different, and there are both physical and chemical processes in the discharge process, which affect each other. It is difficult to determine the specific process in the middle of the final result.

15.6 ADVANTAGES AND ISSUES OF PLASMA ASSIST

Plasma gasification technology has been proven to be an effective, environmentally friendly and sustainable technology for high-value treatment of waste. Under the action of plasma, it can provide extremely high-temperature conditions and efficient conversion of solid waste into syngas and inorganic slag. Because the traditional incineration process can't provide high-temperature conditions, there will still be a lot of unburned waste. The incineration process can reduce solid waste, but it will inevitably produce a lot of toxic substances such as dioxins and furans. However, under the condition of low oxygen and high temperature, the plasma gasification process can completely split large molecular organic substances into small molecular substances, which avoids the generation of toxic pollutants at a large extent. In addition, in the gasification process, the combustible gas generated by gasification can generate electricity through the gas turbine, and the slag generated by inorganic components can be used for roadbed materials, realizing the harmless and resource utilization of waste. Raw material of plasma gasification can be biomass, municipal solid waste (MSW), waste tire, plastic solid, medical waste, oil and natural gas and so on to produce syngas or recovery energy (Messerle et al., 2018). Figure 15.8 shows the treatment of municipal waste by hot plasma process. Plasma gasification can clean up the material without leaving dust behind. Hydrocarbon plasma-assisted hydrogen purification has the advantages of rapid reaction, low reaction temperature and flexible parameter control. Compared with the thermochemical method, its device has the advantages of small size, fast start-up, low energy consumption and wide range of operating parameters. It is especially suitable for the on-board hydrogen production system using natural gas as raw material and small distributed hydrogen production system. And the plasma gasification device adopts automatic technology, showing good flexibility (Yayalık et al., 2020).

Munir et al. (2019) investigated the thermochemical process of plasma co-gasification of MSW and plastic solid waste (PSW) for hydrogen purification. Using air as the primary gas and oxygen or steam as the secondary flow, the components of different plasma formation gases are further investigated. The gasification of MSW has higher overall performance using pure air as a plasma medium. It is found that all waste mixtures can be steadily improved with the increase of oxygen content in plasma medium. As the steam is injected into the plasma medium, the carbon conversion rate of the waste reaches its maximum. In this case, the maximum conversion efficiency is 21.7% at a steam ratio (σ) of 34% in the plasma gas and an equal ratio of MSW to PSW in the waste mixture. Fathollahi et al. (2021) investigated the influence of plasma power and oxygen flow rate on methane conversion and methanol selectivity in a DBD reactor. Their findings demonstrate that increasing power at a constant gas flow rate linearly enhances methane conversion, while methanol selectivity initially improves and then declines. Additionally, maintaining a constant methane flow rate of 100 sccm, increasing oxygen flow rate from 40 sccm to 75 sccm resulted in

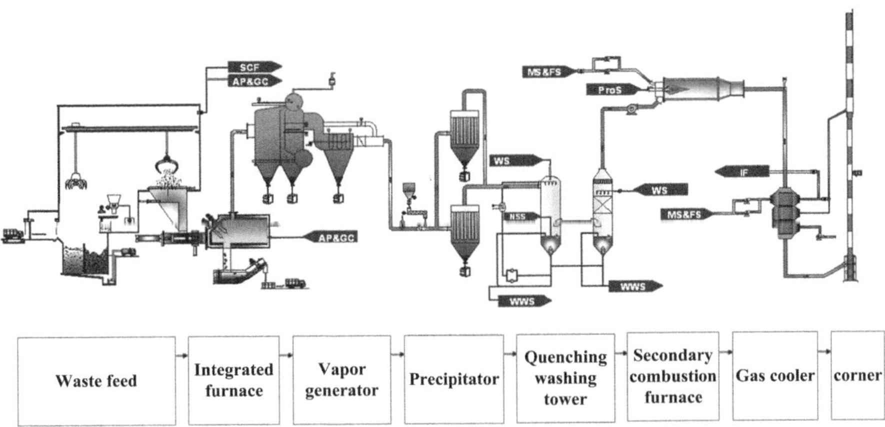

Waste feed	Integrated furnace	Vapor generator	Precipitator	Quenching washing tower	Secondary combustion furnace	Gas cooler	corner

FIGURE 15.8 Flowchart of municipal waste treatment by hot plasma process (Byun et al. 2010).

enhanced methane conversion and methanol selectivity. The study highlights that the formation of copper oxide on the electrode surface effectively promotes high methanol yields. Zeng et al. (2019) explored the possibility of using slag after plasma gasification of different medical wastes as catalyst. At the optimum active temperature of 250°C, the co-treated slag removed 64.4% nitric oxide (1,000 ppm), which was 23.5% higher than that of the slag treated alone. Almost all zinc (15.85–16.99 mg/g) and lead (0.72–1.16 mg/g) were enriched in secondary fly ash. Therefore, if secondary fly ash was treated by plasma gasification again, these heavy metals will be further enriched. Thus, transforming secondary fly ash has potential in metallurgical resources. The main results of this work showed that the appropriate combined treatment of hazardous wastes was carried out through plasma gasification to produce high-value products. Qi et al. (2021) studied the process of MSW into hydrogen purification by plasma gasification and improved the process to solve the problems that the municipal sludge incineration and landfill technology would destroy the environment and occupy a large amount of land. The economic benefit, environmental benefit and exergy efficiency of the two processes were evaluated comprehensively. Compared with the original process, the new process has higher energy conversion efficiency and the ability of plasma gasification to produce hydrogen. The exergy efficiency and economic benefit of the original process are 6.98% and 4.59% respectively, which are lower than that of the new process. The proposed new process provided a promising method to solve the problem of municipal sludge treatment.

15.7 APPLICATION AND CHALLENGE

Plasma solid waste treatment technology originated in the 1960s but limited to the difficult technology and high cost of equipment and other reasons; it is mostly used in the treatment of low-level radioactive waste and medical waste, and other hazardous waste. Since the 1990s, with the continuous improvement of technology and the reduction of equipment costs, the treatment of other solid wastes has been gradually involved. At present, plasma waste treatment technology abroad has made great progress; some begin to operate commercially and some are in the stage of industrialization.

There are several well-established commercial plasma technology companies in the United States, such as Westinghouse Plasma, Phoenix Solutions and Startech (Phoenix Fabry et al., 2013; Sanito et al., 2020). Among them, Westinghouse Plasma Company (later acquired by Alter Corporation of Canada) is the most representative. The company has been engaged in the treatment of household waste, sludge and waste materials for decades and has rich experience in plasma gasification waste. Since 2000, Westinghouse Plasma has built a 220 t/d MSW plasma treatment plant in Japan. The company's gasification technology to treat a wide range of solid waste without any pre-sorting can be directly solid waste treatment. Second, the syngas produced by gasification can be discharged directly after purification. The higher gasification temperature and the anoxic environment avoid the generation of toxic substances such as dioxins and furans. The research on plasma gasification of solid waste is also developing in Korea. The Korea Institute of Nuclear Environmental Technology has been treating low-level radioactive waste using plasma technology since the early 1990s and has developed a commercial radioactive waste treatment facility (Choi et al., 2000). The device can handle radioactive soil, metal and concrete without sorting and compressing the waste, and it can be used for food and glass and other urban waste treatment after hot plasma melting treatment of waste into non-toxic slag; the tail gas after cleaning treatment does not contain nitrogen oxides and sulfide and other pollution gases. The Institute of Electrophysics and Electric Power of the Russian Academy of Sciences has carried out in-depth research in the field of plasma gasification and developed a variety of AC plasma torches that can last several hours at work. These devices have a power of 5–500 kW and a thermal efficiency of 90%–95%. They are mainly used in the treatment of plastic waste, coal and wood. Bellwether, a German company, built a 12 t/h plasma waste treatment plant in Romania in 2007, which is mainly used for the treatment of municipal domestic and industrial waste. It can achieve 80%–85% gasification efficiency and the syngas generated during gasification is sent to the local power plant. The syngas can replace 28,000 tons of coal, greatly easing the local energy crisis. Figure 15.9 shows the common types of plasma

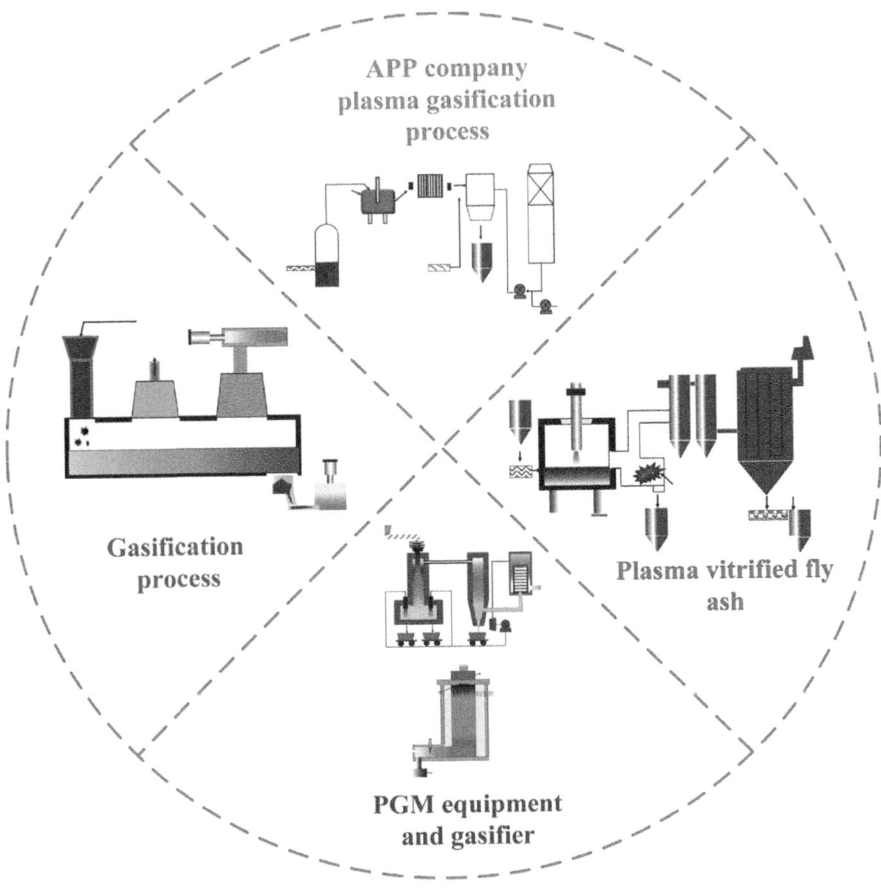

FIGURE 15.9 Schematic diagram of common plasma gasification equipment (Zhang et al., 2011).

gasification equipment. Europlasma has built a demonstration plasma waste treatment facility in France using the CHO-Power multistage gasification process, which uses non-transfer-arc plasma torches and is designed to handle industrial waste and biomass (Zhang et al., 2013).

Since the 1990s, the Institute of Mechanics of the Chinese Academy of Sciences has been engaged in the research of plasma gasification of medical waste and MSW. At present, it has built a laboratory-scale simulated medical waste treatment device and then cooperated with enterprises to build two hazardous waste treatment equipment, the treatment capacity of which reached 5–10 t/d. The process includes waste pretreatment, plasma gasification, gas cooling, deacidification and dust removal and power generation (Zhao et al., 2020). The Taiwan Institute of Nuclear Energy started to develop plasma torches on its own in 1993. After making a breakthrough in this key technology, it developed a 100 kW non-transfer-arc plasma torch device in 1996. The reactor is specially designed to achieve optimal temperature distribution, and the waste is covered by a crucible to avoid spillage or diffusion during treatment. The Taiwan Institute of Atomic Energy uses a hot plasma melting system for harmless treatment of fly ash. The conversion of ash from plasma incinerators into harmless slag that can be used as a building and decorative material solves the problem of having nowhere to store fly ash (Chu et al., 1998).

Plasma gasification technology requires large costs in process design, special materials, hot plasma source, as well as expenses for the replacement of plasma source system and some components in the process of use. Plasma gasification processes need to remain more advanced and efficient in terms of technology and equipment to compete with traditional technologies in the field of solid waste treatment. It is difficult to reduce their initial costs, but efficient and functional plasma waste treatment processes can be established through operation cost. In addition, the plasma gasification process can convert solid waste into high-value products, such as high purity hydrogen, syngas and other valuable compounds, which can subsidize the high operating costs of plasma gasification processes. Plasma gasification is an energy-intensive process that consumes a lot of electricity and places an additional capacity load on the grid. Energy efficiency can be improved through superior process design. Existing processes can be optimized to improve their economy, such as by using insulation to avoid energy loss. It can also be combined with other processes to produce additional products or to recover waste heat, such as integrating the gasification process with a steam turbine to recover additional heat. In addition, switching the plasma gasification process from batch to continuous reduces heating and cooling times and improves feedstock loading and handling capacity, which improves waste disposal efficiency. The commercial application of plasma gasification technology needs the support of corresponding supporting facilities, such as waste treatment system and waste gas cleaning equipment. These problems are solved by making solid waste into refuse derived fuel (RDF). An ideal plasma gasification plant can more fully convert similar wastes into combustible components with HHV. Excess metal and glass residue can reduce the heat value of the RDF and may cause operational problems. Therefore, it can be equipped with perfect waste pretreatment device to optimize waste sorting, which may improve the quality of plasma gasification raw materials and the gasification process.

15.8 CONCLUSION

This chapter summarizes the advantages, disadvantages and development status of traditional hydrogen production technology and novel hydrogen production technology in industry, and it also describes the basic principle of plasma-assisted hydrogen purification technology, commonly used plasma media and plasma generator in detail. Plasma-assisted technology for hydrogen purification does not have high requirements on the feed type, but different medium and electrode structures need to be selected according to product requirements and energy consumption costs. As the development direction of hydrogen purification in the future, the following aspects need to be considered: (a) the energy consumption of plasma generator; (b) how to screen out the best plasma medium; (c) the capture of acid gas and follow-up treatment to reduce the impact on the environment as much as possible and (d) economic analysis of the overall process.

ACRONYMS

ABS	Acrylonitrile butadiene styrene plastic
DBDs	Dielectric barrier discharges
DME	Dimethyl ether
HHV	High heat value
MSW	Municipal solid waste
PSW	Plastic solid waste
PP	Polypropylene
PE	Polyethylene
PVC	Polyvinyl chloride
RDF	Refuse derived fuel
RF	Radio frequency

REFERENCES

Aleknaviciute, I., Karayiannis, T. G., Collins, M. W., Xanthos, C., 2013. Methane decomposition under a corona discharge to generate CO_x-free hydrogen. *Energy* 59, 432–439.

Arkhipenko, V. I., Kirillov, A. A., Simonchik, L. V., Kazak, A. V., Chernukho, A. P., Migoun, A. N., 2016. Ethanol conversion in a DC atmospheric pressure glow discharge. *Int. J. Hydrogen. Energy* 41(41), 18320–18328.

Bednarczyk, J. L., Brzozowska-Rup, K., Luściński, S., 2022. Opportunities and limitations of hydrogen energy in Poland against the background of the European Union energy policy. *Energies* 15(15), 5503.

Blanco, H., Leaver, J., Dodds, P. E., Dickinson, R., García-Gusano, D., Iribarren, D., Baumann, M., 2022. A taxonomy of models for investigating hydrogen energy systems. *Renew. Sust. Energ. Rev.* 167, 112698.

Bogaerts, A., Neyts, E., Gijbels, R., Mullen, J., 2002. Gas discharge plasmas and their applications. *Spectrochim. Acta Part B: At. Spectros.* 57(4), 609–658.

Byun, Y., Namkung, W., Cho, M., Chung, J. W., Kim, Y., Lee, J., Lee, C., Hwang, S., 2010. Demonstration of thermal plasma gasification/vitrification for municipal solid waste treatment. *Environ. Sci. Technol.* 44(17), 6680–6684.

Charles, C., 2014. Grand challenges in low-temperature plasma physics. *Front. Phys.* 2(39), 1–5.

Chen, F. Q., Huang, X. Y., Cheng, D. G., Zhan, X. L., 2014. Hydrogen production from alcohols and ethers via cold plasma: A review. *Int. J. Hydrogen Energy* 39(17), 9036–9046.

Chen, W., Li, T., Ren, Y., Wang, J., Chen, H., Wang, Q., 2022. Biological hydrogen with industrial potential: Improvement and prospection in biohydrogen production. *J. Clean. Prod.* 387, 135777.

Cho, H. H., Strezov, V., Evans, T. J., 2023. A review on global warming potential, challenges and opportunities of renewable hydrogen production technologies. *Sustain. Mater. Technol.* 35, e00567.

Choi, K., Sheng, J. W., Lee, M. C., Song, M. J., 2000. Utilizing the KEP-A glass frit to vitrify low-level radioactive waste from Korean NPPs. *Waste Manag.* 20, 575.

Chu, J. P., Hwang, I., Tzeng, C. C., Kuo, Y. Y., Yu, Y. J., 1998. Characterization of vitrified slag from mixed medical waste surrogates treated by a thermal plasma system. *J. Hazard. Mater.* 58, 179.

Chung, K., Jeong, S., Kim, B. J., Kim, J. S., Park, Y. K., Jung, S. C., 2018. Development of hydrogen production by liquid phase plasma process of water with NiTiO2/carbon nanotube photocatalysts. *Int. J. Hydrogen. Energy* 43(11), 5873–5880.

Czylkowski, D., Hrycak, B., Miotk, R., Jasiński, M., Dors, M., Mizeraczyk, J., 2015. Hydrogen production by conversion of ethanol using atmospheric pressure microwave plasmas. *Int. J. Hydrogen Energy* 40(40), 14039–14044.

Das, A., Peu, S. D., 2022. A comprehensive review on recent advancements in thermochemical processes for clean hydrogen production to decarbonize the energy sector. *Sustainability-Basel* 14(18), 11206.

Dou, B., Zhao, L., Zhang, H., Wu, K., Zhang, H., 2021. Renewable hydrogen production from chemical looping steam reforming of biodiesel byproduct glycerol by mesoporous oxygen carriers. *Chem. Eng. J.* 416, 127612.

Du, C.M., Wu, J., Huang, Y.N., 2016. Technical analysis of hydrogen production from organic waste by plasma pyrolysis and gasification. *China Environ. Sci.* 36(11), 3429–3440.

Fabry, F., Rehmet, C., Rohani, V., Fulcheri, L., 2013. Waste gasification by thermal plasma: A review. *Waste. Biomass. Valori.* 4, 421.

Fathollahi, P., Farahani, M., Rad, R.H., Khani, M.R., Asadi, A., Shafiei, M., Shokri, B., 2021. Selective oxidation of methane to methanol by NTP plasma: The effect of power and oxygen on conversion and selectivity. *J. Electrost.* 112, 103594.

Filippov, S. P., Yaroslavtsev, A. B., 2021. Hydrogen energy: Development prospects and materials. *Russ. Chem. Rev.* 90(6), 627.

Ghorbanzadeh, A., Lotfalipour, R., Rezaei, S., 2009. Carbon dioxide reforming of methane at near room temperature in low energy pulsed plasma. *Int. J. Hydrogen Energy* 34(1), 293–298.

Guo, W., Zheng, X. F., Guo, Q. J., 2022. Hydrogen production from ethanol by rotating arc discharge plasma. *Chem. Eng. Oil Gas* 51(01), 46–52.

Guo, Y., Ye, D., Tian, Y., Chen, K., 2006. Humidity effect on toluene decomposition in a wire-plate dielectric barrier discharge reactor. *Plasma Chem. Plasma Process.* 26, 237–249.

Huang, H., Tang, L., 2009. Pyrolysis treatment of waste tire powder in a capacitively coupled RF plasma reactor. *Energy Convers. Manag.* 50(3), 611–617.

Huang, J. J., Guo, W. K., Xu, P., 2006. Thermodynamic study of water-steam plasma pyrolysis of medical waste for recovery of CO and H2. *Plasma Sci. Technol.* 7(6), 3148–3150.

Ivan, B. G., Boris, I. M., 1992. Some general conclusions from the results of studies on solid fuel steam plasma gasification. *Fuel* 71(8), 895–901.

Jian, H., Qing, G. Z., 2016. A simple method for laser-induced plasma diagnostics under condition of optically thin. *Mod. Phys. Lett. B.* 30(16), 1650197.

Jiang, S., Qiu, L.W., Li, Z., Zhang, L., Rao, J.F., 2020. A new all-solid-state bipolar high-voltage multilevel generator for dielectric barrier discharge. *IEEE Trans. Plasma Sci.* 48(4), 1076–1081.

Khadir, N., Khodja, K., Belasri, A., 2017. Methane conversion using a dielectric barrier discharge reactor at atmospheric pressure for hydrogen production. *Plasma Sci Technol.* 19(9), 95502.

Krasheninnikov, S. I., Kukushkin, A. S., Pshenov, A. A., 2022. Divertor plasma detachment: Roles of plasma momentum, energy, and particle balances. *Plasma Phys. Contr. Fusion* 64, 125011.

Kugai, J., Subramani, V., Song, C. S., 2006. Effects of nanocrystalline CeO2 supports on the properties and performance of Ni-Rh bimetallic catalyst for oxidative steam reforming of ethanol. *Chin. J. Catal.* 238(2), 430–440.

Labanca, A. R., 2020. Carbon black and hydrogen production process analysis. *Int. J. Hydrogen Energy* 45(47), 25698–25707.

Lee, Y.R., Kim, D.Y., Kim, J.Y., Lee, D.H., Bae, G.T., Jang, H., Park, J.Y., Jung, S., Jung, E.Y., Park, C., Lee, H., Tae, H., 2023. Effects of dielectric barrier on water activation and phosphorus compound digestion in gas-liquid discharges. *Nanomaterials* 14(1), 40.

Li, H., Ma, C., Zou, X., Li, A., Huang, Z., Zhu, L., 2021a. On-board methanol catalytic reforming for hydrogen production: A review. *Int. J. Hydrogen Energy* 46(43), 22303–22327.

Li, J., Shang, K.F., Wu, Y., Wang, N.H., Zhang, Y., 2007. The experimental research on electrode configuration and discharge characteristics of pulse discharge. *J. Electrost.* 65(4), 228–232.

Li, Z., Zhang, R., Sun, H. X., Zhang, W. D., Mei, C. X., 2021b. Review on key technologies of hydrogen generation, storage and transportation based on multi-energy complementary renewable energy. *Trans. China Electrotech. Soc.* 36(3), 445–462.

Liu, P., Song, Y.X., Zhang, Z.T., 2021a. A novel dielectric barrier discharge (DBD) reactor with streamer and glow corona discharge for improved ozone generation at atmospheric pressure. *Micromachines* 12(11), 1287.

Liu, Y.W., Wang, S., Zhou, R.S., Fang, Z., Ostrikov, K., 2021b. Development of a battery-operated floating-electrode dielectric barrier discharge plasma device and its characteristics. *Plasma Sci. Technol.* 23, 64008.

Locke, B. R., Sato, M., Sunka, P., Hoffmann, M. R., Chang, J. S., 2006. Electrohydraulic discharge and nonthermal plasma for water treatment. *Ind. Eng. Chem. Res.* 45(3), 882–905.

Lytle, S., Siddiqui, O., Chehade, G., Dincer, I., 2019. Analysis and modelling of microwave plasma hydrogen production utilizing water vapor and tungsten electrodes. *Int. J. Hydrogen Energy* 44(47), 25319–25334.

Mattos, L. V., Noronha, F. B., 2005. Partial oxidation of ethanol on supported Pt catalysts. *J. Power Sources* 145(1), 10–15.

Messerle, V. E., Mosse, A. L., Ustimenko, A. B., 2018. Processing of biomedical waste in plasma gasifier. *Waste Manag.* 79, 791–799.

Mohsenian, S., Esmaili, M. S., Shokri, B., 2015. Physical characteristics of twin DC thermal plasma torch applied to polymer waste treatment. *J. Electrostat.* 76, 231–237.

Moshrefi, M. M., Rashidi, F., 2014. Hydrogen production from methane by DC spark discharge: Effect of current and voltage. *J. Nat. Gas Sci. Eng.* 15: 85–89.

Munir, M. T., Mardon, I., Al-Zuhair, S., Shawabkeh, A., Saqib, N. U., 2019. Plasma gasification of municipal solid waste for waste-to-value processing. *Renew. Sustain. Energy Rev.* 116, 109461.

Neeraj, B., Amit, P., Mishra, R. S., 2022 Plasma reforming for hydrogen production: Pathways, reactors and storage. *Int. J. Hydrogen Energy* 48(7), 2467–2482.

Neretti, G., Popoli, A., Scaltriti, S.G., Cristofolini, A., 2022. Real time power control in a high voltage power supply for dielectric barrier discharge reactors: Implementation strategy and load thermal analysis. *Electronics* 11(10), 1536.

Nishikawa, H., Ibe, M., Tanaka, M., 2004. A treatment of carbonaceous wastes using thermal plasma with steam. *Vacuum* 73(3/4), 589–593.

Oliveira-Vigier, K. D., Abatzoglou, N., Gitzhofer, F., 2008. Dry-reforming of ethanol in the presence of a 316 stainless steel catalyst. *Can. J. Chem. Eng.* 83(6), 978–984.

Ongena, J., Ogawa, Y., 2016. Nuclear fusion: Status report and future prospects. *Energy Policy* 96, 770–778.

Oost, G. V., Hrabovsky, M., Kopecky, V., 2006. Pyrolysis of waste using a hybrid argon-water stabilized torch. *Vacuum* 80(11), 1132–1137.

Orhan, M. F., Kahraman, H., Babu, B. S., 2017. Approaches for integrated hydrogen production based on nuclear and renewable energy sources: Energy and exergy assessments of nuclear and solar energy sources in the United Arab Emirates. *Int. J. Hydrogen Energy* 42(4), 2601–2616.

Pu, G., Zhou, H., Hao, G., 2013. Study on pine biomass air and oxygen/steam gasification in the fixed bed gasifier. *Int. J. Hydrogen Energy* 38, 15757–15763.

Qi, H., Cui, P., Liu, Z., Xu, Z., Yao, D., Wang, Y., Zhu, Z., Yang, S., 2021. Conceptual design and comprehensive analysis for novel municipal sludge gasification-based hydrogen production via plasma gasifier. *Energy Convers. Manag.* 245, 114635.

Ran, J.X., Li, C.X., Ma, D., Luo, H.Y., Li, X.W., 2018. Homogeneous dielectric barrier discharges in atmospheric air and its influencing factor. *Phys. Plasmas* 25, 33511.

Ricard, A., Sarrette, J., Oh, S., Kim, Y.K., 2016. Comparison of the active species in the RF and microwave flowing discharges of N2 and Ar-20%N2. *Plasma Chem. Plasma Process.* 36, 1559–1570.

Richard, G., Tilmatine, A., Benabderrahmane, A., Zeghloul, T., Medles, K., Dascalescu, L., 2019. Factors that affect the tribocharging of polyethylene (PE) granules after exposure to a dielectric barrier discharge (DBD). *IEEE Trans. Ind. Appl.* 55(1), 812–816.

Riera, J. A., Lima, R. M., Knio, O. M., 2023. A review of hydrogen production and supply chain modeling and optimization. *Int. J. Hydrogen Energy* 48(37), 13731–13755.

Rincón, R., Marinas, A., Muñoz, J., Calzada, M. D., 2014. Hydrogen production from ethanol decomposition by microwave plasma TIAGO torch. *Int. J. Hydrogen Energy* 39(22), 11441–11453.

Romat, H., Zastawny, H., Chang, J., 2007. Pulsed arc discharges for water treatment and disinfection. Electrost. Ser.: Inst. Phys. Conf. Ser., 178, 325–330.

Sanito, R. C., You, S. J., Chang, T. J., Wang, Y. F., 2020. Economic and environmental evaluation of flux agents in the vitrification of resin waste: A SWOT analysis. *J. Environ. Manag.* 270, 110910.

Sarmiento, B., Brey, J. J., Viera, I. G., González-Elipe, A. R., Cotrino, J., Rico, V. J., 2007. Hydrogen production by reforming of hydrocarbons and alcohols in a dielectric barrier discharge. *J. Power Sources* 169(1), 140–143.

Schumacher, U., 2007. Similarities and particularities of low- and high-temperature plasma technologies. *Plasma Process. Polym.* 4(3), 318–325.

Seelam, P. K., Linguori, S., Iulianelli, A., 2012. Hydrogen production from bio-ethanol steam reforming reaction in a Pd/PSS membrane reactor. *Catal. Today* 193(1), 42–48.

Sreethawong, T., Thakonpatthanakun, P., Chavadej, S., 2007. Partial oxidation of methane with air for synthesis gas production in a multistage gliding arc discharge system. *Int. J. Hydrogen Energy* 32(8), 1067–1079.

Sun, B., 2013. *Liquid Phase Discharge Plasma and Its Application.* Beijing Science and Technology Press, China.

Sun, B, Zhao, X. T., Xin, Y. B., Zhu, X. M., 2017. Large capacity hydrogen production by microwave discharge plasma in liquid fuels ethanol. *Int. J. Hydrogen Energy* 42(38), 24047–24054.

Tendler, M., Rutberg, P., Oost, G. V., 2005. Plasma based waste treatment and energy production. *Plasma Phys. Control. Fusion* 47(5), 219–230.

Thirumdas, R., Kadam, D., Annapure, U. S., 2017. Cold plasma: An alternative technology for the starch modification. *Food Biophys.* 12, 129–139.

Valentino, C., Riccardo, S., Michele, B., Carlo, N. G., Massimo, A. D R., Douglas, B., Laxminarayan, L. R., 2020. Multidimensional modeling of non-equilibrium plasma generated by a radio-frequency corona discharge. *Plasma Sources Sci. Technol.* 29, 115013.

Vasilev, M., Conlon, P., Bohl, D., Thagard, S. M., 2022. The effect of discharge frequency of a gas-liquid plasma reactor on bulk liquid transport and removal of organic contaminants. *Plasma Chem. Plasma Process.* 42, 759–783.

Venkatramani N., 1995. Thermal plasmas in material processing. *Bull. Mater. Sci.* 18, 741–754.

Wang, M., 2018b. Development status and future trends of new energy hydrogen production at home and abroad. *Chem. Ind. Chem. Eng. Q.* 36(6), 13–18.

Wang, K. P., Shariful, I. B., Md, A. H. B., Jamie, K., Christopher, C., Howard, J., David, S., 2012b. Electric fuel conversion with hydrogen production by multiphase plasma at ambient pressure. *Chem. Eng. J.* 433, 133660.

Wang, B. W., Sun, Q. M., Lv, Y. J., Yang, M. L., Yan, W. J., 2014. Steam reforming of dimethyl ether by gliding arc gas discharge plasma for hydrogen production. *Chin. J. Chem. Eng.* 22(1), 104–112.

Wang, L. L., Wang, X. Q., Ning, P., Cheng, C., Ma, Y. X., Zhang, R., 2018a. Simultaneous removal of COS, H2S, and dust in industrial exhaust gas by DC corona discharge plasma. *Ind. Eng. Chem. Res.* 57(18), 6568–6575.

Wang, C. Q., Zhang, G. X., Wang, X. X., 2012a. Comparisons of discharge characteristics of a dielectric barrier discharge with different electrode structures. *Vacuum* 86, 960–964.

Xin, Y. B., 2018. *Characteristics and Mechanism of Hydrogen Produced by Pulsed Discharge Plasma in Liquid.* Dalian Maritime University, China.

Xiang, H., Lei, B., Yuan, X., Lv, Q., Zhang, Q., 2019. Design and simulation of new type reactor in the wastewater treatment system based on discharge plasma. *IEEE Trans. Plasma Sci.* 47(1), 952–957.

Yan, J. H., Du, C. M., 2017. *Hydrogen Generation from Ethanol Using Plasma Reforming Technology.* Singapore: Springer.

Yang, J., Chen, Z.Q., Dai, T., Chen, S., Wang, W.Y., Su, Q.C., Zhang, L., Wang, B., Zhou, Y.M., Lu, X.P., 2023. Discharge characteristics of atmospheric pressure pulsed microwave He/N2 plasma jet. *Plasma Process. Polym.* 25(5), 2200203.

Yayalık, İ., Koyun, A., Akgün, M., 2020. Gasification of MSW in plasma arc medium. *Plasma Chem. Plasma Process.* 40, 1401–1416.

Zeng, J., Yue, Y., Gao, Q., Zhang, J., Zhou, J., Pan, Y., Qian, G., Tang, J., Ruan, J., 2019. Co-treatment of hazardous wastes by the thermal plasma to produce an effective catalyst. *J. Clean. Prod.* 208, 243–251.

Zhang, Q. L., Dor, L., Fenigshtein, D., Yang, W. H., Blasiak, W., 2011. Gasification of municipal solid waste in the plasma gasification melting process. *Appl. Energy* 90 (1), 106–112.

Zhang, Q. L., Wu, Y. S., Dor, L. R., Yang, W. H., Blasiak, W., 2013. A thermodynamic analysis of solid waste gasification in the plasma gasification melting process. *Appl. Energy* 112, 405.

Zhang, H., Zhu, F. S., Li, X. D., Cen, K. F., Du, C. M., Tu, X., 2016. Enhanced hydrogen production by methanol decomposition using a novel rotating gliding arc discharge plasma. *RSC Adv.* 6, 12770–12781.

Zhao, W., Ma, J., Wang, Z., Li, Y., Zhang, W., 2022. Potential hydrogen market: Value-added services increase economic efficiency for hydrogen energy suppliers. *Sustainability-Basel* 14(8), 4804.

Zhao, P., Ni, G. H., Jiang, Y. M., Chen, L. W., Chen, M. Z., Meng, Y. D., 2020. Destruction of inorganic municipal solid waste incinerator fly ash in a DC arc plasma furnace. *J. Hazard. Mater.* 58, 580.

Zou, J., Zhang, Y., Liu, C., 2007. Hydrogen production from partial oxidation of dimethyl ether using corona discharge plasma. *Int. J. Hydrogen Energy* 32(8), 958–964.

Index

Note: **Bold** page numbers refer to tables; *italic* page numbers refer to figures.